Semiconductor Lasers II

Materials and Structures

OPTICS AND PHOTONICS

(formerly Quantum Electronics)

SERIES EDITORS

PAUL L. KELLEY

Tufts University
Medford, Massachusetts

IVAN P. KAMINOW

AT&T Bell Laboratories
Holmdel, New Jersey

GOVIND P. AGRAWAL

University of Rochester
Rochester, New York

CONTRIBUTORS

G.A. Acket
Markus-Christian Amann
P.G. Eliseev
Kenichi Iga
Fumio Koyama
David G. Mehuys
A. Valster
C.J. van der Poel

A complete list of titles in this series appears at the end of this volume.

Semiconductor Lasers II

Materials and Structures

Edited by Eli Kapon

Institute of Micro and Optoelectronics
Department of Physics
Swiss Federal Institute of Technology, Lausanne

OPTICS AND PHOTONICS

ACADEMIC PRESS

San Diego   London   Boston
New York   Sydney   Tokyo   Toronto

ACADEMIC PRESS
a division of Harcourt Brace & Company
525 B Street, Suite 1900, San Diego, CA 92101-4495, USA
http://www.apnet.com

ACADEMIC PRESS
24–28 Oval Road, London NW1 7DX, UK
http://www.hbuk.co.uk/ap/

Library of Congress Cataloging-in-Publication Data
Semiconductor lasers : optics and photonics / edited by Eli Kapon.
        p.   cm.
    Includes indexes.
    ISBN 0-12-397630-8 (v. 1). — ISBN 0-12-397631-6 (v. 2)
    1. Semiconductor lasers.   I. Kapon, Eli.
TA1700.S453   1998
621.36'6—dc21   98-18270
                                                              CIP

Printed and bound in the United Kingdom
Transferred to Digital Printing, 2011

Contents

Preface

More than three decades have passed since lasing in semiconductors was first observed in several laboratories in 1962 (Hall *et al.*, 1962; Holonyak, Jr. *et al.*, 1962; Nathan *et al.*, 1962; Quist *et al.*, 1962). Although it was one of the first lasers to be demonstrated, the semiconductor laser had to await several important developments, both technological and those related to the understanding of its device physics, before it became fit for applications. Most notably, it was the introduction of heterostructures for achieving charge carrier and photon confinement in the late sixties and the understanding of device degradation mechanisms in the seventies that made possible the fabrication of reliable diode lasers operating with sufficiently low currents at room temperature. In parallel, progress in the technology of low loss optical fibers for optical communication applications has boosted the development of diode lasers for use in such systems. Several unique features of these devices, namely the low power consumption, the possibility of direct output modulation and the compatibility with mass production that they offer, have played a key role in this development. In addition, the prospects for integration of diode lasers with other optical and electronic elements in optoelectronic integrated circuits (OEICs) served as a longer term motivation for their advancement.

The next developments that made semiconductor lasers truly ubiquitous took place during the eighties and the early nineties. In the eighties, applications of diode lasers in compact disc players and bar-code readers have benefited from their mass-production capabilities and drastically reduced the prices of their simplest versions. In parallel, more sophisticated devices were developed as the technology matured. Important examples are high power lasers exhibiting very high electrical to optical power conversion efficiency, most notably for solid state laser pumping and medical applications, and high modulation speed, single frequency, distributed feedback lasers for use in long-haul optical communication systems.

Moreover, progress in engineering of new diode laser materials covering emission wavelengths from the blue to the mid-infrared has motivated the replacement of many types of gas and solid state lasers by these compact and efficient devices in many applications.

The early nineties witnessed the maturing of yet another important diode laser technology, namely that utilizing quantum well heterostructures. Diode lasers incorporating quantum well active regions, particularly strained structures, made possible still higher efficiencies and further reduction in threshold currents. Quantum well diode lasers operating with sub-mA threshold currents have been demonstrated in many laboratories. Better understanding of the gain mechanisms in these lasers has also made possible their application in lasers with multi-GHz modulations speeds. Vertical cavity surface emitting lasers, utilizing a cavity configuration totally different than the traditional cleaved cavity, compatible with wafer-level production and high coupling efficiency with single mode optical fibers, have progressed significantly owing to continuous refinements in epitaxial technologies. Advances in cavity schemes for frequency control and linewidth reduction have yielded lasers with extremely low (kHz) linewidths and wide tuning ranges. Many of these recent developments have been driven by the information revolution we are experiencing. A major role in this revolution is likely to be played by dense arrays of high speed, low power diode lasers serving as light sources in computer data links and other mass-information transmission systems. Tunable diode lasers are developed mainly for use in wavelength division multiplexing communication systems in local area networks.

In spite of being a well established commercial device already used in many applications, the diode laser is still a subject for intensive research and development efforts in many laboratories. The development efforts are driven by the need to improve almost all characteristics of these devices in order to make them useful in new applications. The more basic research activities are also driven by the desire to better understand the fundamental mechanisms of lasing in semiconductors and by attempts to seek the ultimate limits of laser operation. An important current topic concerns the control of photon and carrier states and their interaction using micro- and nano-structures such as microcavities, photonic bandgap crystals, quantum wires, and quantum dots. Laser structures incorporating such novel cavity and heterostructure configurations are expected to show improved noise and high speed modulation properties and higher efficiency. Novel diode laser structures based on intersubband quantum-cascade transitions are explored for achieving efficient lasing in the mid infrared range. And new III nitride compounds are developed for extending the emission wavelength range to the blue and ultraviolet regime.

The increasing importance of semiconductor lasers as useful, mature device technology and, at the same time, the vitality of the research field related to these devices, make an up-to-date summary of their science and technology highly desirable. The purpose of this volume, and its companion volume *Semiconductor Lasers: Fundamentals*, is to bring such a summary to the broad audience of students, teachers, engineers, and researchers working with or on semiconductor lasers.

While the companion volume treats the fundamentals of diode lasers, the present one concentrates on several important issues related to their material systems and structures. In particular, it discusses several material systems employed to fabricate diode lasers emitting over a wide range of the electromagnetic spectrum, structures designed for controlling the emission wavelength and linewidth or for generating very high output powers, and vertical cavity lasers optimized for large array and extremely low power operation. Both volumes are organized in a way that facilitates the introduction of readers without a background in semiconductor lasers to this field. This is attempted by devoting the first section (or sections) in each chapter to a basic introduction to one of the aspects of the physics and technology of these devices. Subsequent sections deal with details of the topics under consideration.

We begin in Chapter 1, by G. A. Acket, A. Valster, and C. J. van der Poel, by discussing diode lasers emitting in the visible wavelength range. Several lattice matched and strained semiconductor systems used in such lasers are described, including red-emitting AlGaInP compounds and blue-emitting II–VI compounds. Emphasis is on materials and design issues related to room-temperature, reliable operation of such devices. Recent developments in III–Nitride compounds for blue and ultraviolet emission are also briefly discussed.

Chapter 2, by P. G. Eliseev, introduces compound semiconductors employed in long wavelength ($>2\mu$m) diode lasers. It discusses the lasing mechanisms in more traditional, narrow bandgap semiconductors utilized for this purpose, as well as more recent developments in unipolar quantum cascade lasers emitting in the mid-infrared region. Device issues relevant to applications of such lasers in infrared spectroscopy applications are mentioned.

The mechanisms and most important cavity structures used for controlling the diode laser emission wavelength and linewidth are described in Chapter 3, by Markus Christian Amann. InGaAsP/InP lasers emitting in the 1.3- to 1.55-μm wavelength range are reviewed here because of the importance of these issues in optical communication applications. Several cavity configurations designed for stable, single longitudinal mode operation and more complex ones allowing wide range wavelength tuning are analyzed, and tradeoffs between tunability and wavelength selectivity are examined.

Chapter 4, by David G. Mehuys, deals with high power diode lasers. The main design parameters important for achieving high power and high efficiency, reliable operation are introduced. Several device structures designed for high power (several 100mW), spatially coherent emission are discussed. Integrated laser-amplifier structures useful for increasing such output power to several watts are then described and analyzed. Very large array structures capable of emitting in the kilowatt range under pulsed operation are finally mentioned.

We conclude with Chapter 5, by Kenichi Iga, which discusses surface emitting diode lasers. Several laser cavity designs, different from the traditional edge emitting lasers treated in most other chapters, are described, particularly vertical cavity surface emitting lasers. The development of such vertical cavity lasers using different material systems for the active regions and various multilayer Bragg reflector designs is outlined. The current level of device performance and prospects for further future progress are assessed, with particular consideration of applications in microoptics array applications.

While it is difficult to include all aspects of this very broad field in two volumes, we have attempted to include contributions by experienced persons in this area that cover the most important basic and practical facets of these fascinating devices. We hope that the readers will find this book useful.

References

Hall, R. N., Fenner, G. E., Kingsley, J. D., Soltys, T. J., and Carlson, R. O. (1962). *Phys. Rev. Lett.*, **9**, 366.

Holonyak, N. Jr., and Bevacqua, S. F. (1962). *Appl. Phys. Lett.*, **1**, 82.

Nathan, M. I., Dumke, W. P., Burns, G., Dill, F. H. Jr., and Lasher, G. (1962). *Appl. Phys. Lett.*, **1**, 62.

Quist, T. M., Rediker, R. II, Keyes, R. J., Krag, W. E., Lax, B., McWhorter, A. L., and Zeigler, H. J. (1962). *Appl. Phys. Lett.*, **1**, 91.

Chapter 1

Visible-Wavelength Laser Diodes

G. A. Acket, A. Valster and C. J. van der Poel

Philips Optoelectronics, Prof. Holstlaan 4, 5656 AA Eindhoven, The Netherlands

1.1 Introduction

1.1.1 Relevance of visible wavelength lasers

AlGaAs, heterojunction laser diodes, and lead salt laser diodes (emitting at wavelengths of about 0.8 μm and 4–5 μm respectively) were developed at about the same time in the early 1970s. Thereafter, developments were directed toward wavelengths of 1–2 μm using materials like InGaAsP/InP and AlGaAsSb/GaSb; these developments were stimulated by the rapid progress in silica optical fiber technology, which started in the late 1970s. At shorter wavelengths, the first visible red InGaAlP laser diodes appeared by the mid-1980s and have shown a marked progress since then, but this field has not yet reached the maturity of the AlGaAs and InGaAsP. Most recently, developments into blue–green laser diodes have shown considerable progress. This progress has taken place mainly in the MgZnSSe system, but the GaAlInN system is coming up strongly. In view of all these developments, a review of the present status of visible laser

1

diodes seems appropriate at present. Besides that, it should be noted that in recent years quite some effort has been devoted to obtain blue laser radiation by frequency doubling of AlGaAs lasers in the 850 nm wavelength range by using optically nonlinear materials such as $KNbO_3$, $LiNbO_3$, or KTP. The latter subject will not be treated in detail here. In the present chapter, the emphasis will be on the InGaAlP and the II–VI, systems, but a good measure of the state-of-the-art in frequency doubled devices can be found in Jongerius (1995).

There is a strong interest from industry in visible semiconductor lasers. Of course, the availability of laser diodes at all colors would stimulate dreams such as laser diode TV displays (Glenn, 1993), but many applications are already present today. This is evidently the case for red laser diodes. They are replacing HeNe lasers in bar-code scanning and optical printing. In both applications, rotating polygon mirrors are frequently used for scanning of the beam. In printing, the current modulated laser diode replaces the combination of a HeNe laser with an external modulator (electro-optic or acoustic-optic). The first visible AlGaInP lasers emitted light around 670 nm, a wavelength where most red laser diodes still operate. However, around 630 nm, the sensitivity of the human eye is about a factor of seven greater than that near 670 nm. This improved visibility may also be an important safety aspect, for instance for use in hand-held bar-code scanners or in optical levellers. Fig. 1.1 shows the variation of eye sensitivity with wavelength, demonstrating the rapid decrease in eye sensitivity toward 700 nm.

Another application in which much use is made of shorter-wavelength diode lasers is optical recording. The principles of this technology have been described rather fully in the book by Bouwhuis *et al.* (1985). A more recent monograph by Marchant (1990) describes various types of practical systems developed up to 1990. The resolution the focused optical spot achieves on the disc surface is nearly diffraction limited and thus corresponds to $\lambda/2NA$ where λ is the free-space wavelength and NA the numerical aperture of the focusing objective. Because in practical systems NA = 0.50, the area of written or detected mark roughly equals λ^2. The emerging need for future storage of digital TV and HDTV signals on optical discs gives a strong impetus to high-density optical recording and hence shorter wavelengths for the laser diodes. In fact, the new Digital Video Disc (DVD) standard is based on the use of red laser diodes emitting in the 640–650 nm wavelength range.

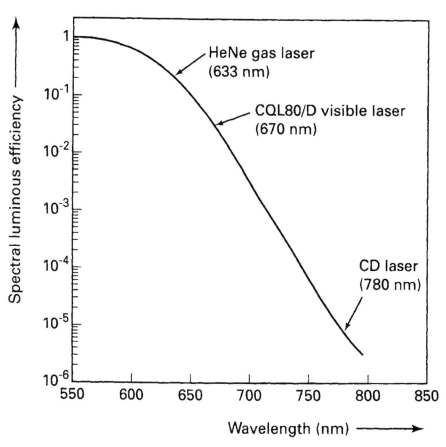

Figure 1.1: Variation of eye sensitivity with wavelength.

1.1.2 Thermal aspects; the temperature-dependence of threshold current

Of great importance with respect to laser diodes at visible wavelengths is the maximum temperature, T_{max}, at which c.w. operation can still be achieved. T_{max} is determined by the heat resistance R_{th} and the dissipation on the one hand and by the temperature dependence of threshold current J_{th} on the other. Early work by Gooch (1969) yielded T_{max} in terms of these quantities (at that time the lasers considered were still simple homojunction GaAs lasers, but this has no effect on the relevance of the argument). If the heat sink temperature equals T and the junction temperature is higher by an amount ΔT, then

$$\Delta T = R_{th} \{V_j + I_{th}(T + \Delta T) \cdot R_s\} I_{th}(T + \Delta T) \tag{1.1}$$

where V_j is the junction voltage and R_s the series resistance. Gooch used a T^3 variation for the temperature dependence of the threshold current $I_{th}(T)$ instead of the now more commonly used $\exp(T/T_0)$. He presented criteria for c.w. operation that read, when translated using $I_{th} \propto \exp(T/T_0)$:

$$I_{th}V_jR_{th}/T_0 \leq 1/e \qquad (e = 2.716\ldots) \tag{1.2A}$$

for the case $I_{th}R_s \ll V_j$ and

$$I_{th}{}^2R_sR_{th}/T_0 \leq 1/2e \tag{1.2B}$$

for the case $I_{th}R_s \gg V_j$.

Fortunately, most present-day laser diodes are in the category of eq. (1.2A), where the voltage drop across the series resistance is at most 20% of the junction voltage.

Equation (1.2) shows the strong need to reduce I_{th} as much as possible, especially if at shorter wavelengths T_0 is reduced (which in itself always leads to an additional increase of I_{th}). R_{th} should be also reduced as much as possible. That can be done by means of soldering the laser chips to well-conducting heat sinks, preferably with the epitaxial layer downside.

The characteristic temperature T_0 is not a real physical quantity but rather a phenomenological parameter. For AlGaAs lasers, $T_0 = 150$ K ; for 670 nm InGaAlP lasers $T_0 = 110$ K (Valster *et al.*, 1990b, and Hagen *et al.*, 1990), but for 630 nm $T_0 = 50$–60 K. In InGaAsP/InP lasers, $T_0 = 60$ K, which has led to the development of strongly index-guided lasers in order to reduce I_{th}, in agreement with eq. (1.2A). Generally, if within a given material system the wavelength is decreased, T_0 also decreases markedly. In AlGaAs this decrease takes place at emission wavelengths below 750 nm (Yamamoto *et al.*, 1983); in InGaAlP below 650 nm (see Sec. 5.). As will be discussed below, one of the causes of this is the electron leakage out of the active layer into the *p*-type cladding layer.

Generally, one makes use of the quantum shift of the confinement energies of the electrons and holes in order to shorten the wavelength of the laser. In quantum-well lasers (see the contribution by Zhao and Yariv, Chapter 1 in Volume 1), the quantum wells are usually surrounded by barrier material having a lower band gap than that of the cladding layers and hence a higher refractive index, so that a waveguide surrounding the quantum wells results. Such a configuration, which is very suitable for achieving low threshold current densities, may lead to some increase of the temperature dependence of threshold due to the thermal leakage of electrons out of the quantum wells into the barrier material followed by nonradiative recombination in the barrier material (see Blood *et al.*, 1988, and Bour *et al.*, 1989). In conventional lasers, where the wavelength is shortened by adjusting the composition of the active layer, the indirect and edge minima (*e.g.* the X minima) will come energetically so close to the Γ minimum that thermal excitation followed by nonradiative recombination from the higher minima becomes an important competing process. In the case of electron leakage toward the p-type cladding layer, recombination may also take place via the indirect minima there.

In the aforementioned case, the temperature dependence of the threshold current can be expressed as follows:

$$I_{\text{th}}(T) = I_{\text{th }0}(T) + f(T)\exp(- \Delta E/kT), \qquad (1.3)$$

where $I_{\text{th }0}(T)$ describes the temperature dependence of the threshold in the absence of carrier leakage, $f(T)$ involves the densities of states of the band edges involved, k is Boltzmann's constant, T is the absolute temperature, and ΔE is the thermal energy difference to be surmounted by the electron in order to escape, either to the barrier, to the higher energy minima, or to the conduction band of the p-type cladding material. For the case of electron leakage toward the p-type cladding of a normal DH laser ΔE can be expressed in terms of the band gap difference and the positions of the electron and hole quasi-Fermi levels. It is found that (Hagen *et al.*, 1990):

$$\Delta E = \Delta E_{\text{g}} - E_{\text{fe act}} + E_{\text{fh act}} - E_{\text{fh clad}} \qquad (1.4)$$

where ΔE_{g} equals the band gap difference between the active layer and the cladding layers and E_{fe} and E_{fh} stand for the electron and hole quasi-Fermi levels. Generally, when the wavelength is reduced, either by increasing the band gap of the active layer or by the use of thinner quantum wells, ΔE_{g} as well as ΔE decrease, which leads to an increase of the leakage contribution. Therefore, ΔE_{g} should be made as high as possible to achieve short wavelength operation. It can also be seen from eq. (4) that it is advantageous to use laser chips that are longer than normal, since this decreases the value of $E_{\text{fe act}}$ required to overcome the mirror losses. In eq. (3), the temperature dependence of I_{th0} can be described by $I_{\text{th0}}(T) \propto$

$\exp(T/T_0)$, but the total threshold current according to eq. (3) cannot. This leads to a strong apparent decrease in T_0 toward higher temperatures. As a consequence, at higher temperatures, the numerator of eq. (2) increases whereas the denominator decreases, which may lead to a rather fast transition to thermal runaway. From the above it follows that appropriate heatsinking, increase of the band gap difference between the active layer and the cladding layers, and the use of long cavities help to raise the maximum operating temperature at a given wavelength or to lower the minimum wavelength at a given temperature. It can be seen from eq. (4) that it is advantageous to make the doping level in the p-type cladding layer as high as possible because this reduces the value of $E_{\text{fh clad}}$ and thus leads to an increase of ΔE.

Several binary, ternary, and quaternary compounds have been mentioned up to now. Some of their properties have been presented in Casey and Parnish (1978), where special attention has also been given to the composition ranges where the band gap is direct so that laser operation is possible.

1.1.3   Beam properties, astigmatism

In many applications, for example optical recording, the near field of the laser is focused to an almost diffraction-limited spot. Many types of laser diodes have some astigmatism in the emitted beam. This is due to the presence of an optical loss at both sides of the laser stripe combined with the fact that the injected charge carriers in the stripe region contribute negatively to the refractive index (Thompson, 1980). The magnitude of the astigmatic distance, D, depends on the wavelength; the radius of curvature of the wavefront, R; and the width of the nearfield at the laser mirror, w. The relation between these quantities can be derived from the formulae presented by Cook and Nash (1975) for Gaussian beam propagation and is given by

$$D = R \left[1 + (\lambda R / \pi w^2)^2 \right]^{-1} \tag{1.5}$$

For gain-guided lasers, the dimensionless quantity $\lambda R / \pi w^2$ is comparable to or smaller than unity; for index-guided lasers, it is much larger than unity. R is usually of the order of 20–80 microns. It can be inferred from eq. (5) that for gain-guided lasers, $D \approx R$, whereas for index-guided lasers, $D \ll R$. The question of which values of D are tolerable depends strongly on the numerical aperture used (Born and Wolf, 1975) and hence on the type of application.

1.1.4   High-power operation

In applications such as optical recording, high-power lasers should operate in a stable lowest-order waveguide mode, preferably with only a small astigmatism. Lateral refractive index differences are usually of the order of 10^{-3} or higher, which limits the width of the near field to values below 5 microns. This limits practical power levels to values of about 100 mW. This type of laser power is referred to here as diffraction-limited laser diode power.

The use of quantum wells in AlGaAs and AlInGaP lasers has reduced the threshold current densities to values of the order of 200 A/cm^2, so that wide-stripe laser diodes can also easily operate c.w. as can be seen from eq. (1.2A). Using broad-stripe lasers with a width of approximately 50 microns, c.w. powers of the order of 1W can be generated. However, for such broad stripes the lateral waveguide becomes multimoded, and the resulting beam will generally consist of a mixture of higher-order modes; the beam will no longer be diffraction limited.

A somewhat similar situation arises if a phase-coupled array is used instead of a wide-stripe laser. This also leads to an effective enlargement of the emitter facet area, but it is difficult to ensure operation in the lowest-order lateral supermode (Kapon *et al.*, 1984), even though various schemes have been developed to favor lowest-order supermode operation (Streifer *et al*, 1986). The use of phase-coupled arrays has also led to the generation of total c.w. output powers in the order of several Watts. The applications determine whether a diffraction-limited zero-order mode is required. Even higher powers can be obtained for arrays without phase coupling because the spacing between the individual elements in the array can be greater. Powers as high as 100 W quasi-c.w. can thus be obtained (Serreze and Harding, 1992).

1.1.5   Organization of this Chapter

The remainder of this chapter is organized as follows. In Sect. 1.2, the materials systems relevant to short-wavelength lasers will be described.

Here the focus will be mainly on essential parameters such as band gap versus composition, band offset, and refractive index. The actual preparation and in particular the epitaxial growth processes will be described for these material systems in Sect. 1.3. In Sect. 1.4, an illustrative discussion of the importance of quantum wells and strain will be presented, which should be complementary to the more rigorous discussions presented elsewhere in this volume by Zhao and Yariv (Chapter 1 in Volume 1) on quantum wells and by Adams and O'Reilly on strain (Chapter 2 in Volume 1). Section 1.5 describes the present status of InGaAlP red visible diode lasers. Section 1.6 presents recent developments in the area of blue-green and blue semiconductor lasers, whereas Sect. 1.7 ends the chapter with conclusions and a glimpse into the future.

1.2   Material systems for visible lasers

1.2.1   Introduction

In this section we will describe the material systems that are suitable for use in visible diode lasers. Two of these systems, namely the InGaAlP system and the MgZnSSe system, are already in full use. The AlGaInN system is at present already used for blue LEDs, and its use for blue lasers has just started. The fourth system is the chalcopyrites; relatively little work has been done up to now, but this system will be mentioned briefly here for the sake of completeness.

1.2.2   The InGaAlP material system

This system was first discussed in 1978 (Casey and Panish, 1978, Part B, Section 5.3). The ternary $In_yGa_{(1-y)}P$ system has a direct energy gap over most of its composition range. The crossover with the X-minima lies at about $y = 0.70$. Because of the appreciable difference in lattice constants between GaP ($a=5.451$ Å) and InP ($a = 5.869$ Å), the lattice constant varies strongly with composition. At $y = 0.48$ the material is lattice-matched to GaAs ($a=5.653$ Å), which is therefore the preferred substrate. In the case of these covalent semiconductors, the substitution of Ga by Al produces hardly any change of the lattice constant, as is also observed with the AlGaAs system. Hence, with the indium constant fixed at 0.48, the Ga may partially be replaced by Al to provide various band gaps while maintaining a fixed lattice constant that matches the GaAs substrate. If we now denote the fraction of Ga replaced by Al by x, the system used has the composition $In_{0.48}(Ga_{(1-x)}Al_x)_{0.52}P$. For aluminum fractions x up to about 65% (*i.e.*, an aluminum content of about 32.5%), the energy gap is direct; the crossover with the X-minima takes place at $x = 0.65$. Details about the band gap versus composition in this system will be presented below.

Like most other III–V semiconductor materials, InGaAlP crystallizes into a cubic zinc blend structure. However, an important property of the InGaP and InGaAlP materials system is the presence of a strong tendency to order the In atoms on the one hand and the Ga and Al atoms on the other on alternate (111)B planes, resulting in a CuPt-like structure. The ordering has important influences on the electro-optical properties of the material namely:

1. An anisotropy is present for example in the optical gain.

2. The bandgap of the (partially) ordered phase is lower compared to that of the disordered phase by about 90 meV.

The degree of ordering depends strongly on the epitaxial growth conditions and the substrate orientation (more details will be presented in Sect. 1.3). It has been extensively studied and discussed by Suzuki and Gomyo (1993).

The quantities that are most relevant for the design of semiconductor lasers are the bandgap, the band offsets, and the refractive index. The direct and indirect energy gaps have been determined by Liedenbaum and Valster, (unpublished) from photoluminescence measurements for a number of compositions, both for ordered and for disordered material. The results are shown in Fig. 1.2a. Both the direct transitions at lower aluminum content as well as the indirect transitions at higher aluminum contents are observed. The aforementioned shift of the direct transition toward higher photon energy for disordered material is clearly seen. Of special interest is the fact that also the indirect transition increases toward higher photon energy by the same amount. This indicates that the ordering/disordering mainly affects the valence band, which is understandable, because the ordering reduces the symmetry just as biaxial strain does, and this is known to lead to a splitting between the light-

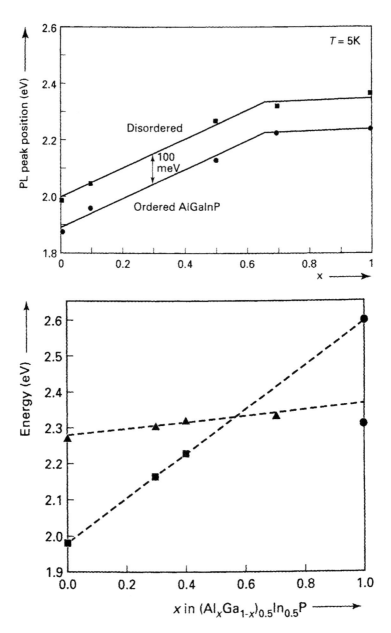

Figure 1.2: (a) Bandgap of InGaAlP lattice matched to GaAs as a function of the Al content as determined from PL data (5K) for ordered and disordered material. The positions of both the direct as well as the indirect conduction band minima are apparent from the measurements. (b) Similar data for disordered material from PL measurements (2K) after Meney *et al.* (1995). © 1995 IEEE.

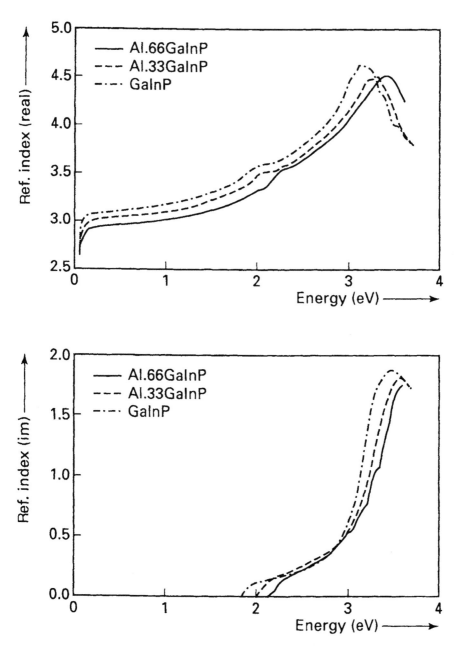

Figure 1.3: Refractive index as a function of photon energy for ternary InGaP and quaternary InGaAlP after Moser *et al.*, (1994). © 1994 American Institute of Physics.

hole and heavy-hole valence subbands. Prins *et al.* (1994a) and Meney *et al.*, (1995) also deduced the energy band structure from photoluminescence measurements for disordered material of various compositions. The latter results are displayed in Fig. 1.2b. Although these results do not agree exactly with those of Fig. 1.2a (notably the crossover is found at a somewhat lower Al content), the data agree sufficiently to state that by now the energy states of the InGaAlP system lattice-matched to GaAs have been well established. The band offset has been determined by Liedenbaum *et al.* (1990) by means of Photo Luminescence Excitation Spectroscopy (PLE) and has been confirmed by Dawson and Duggan (1993). Prins *et al.* (1994b) determined the band offset in InGaAlP/InGaP heterostructures from photoluminescence measurements at high pressures.

Moser *et al.* (1994) have reported the refractive indices of the InGaAlP system lattice matched to GaAs for different compositions. The results are presented in Fig. 1.3. The authors observed that, leaving aside the shift of the bandgap, the occurrence of ordering/disordering does not affect the magnitude of the refractive index appreciably. They also present a theoretical fit for their experimental refractive index data with various band gaps as the essential parameters.

1.2.3 The MgZnSSe material system

The ZnSe/ZnS binary system and the ternary ZnSSe have for many years been considered as possible materials for blue lasers. Optically pumped laser action and electron beam pumped laser action were demonstrated long ago. Practical diode lasers were for a long time not possible because it was impossible to dope the material *p*-type sufficiently. If we first consider the optical properties of the ternary ZnSSe, the material has a direct band gap over the whole composition range, but the lattice constant varies considerably (by about 4%) with S content. If the sulphur content is not too high, the cubic sphalerite or zinc blend structure is the more stable one. The lattice constant a for ZnSe equals 5.669 Å, *i.e.* 0.27% greater than that of GaAs, which makes GaAs a logical and in fact preferred choice as the substrate. That of ZnS equals about 5.409 Å. This implies that an exact lattice match of ZnSSe onto GaAs can be achieved at a sulphur content of about 6%. The corresponding band gap values are 2.77 eV for ZnSe and 3.77 eV for ZnS, and therefore the band gap difference between ZnSe and ZnSSe matched to GaAs is only about 30–40 meV. A plot of band gaps and lattice constants of some important II–VI semiconductors is presented in Fig. 1.4.

The heterojunction system considered in the past was, not surprisingly, the ZnSe/ZnSSe matched to GaAs. It turns out that the properties to be expected from this system are unfavorable. As already mentioned, the band gap difference between the binary and the ternary is only relatively small. In addition, according to the common-cation rule (which states that for the more ionic semiconductors, the valence band should be mainly related to the states of the anion and the conduction band states should be more related to the cation) predicts that this band gap difference should be mainly in the valence band, leaving a step in the conduction band of only 5 meV or the like. Since the electrons are expected to be the more "leaky " carrier because of their lighter mass and higher mobility, a very high thermal leakage current is expected from such a structure even though the electrostatic forces between the confined holes and the electrons will try to prevent them from escaping. Another disad-

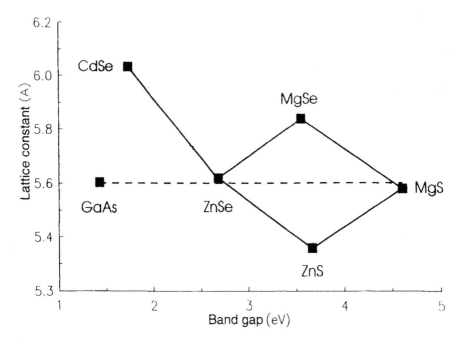

Figure 1.4: Plot of band gaps and lattice constants of some important II–VI semiconductors after Gaines (1995). © 1995 *Philips J. of Research.*

vantage of this structure is that the ZnSe active layer is not matched to the other parts of the structure.

In this structure, the problem of the thermal escape of charge carriers into the cladding can be alleviated by using a longer lasing wavelength. This is usually accomplished by inserting a thin ZnCdSe active layer, with a thickness of about 65 Å within the ZnSe intermediate layer. CdSe has a considerably lower bandgap than ZnSe and a considerably larger lattice constant (about 1.70 eV and 6.04 Å, respectively) (Gaines, 1995). Taking a Cd content of about 20% yields an emission wavelength of about 490 nm at liquid nitrogen temperature. In fact, this value is also somewhat influenced by the compressive strain in the active layer and some quantum shift due to the small thickness (see Sect. 1.4).

Another material that is vital for the design of II–VI lasers, especially for use as a wider-gap cladding material, is MgZnSSe. It is known that the addition of magnesium increases both the band gap and the lattice constant. This can be combined with the tendency of sulphur to increase the band gap and decrease the lattice constant to realize quaternary compounds with a band gap up to about 3.5 eV, lattice-matched either to GaAs or the ZnSe itself. The growth and doping properties of this material will be discussed in Sect. 1.3.

Finally, an important parameter for the design of semiconductor lasers is the refractive index and notably difference of the refractive index. Due to the higher ionicity, the refractive index is generally lower than that of the commonly used III–V semiconductors. Some typical data as a function of wavelength are presented in Fig. 1.5 (Gaines and Petruzello, unpublished). It can be seen that in the relevant wavelength range of the CdZnSe emission (490 nm), the refractive index of ZnSe is around 2.80 and that of the quaternary cladding material about 2.63. In addition, it can be seen that the refractive index difference between ZnSSe lattice-matched to GaAs and ZnSe is rather small, so that the aforementioned ZnSSe/ZnSe/ZnSSe waveguide is, in fact, rather weak (for more details, see also Gaines *et al.* [1993]), the more so since the layer thickness of the ZnSe layer is limited to values of the order of 10 nm or lower due to the lattice mismatch.

1.2.4   The AlInGaN system

The basic binaries of this general III–V nitride system, GaN and AlN, are direct semiconductors of very large band gaps of 3.40 eV and 6.2 eV,

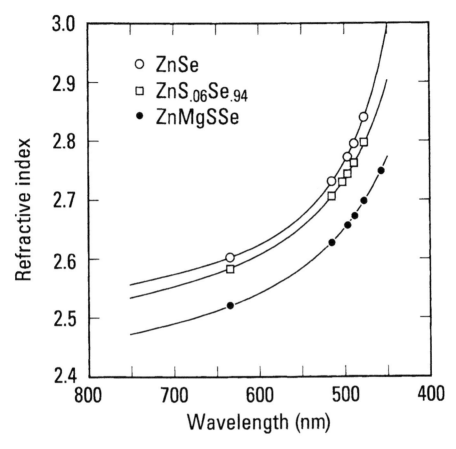

Figure 1.5: Refractive index data of some selected II–VI semiconductors (after Gaines and Petruzello, unpublished).

respectively. The materials crystallize into either the cubic zinc blend structure or the thermodynamically stable hexagonal wurtzite structure. An excellent review covering the research done into the materials up to 1993 can be found in the paper by Strite *et al.* (1993). As already mentioned, the growth, doping, and other technological aspects will be discussed separately in Sect. 1.3, but it may already be mentioned here that, as with ZnSe, *p*-type doping remained impossible for quite some time. In addition, the lack of a well-fitting substrate hampered technological progress in the past. Unlike in the AlGaAs and InGaAlP systems, the incorporation of aluminum into the GaN influences the lattice constant appreciably. For the wurtzite structure a = 3.189 Å for GaN and for AlN 3.112 Å, which corresponds to a mismatch of about 2%. As the emission of GaN is in the UV around 365 nm, often InGaN is used in order to shift the emission into the visible. The addition of In results in a decrease of the band gap and an increase of the lattice constant, just as in the zinc blend III–Vs. The band gap of hexagonal InN is about 1.9 eV, and its lattice constant is about 3.548 Å. Hence, the total wavelength span of the AlGaInN system ranges from 360 nm up to about 630 nm.

Because of the high ionicity of the nitrides, their refractive indexes are even smaller than those of the sulpho-selenides, namely about 2.3 at a wavelength of about 400 nm. Here too, appreciable refractive index differences can be realized by changing the band gap by adjusting the composition, as a result of which the realization of waveguiding as required for the design of semiconductor lasers will be no problem. Etching technology will also be discussed in Sect. 3. Device results, as far as they are available, will be presented in Sect. 4.

A favorable property of the nitrides for the future c.w. operation of laser diodes is their relatively high thermal conductivities. The thermal conductivity, expressed in W/cm.K, are, for GaN, 1.3, and for AlN, 2.0, compared to the values for GaAs (0.5) and Si (1.5). The thermal conductivities of the substrates used in GaN technology (see Sect. 1.3) are sapphire, 3.5, and 6H-SiC, 4.9.

1.2.5   The chalcopyrites

The chalcopyrites are interesting because they are related to the II–VI compounds, but to some extent they also resemble the III–V compounds. Essentially, they are I–III–VI$_2$ semiconductors, where the I is from the first B column of the periodic system, the III from the third column as with the III–Vs and the VIs from the sixth column. For the I-element Cu has been chosen; the group III element is Ga, partially substituted by aluminum; and the VIs are a mixture of sulphur and selenium. Double heterostructures of the Cu(Al, Ga)SSe system can be grown on GaP substrates and possibly also on Si substrates. Details of the system can be found in Shay and Wernick (1975). The band gap of CuAl$_x$Ga$_{1-x}$Se$_2$ is direct over the whole composition range and varies from 1.63 eV up to 2.52 eV. The chalcopyrite structure is tetragonal, but it can be regarded as two zinc blend unit cells stacked on top of each other in the c-direction. LEDs have been realised, and optically pumped laser action has been observed on single-layer structures. Some high-quality growths by MOVPE have been reported (Hara *et al.*, 1991). As the progress with the II–VI and the nitrides advances, the work on this system is expected to diminish because the wavelength range to be expected from this system will not extend below 500 nm.

1.3 Epitaxial growth and technology

1.3.1 The AlGaInP system

OMVPE is at present the dominant growth technology for the InGaAlP system. In the past, successful growths have been reported of InGaP by LPE and conventional MBE, but the growth of the quaternaries by these technologies is complicated due to the high segregation constant of aluminum and the high vapor pressure of phosphorus, respectively. OMVPE of InGaAlP has been described, *e.g.* by Valster *et al.* (1990b). Usually, trimethyl-indium (TMI), trimethyl-gallium (TMG), and trimethyl-aluminum (TMA), are used as the group III precursors and phosphine (PH_3) for the group V. Growth usually takes place at reduced pressures (around 50 millibar), at temperatures of around 650—750 °C. Due to the low sticking of the phosphorus at these high temperatures, very high V/III ratios of the order of 300 are required. For *n*-type doping of the material, Si is used from SiH_4 as the precursor; and generally for the *p*-type dopant, Zn is used. The precursor is dimethyl-zinc (DMZn). For the growth of *p*-type InAlP, magnesium acceptors using Cp_2Mg (bis-cyclopentadienyl-magnesium) are also used. In our laboratories we have grown InGaAlP both in a horizontal single-wafer reactor (AIXTRON AIX-200) and in a multiwafer reactor (AIXTRON AIX-2000) (Ambrosius, to be published) with rotating substrates (Frijlink, 1988). Typical growth rates are in the order of 1 μm/min. Photoluminescence is generally used to check the composition of the various layers; the lattice mismatch with respect to the substrates is usually determined with the help of double crystal X-ray diffractometry (Bartels, 1983). Usually GaAs substrates with an orientation of approximately 001 are used. If the material has to be disordered (see Sect. 2), many authors use misoriented substrates (see the following discussion).

Using disordered InGaAlP material for semiconductor lasers presents some advantages. First (see Sect. 4), lasers emitting at the shorter wavelengths are more easily obtained by using disordered material, because the emission wavelength is about 20 nm shorter than that of ordered material so that, especially for 630 nm-band lasers, the remaining shift toward shorter wavelengths can more easily be accomplished by the use of quantum wells. Secondly, the internal optical loss of disordered material is significantly lower compared to that of the ordered one (van der Poel *et al.*, 1992). A comparison of the optical losses deduced from η^{-1} versus L measurements is shown in Fig. 1.6. Furthermore, the gain is anisotropic in ordered material leading to a threshold current density that is anisotropic, the threshold current density of lasers having the stripes oriented in the $\langle -110 \rangle$ direction being lower by several tens of percents compared to that of lasers with stripes along the $\langle 110 \rangle$ (Forstmann *et al.*, 1994). Thus, lasers that are to be made from ordered material must always have their stripes oriented along the former direction.

Whether the material grown will be ordered or not depends essentially on the OMVPE growth conditions and the orientation of the substrate. Suzuki and Gomyo (1993) conducted a detailed study into the nature of the ordering and the effect of substrate misorientation with respect to the 001 plane. They found that the ordering starts within the layer due

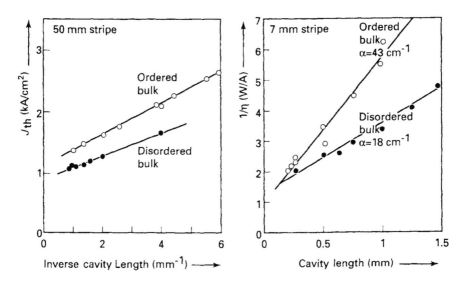

Figure 1.6: Plots of threshold current density versus reciprocal cavity length and reciprocal differential efficiency versus cavity length L for AlGaInP lasers made from ordered and disordered material.

to the large difference in size of the Ga and In atoms and is continued in the next plane holding a phase relation so that alternating Ga planes and In planes in the [111]B directions result. This mechanism was explained by the presence of B-steps together with the assumption of dimer formation of the P-atoms leading to a 2×4 surface reconstruction.

If index-guided structures are made from wafers with some degree of misorientation with respect to the 001 plane, care should be taken that the laser stripe be oriented approximately perpendicular to the direction of misorientation; otherwise the cleaved facet mirrors become strongly tilted with respect to the plane of the layers. A disadvantage of the use of substrate misorientation is that the orientation dependence of the wet-chemical etching processes used, *e.g.*, for ridge etching, will generally lead to a lateral asymmetry of the waveguide, which has an adverse effect on the electro-optical properties of the lasers, especially at high output powers.

1.3.2   The MgZnSSe system

1.3.2.1   Introduction

Some characteristics of the MgZnSSe-system have already been discussed in Sect. 2. As far as preparation and technological aspects are concerned, several comments may be made. P-type doping and especially p-doping to a high hole concentration has always been difficult, whereas n-type doping has never been a problem. The only p-dopants that have proven to be successful up to now are Li and atomic N. Li has only partially been successful in that hole concentrations only up into the high 10^{16} cm^{-3} level have been achieved, whereas n-doping has allowed nitrogen-concentrations that are more than an order of magnitude higher. The two technologies currently used in epitaxial growth at present are Molecular Beam Epitaxy (MBE) and Organo-Metallic Vapor Phase Epitaxy (OMVPE). These two technologies will be discussed in the following paragraphs. Another aspect, which always created difficulties in the past, is the achievement of low-resistance Ohmic contacts. Since n-type GaAs substrates are used in practice, the n-type contact is no problem at all; but it turns out that Ohmic contacts to the p-type materials, even to highly p-type doped ZnSe, are very difficult to achieve. In principle, there could also be a problem at the n-GaAs/n-ZnSe interface, but as the n-ZnSe can be doped high enough, no problem is encountered in practice there.

1.3.2.2   MBE

The MBE growth of II–VI materials has been discussed by A. O. Haase *et al.* (1991) and Gaines (1995). ZnSe, ZnSSe lattice matched to GaAs, ZnTe, ZnCdSe, and MgZnSSe can be successfully grown using this technique. Moreover, at present MBE is the only technique that is capable of producing high *p*-type doping, and hence it is the preferred technique for growing laser wafers. The doping is done by using atomic N produced in an N_2 plasma. Apart from this dopant, the sources for the materials are Zn, Mg, ZnS, and Se. *N*-type doping is usually effected with Cl, mostly by using $ZnCl_2$ as the source material. The active layer of II–VI lasers generally consists of CdZnSe having a Cd content of about 20%. Metallic cadmium can also be used as a source in the MBE-process. In the MBE technique, Reflected High Energy Electron Diffraction (RHEED) can be applied *in situ* and is an important tool for monitoring and controlling the growth. For MgZnSSe the *p*-type doping concentrations that can be achieved become smaller at higher Mg and S contents, thus practically limiting the useful range of energy gaps to values below 3.2 eV. The presence of the gas discharge for the nitrogen dissociation produces a gas inflow into the vacuum chamber so that pressures in the MBE reactor are higher than, for example, in the normal AlGaAs MBE. Moreover, the vapor pressure of the material limits the growth temperature to about 300 °C.

1.3.2.3   OMVPE

In principle, OMVPE presents the advantages over MBE that high vapor pressure elements such as sulfur can more easily be handled by OMVPE and that simultaneous growth on several substrates is also more readily achieved. In OMVPE normal precursors such as dimethyl zinc or diethyl zinc can be used as the Zn source. In principle, the hydrides (H_2S and H_2Se) can be used as the group VI sources. However, since prereactions between the hydrides and the zinc precursor may take place, often use is made of organic precursors for the group VI elements (see the review by Heuken, 1995). Examples are diethylselenide (DESe) and diisopropylselenide (DIPSe) or di-tertiary-butylselenide (DTBSe) for Se. In atmospheric OMVPE, higher-growth temperatures can in principle be used, even up to 600 °C. However, lower growth temperatures are preferred, because at lower temperatures the sticking of nitrogen, to be used as the acceptor, is higher. As in MBE, atomic N can be produced by means of plasma excitation of nitrogen gas. However, the high hole concentrations reported for MBE-grown II–VI materials have not yet been achieved, but further study may improve this situation. Compensation mechanisms related to N-H complexes or N-C complexes may be responsible for the higher degree of compensation observed in OMVPE material compared to that grown by MBE. For *n*-type doping, group VII elements such as Cl can be used.

1.3.2.4 Ohmic contacts to *p*-type ZnSe

It has been found that the work function of *p*-type ZnSe is larger than that of any known metal. However, there are other II–VI semiconductors that show much better properties. Examples are *p*-type CdTe and *p*-type ZnTe. Because of the common cation, the latter is preferred for making contact to *p*-type ZnSe. Graded layers or graded ZnTe-ZnSe superlattices built between the Ohmic contact metal (Au-Pt-Pd or the like) to be used and the *p*-type ZnSe top layer have been successful in reducing the voltage drop across the *p*-type contact. Early ZnSe-based lasers showed operation voltages of well over 20 V, whereas values of around 6 V or lower are generally reported (Nurmikko and Gunshor, 1994; Ishibashi, 1995; and Drenten *et al.*, 1995).

1.3.3 The AlInGaN system

1.3.3.1 OMVPE/MOMBE

The epitaxial growth of the group III nitrides made great progress during the late 1980s and early 1990s. The most suitable technology so far seems to be OMVPE. Most work concentrates on the hexagonal wurtzite structure because this structure appears to be the more stable one. Papers have been published by Nakamura *et al.*(1991a) describing a vertical reactor and Ito *et al.* (1991), who used a horizontal one. Sapphire was mostly used as the substrate, but growth of the hexagonal wurtzite material on (111)-oriented cubic spinel magnesium aluminate substrates has also been reported (Sun *et al.*, 1996). The precursors applied are usually trimethyl-gallium, trimethyl-aluminum, and NH_3. Mg is used for doping *p*-type material (Nakamura, 1991), the corresponding precursor being Cp_2Mg (*bis*-cyclopentadienyl-magnesium). Post-growth treatment by annealing or irradiation is essential in order to "activate" the Mg-acceptors and to obtain low-resistivity *p*-type material. Recent evidence (Neugebauer and Van de Walle, 1996) indicates that Mg-H complexes are formed and that the post-growth treatment serves to dissociate these complexes. N-type doping can be accomplished with the aid of Si or Ge using hydride sources. Growth is usually performed at very high temperatures (1000–1100 °C), but it has been found that for high-quality growth the use of a GaN or AlN buffer layer grown at a much lower temperature (about 500–600 °C) is essential. These papers were mainly related to the growth of GaN. During the years 1993 and 1994, the growth of AlGaN/ InGaN double heterostructures was reported (Amano *et al.*, 1994). Tri-methyl-indium is generally used as the In-precursor for growing InGaN. An essential feature of the vertical reactor described by Nakamura *et al.* (1991a) is that it has two different gas flows, the main flow that carries the reactant gases and enters the reactor parallel to the plane of the substrate and a subflow that carries only inactive gases ($N_2 + H_2$) and enters from the top of the reactor. Metalorganic molecular beam epitaxy (MOMBE), using triethyl-gallium, trimethyl-indium, and atomic nitrogen from an ECR source, has also been reported (Abernathy, 1993).

1.3.3.2 Processing and contacting

Wet-chemical etching of GaN and related compounds appears to be diffi-cult. Therefore, considerable effort has been devoted to the development of dry-etching processes, notably Reactive Ion Etching (RIE) and Electron Cyclotron Resonance Reactive Ion Etching (ECR-RIE). Usually chlorine or chlorides are used as the etching media (Adesida *et al.*, 1993, and Lin *et al.*, 1994). ECR-RIE at high temperatures of GaN, InN, and AlN using both $Cl_2/ H_2/ CH_4/ Ar$ and $Cl_2/ H_2 / Ar$ plasmas have been reported (Shul *et al.*, 1995). Cleaving GaN-based materials, especially when they are on a sapphire substrate, seems so far rather difficult. Therefore, the produc-tion of laser mirrors by dry etching may well become the favored process. Contacting of n-type and p-type GaN is found to be less difficult than expected. Metals like In-Au are often used (Strite *et al.*, 1993).

1.4 Quantum wells and strain in laser diodes

1.4.1 The effects of quantum wells

The aim of this section is to present an illustrative discussion of some aspects of quantum wells and strain in diode lasers, which will complement the more extensive contributions by Zhao and Yariv, Chapter 1 of Volume I and Adams and O'Reilly, Chapter 2 of Volume I.

A remarkable characteristic of present-day OMVPE and MBE growth techniques is the possibility of controling the layer thicknesses and compositions of III–V compounds with extremely high accuracy. Using the recent designs for OMVPE in rotating wafer systems (Frijlink, 1988), extremely homogeneous layer structures on large, 3-in. diameter wafers can be obtained enabling the reproducible fabrication of sophisticated semiconductor laser structures. A typical example illustrating the possibility of layer thickness control in OMVPE of visible laser structures is shown in Fig. 1.7. In the sample, on a GaAs substrate, $In_{0.48}Ga_{0.52}P$ layers (dark in the inset) of decreasing widths were grown between thick $(Al_{0.6}Ga_{0.4})_{0.52}In_{0.48}P$ cladding material. When laser structures are fabricated with the active layer thickness of less than 100

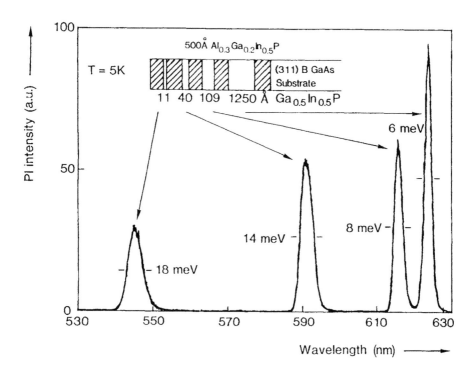

Figure 1.7: Photoluminescence spectrum of a structure containing InGaP quantum wells of various thicknesses between AlGaInP barriers. The dimensions are indicated in the inset. After Valster *et al.* (1991). ©1991 *Journal of Crystal Growth.*

A, quantum confinement effects become important (see the contribution
by Zhao/Yariv in Volume 1) and the effective band gap of the quantum
well (QW) material increases with decreasing QW thickness. Calculated
results (Valster *et al.*, 1991) are compared with experimental data in Fig.
1.8. The finite barrier height of the QWs and the lifting of the valence
band degeneracy by quantum effects have been taken into account. Also
indicated in the figure are confinement energies as obtained from low-
temperature Photo-Luminescence (PL) and Photo Luminescence Excitation (PLE) experiments. The agreement between the model calculation
and experiment is satisfactory and, in addition, enables the experimental
determination of the band offset Q (which is the fraction of the band
gap discontinuity that occurs in the conduction band) in InGaP/InGaAlP
heterojunctions to be $Q = 0.65 +/- 0.05$. From the photoluminescence
line widths and peak positions, it also follows that the interface abruptness
of the QW is less than two monolayers.

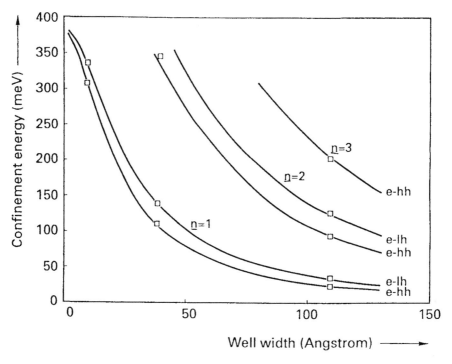

Figure 1.8: Confinement energies as a function of well thickness obtained
from photoluminescence and photoluminescence excitation spectra of the structure of Fig. 7. After Valster *et al.*(1991). ©1991 *Journal of Crystal Growth.*

In visible red semiconductor lasers, quantum confinement effects are employed to shorten the wavelength. When use is made of conventional OMVPE conditions, ordered phase InGaP active layers are grown, and the resulting wavelength for bulk active layers is around $\lambda = 675$ nm. One method for obtaining shorter lasing wavelengths involves the incorporation of AlInGaP compounds into the active layer. Although the higher band gap active layer material leads to shorter wavelengths, the quality of the material deteriorates with an increasing Al content, and the reliability of the lasers becomes unsatisfactory. With optimized OMVPE conditions, disordered phase InGaP active layers can be grown and the lasing wavelength is shifted down to about $\lambda = 655$ nm, without the use of AlInGaP in the active layer. Further wavelength reduction is accomplished by using quantum effects. In Fig. 1.9, we show the resulting wavelengths of three lasers (at about 100 K) in which the layer compositions are similar while only the width of the quantum wells is varied, namely from 50 Å (red), to 30 Å (yellow, $\lambda = 584$ nm), and 13 Å (green, $\lambda = 555$ nm). In these lasers, InGaP QWs have been centered in an $(Al_{0.5}Ga_{0.5})_{0.52}In_{0.48}P$ separate confinement structure between $(Al_{0.7}Ga_{0.3})_{0.52}In_{0.48}P$ cladding layers. Figure 1.9 is a striking and beautiful illustration of the classic particle-in-a-box (carriers in a quantum well) quantum mechanical problem, which is directly visible by the naked eye. Moreover, it is a direct demonstration of the state of the art in epitaxial growth techniques in this material system.

The results of Fig. 1.9 can only be realized at reduced temperatures (around 100 °K). At higher temperatures, thermally activated carrier leakage from the quantum well into the separate confinement and cladding layers becomes a major problem. Unfortunately, the $(Al_xGa_{1-x})_{0.52}In_{0.48}P$ system becomes an indirect band gap semiconductor at $x \geq 0.65$, and a further substantial increase of the band gap of the cladding layer is therefore impossible. Thus c.w. lasing can be achieved only up to 113 °K for the green laser and up to 263 °K for the yellow laser in the case of lattice matched quantum wells. However, lasing at 633 nm can already be achieved far above room temperature (see Sect. 1.5).

1.4.2 The effects of strain

The modification of the band structure of semiconductors by strain used in the active layer of diode lasers is quite complicated and requires extensive theoretical modelling (Adams and O'Reilly, Chapter 2 of this Volume 1). A simplified view that explains the main consequences of strained layers

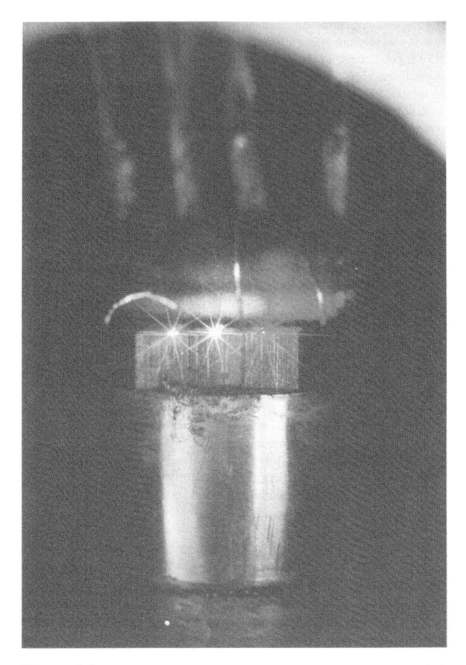

Figure 1.9: Laser emission from three lasers made from structures that differ only in quantum well thickness.

in diode lasers and gives the means to predict trends in laser performance when strain parameters are changed is obtained by using a simple tight binding approach for material with a cubic lattice. The model used is essentially too simple and completely ignores effects due to hybridization and spin-orbit coupling. Nevertheless, even in this simplified model, the effects of strain can be understood qualitatively. We will illustrate the predicted trends with experimental data for strained layer visble red laser structures. The most important orbitals for the III–V compounds considered are the p_x, p_y, p_z, and s orbitals. When interacting in a solid, the p_x, p_y, and p_z combine to give rise to the valence band and the combination of s-orbitals generates the conduction band. The configuration resulting for the valence band for the situation without built-in strain is depicted in Fig. 1.10b. The binding energy and the effective mass for holes moving in the valence band depends on the overlap of the p-wave functions from neighboring atoms. In directions in which the overlap is large, transport becomes easy and the resulting effective mass is low; when the overlap is small, the effective mass becomes large. In unstrained thick active layers, the resulting valence bands for Light Holes (LH) and Heavy Holes (HH) are degenerate at $k = 0$. As can be seen in Fig. 1.10b, we have labeled the bands in accordance with their character in the direction of growth, z.

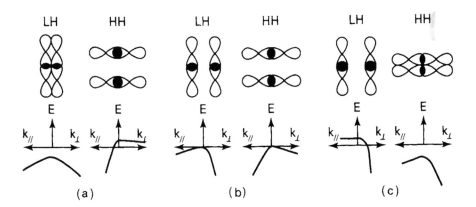

Figure 1.10: The effects of pressure on the p-orbitals corresponding to light - and heavy hole bands, respectively. Fig. 1.10a corresponds to the situation of compressive strain, Fig. 1.10b to the lattice matched situation and Fig. 1.10c to tensile strain.

It should be realized that, due to the critical thickness rule for the product of the strain ε (which is defined as $\varepsilon = \Delta a / a$) and the layer thickness of the strained layer, d is given by $\varepsilon d \leq 100\text{-}200\%$ Å, where ε is in % and d is in Å. This limit is imposed by the necessity of preventing the occurrence of misfit dislocations. Strained layers with a reasonable amount of strain always have thicknesses in the quantum well regime. It turns out that the effective mass perpendicular to the plane of the layers determines the quantum shift and hence the lasing wavelength, whereas the effective mass parallel to the plane of the layers determines the optical gain.

When strain is applied, the unit cell becomes elongated (compressive strain) or shortened (tensile strain) in the growth direction through the Poisson extension/contraction. The deformation of the strained layer materials used in diode lasers is practically elastic, and the volume of the unit cell remains almost identical to that of the unstrained material. The change in the lattice constant along the z-direction is thus about twice as large compared to that in the x-direction. Due to this deformation, the overlap of the p-orbitals is modified as shown in Figs. 1.10a and 1.10c. For compressive strain (as shown in Fig. 1.10a), the unit cell becomes elongated and the overlap between the LH p-orbitals becomes smaller. Near $k = 0$, the sign of the overlap is negative and is observed to become less negative due to compressive strain. For the same orbital, the overlap in the directions perpendicular to the z axis is positive and increases due to compression. Summed over both directions, the overlap increases, and the binding energy is enhanced. The opposite effect is observed for the HH band, and the binding energy becomes smaller. Therefore the deformation due to compressive strain induces an increase in binding energy of the LH band with respect to the HH band, and the degeneracy at $k = 0$ is lifted. The resulting band splitting is indicated in the figure. At the same time, the effective masses of the LH and HH bands change accordingly; for the HH band the parallel effective mass becomes lower than in the unstrained situation. Hence, the band structure is altered in two ways. First, the LH band is shifted downwards and will carry a smaller hole population. Second, the perpendicular effective masses of electrons in the conduction band and holes in the valence band approach more the ideal situation of being equal. As a result, improved laser characteristics can be expected. The situation for tensile strain is sketched in Fig. 1.10c. Here, the situation is somewhat more complex. Following the same lines of reasoning, it can now be concluded that the HH band gains binding energy and that the parallel effective mass of the LH band becomes larger than in the unstrained condition. We saw earlier that that the quantum wells alone tend to split the heavy holes and light holes. At small tensile strains the heavy and light hole subbands tend to coincide again, but more detailed calculations show that at higher tensile strain the light hole band is situated sufficiently above the heavy hole band and that for an appreciable range of k-vectors, the density of states becomes favorable again (*cf.* Chapter 2 by Adams and O'Reilly in this volume). Again, improved laser characteristics are expected at higher tensile strains, but at the smaller tensile strains around the band crossing a considerable increase of threshold current density is observed (see the following).

In PL experiments the shifts of the PL-peaks (both of the conduction band to LH and to HH transitions) can be observed over a certain range of strain values (see Fig. 1.11). Experimentally, the following values for the energy shifts have been experimentally determined

$$\Delta HH = 10.3\varepsilon \qquad \text{and} \qquad \Delta LH = 5.9\varepsilon \qquad (1.6)$$

where ε is the strain, which is found from the difference between the lattice constants of InP and GaP to be:

$$\varepsilon = 3.815. \ 10^{-2} - 7.394. \ 10^{-2} x \qquad (1.7)$$

The energy shifts indicated in eq.(1.6) are in electron-volts.

Lattice match, $\varepsilon = 0$, is obtained for the In-content $x = 0.48$ (see also Sect. 1.3).

It should be noted that with the possibility of introducing strained layers in diode lasers, the designer has now a new degree of freedom that can be used to design lasers over a wider wavelength range than was possible in the past.

Figure 1.12 shows the threshold current densities obtained for (a) wide stripe structures in the InGaP (630 – 650 nm both tensile and compressive), (b) the GaAsP tensile strained (GaAs substrate) and the compressively strained InGaAs (4.1), (c) and the InGaAs on InP substrate systems, both compressive and tensile, as a function of strain in one figure. Data are shown for three temperatures. The peak in the threshold for small tensile strain values is attributable to the aforementioned effects of band crossing of the LH and HH subbands.

Additional information can be obtained by measuring the polarization properties of the spontaneous radiation emitted from the edge of strained layer structures and lasers. An early report on the effect of strain on the

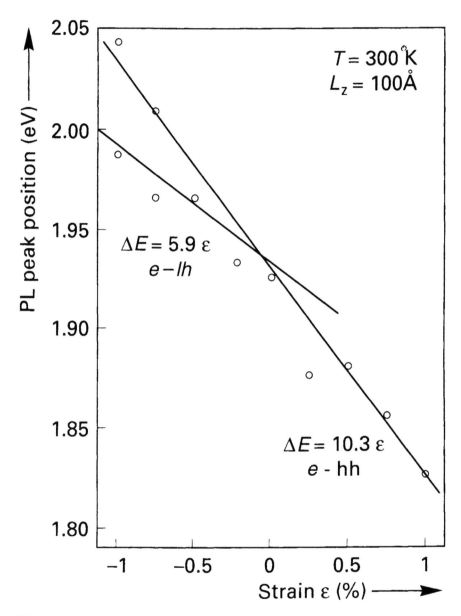

Figure 1.11: Positions of the photoluminescence peaks for InGaP quantum well structures with quantum wells with the same thickness but of different strain. The different values of the shift for the LH and HH-band edges with strain have been derived from this plot (see text).

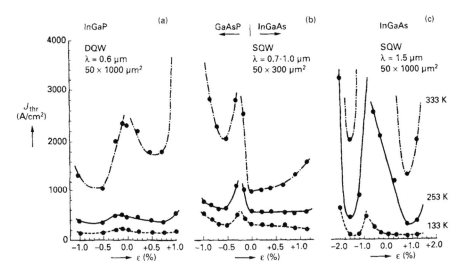

Figure 1.12: Variation of threshold current with strain for red visible lasers (InGaP) (a), near infrared (GaAsP for tensile and InGaAs for compressive strain) (b) and long-wavelength lasers (InGaAs/InP) (c).

polarization properties of InGaAlP lasers was given by Boermans *et al.* (1990). If conditions are such that holes only populate the HH subband, the radiation emitted will be completely TE-polarized, since the matrix element for the c.b.→HH TM transition is zero (Bastard, 1991). On the other hand, for complete population of the LH-subband, a dominance of the TM polarized radiation will be found. In fact, the lasers made from such material will lase TM because that polarization has the highest gain. However, when the spontaneous emission is analyzed, an appreciable TE component is found since the transition probability (which equals the matrix element squared) of the c.b.→ HH transition is not zero but equals one quarter of that for the TM polarization (Bastard, 1991). For experimental evaluation, a suitable method is to measure the so-called ρ parameter of the spontaneous radiation emitted from the side faces as a function of current and temperature. The ρ parameter is defined as

$$\rho = (TE - TM)/(TE + TM), \tag{1.8}$$

where TE and TM indicate the intensities of the TE and TM polarizations, repectively. Figure 1.13 shows the results of the measurement of ρ as a function of strain, both compressive and tensile, at different temperatures

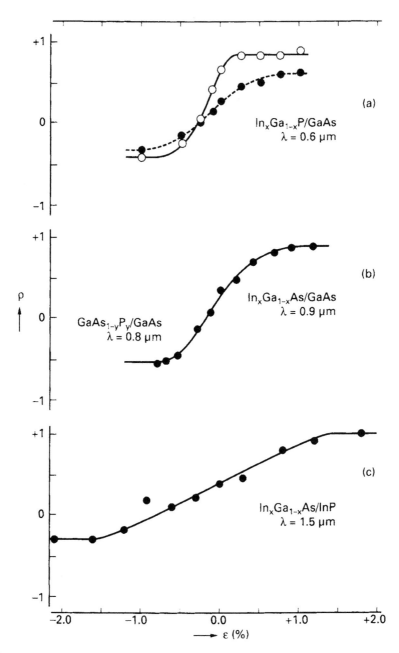

Figure 1.13: Measurement of the parameter ρ (see text for definition) as a function of strain for the lasers shown in Figure 1.12.

(room temperature for the closed circles and 70 K for the open ones), for
the three materials systems of Fig. 1.12. In theory, one would expect ρ to
become nearly equal to 1.0 for complete dominance of the HH subband,
whereas -0.60 would be expected for LH dominance. The behavior clearly
illustrates the dominance of the HH band for compressive strain and of
the LH band for (high enough) tensile strains. Information about the
relative populations of both subbands can be obtained from the variation
at intermediate strain levels. Since the $\rho(\varepsilon)$ curve appears to be universal
for the various material systems at room temperature, a measurement
of ρ could in principle be used to determine the strain ε experimentally.

As expected, at lower temperatures the transition dominance of one
of the bands takes place more abruptly.

1.5   InGaAlP laser structures

As with AlGaAs, several types of laser diodes, both index-guided and
gain-guided, exist. However, the variety of currently available structures
is far less in InGaAlP than it is in AlGaAs. This is due partially to the
fact that the work on InGaAlP is much more recent and partially also to
a less easy lattice-matched growth of InGaAlP on patterned substrates
with several crystallographic planes. We will first discuss the gain-guided
structures and then the index-guided ones. But first we will discuss the
transverse structure, together with the use of strain.

1.5.1   Transverse laser structure with bulk InGaP active layer

The most commonly used transverse structure is shown in Fig. 1.14. On
top of the n-GaAs substrate has been grown a buffer layer, usually also
n-GaAs, followed by a $In_{0.50}(Ga_{0.30}Al_{0.70})_{0.50}P$ cladding layer that is n-
type doped (often by using Si). Then comes the active layer, which in the
case of normal 670 nm lasers is composed of ordered InGaP, but for shorter
wavelengths often consists of a wave guide layer containing quantum
wells. Next comes the p-type cladding, which usually has the same composition as the n-type cladding layer but which is doped with zinc or magnesium. Often an intermediate layer of p-InGaP is used because a p-GaAs/
p-InGaAlP heterojunction may present considerable non-Ohmic behavior
due to the large valence band offset. The top layer is the well-known

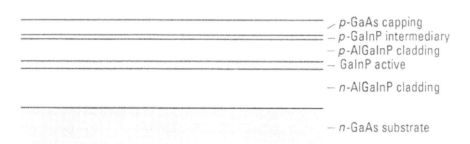

Figure 1.14: Transverse 670 nm InGaAlP laser structure.

p^+-GaAs contacting layer. For lasers emitting at shorter wavelengths (below 650 nm) quantum well structures are preferred (Dallesasse *et al.*, 1988; Valster *et al.*, 1990b; see also Sect. 1.5.8).

1.5.2   InGaAlP strained layer quantum well lasers

As has already been discussed in Sect. 1.4 (see also the Chapter 2 of Volume 1 by Adams and O'Reilly), the use of strained layer quantum wells can result in a considerable reduction of both threshold current densities and optical losses in such lasers. These phenomena have been observed extensively in long wavelength lasers. Here, we will restrict ourselves to the more recent material on the effects of strain on InGaAlP (quantum well) lasers.

Strain is produced by changing the In/Ga ratio of thin InGaP layers. An increased In content leads to compressive strain and an increased Ga content to tensile strain. Usually strain values are of the order of 0.5–1.0%, and thicknesses are smaller or equal to 100 Å, so that indeed the limiting condition mentioned earlier is fulfilled. Often the strained layer (quantum well) is embedded in a separate confinement structure, and a strained layer SQW or MQW laser results. The cladding layers are always lattice-matched to the substrate, but it could be advantageous to give the barrier layers a strain opposite to the strain of the active layer. An example of the L-I characteristics of a compressively strained 670 nm quantum well laser is shown in Fig. 1.15; the data are compared with those obtained on a similar laser with a bulk InGaP active layer.

Although some reports in the literature deal with tensile strain, the majority of the papers refer to the use of compression. It should be men-

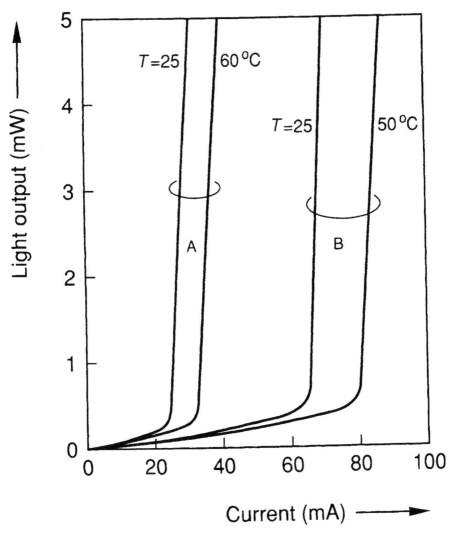

Figure 1.15: Comparison of L-I characteristics of a 670 nm strained layer QW laser at various temperatures (a) with those on a similar laser with bulk InGaP active layer (b).

tioned that the higher In-content leads to a reduction of the band gap and hence to a shift toward longer wavelengths and also to an increase of both the optical and the carrier confinements. Therefore, the strained InGaP active layer results generally refer to a wavelength of about 680 –

690 nm rather than the 670 nm corresponding to the unstrained InGaP. Threshold current densities of about 200–300 A/cm^2 have been reported for compressively strained SQW lasers emitting near 690 nm (Katsuyama *et al.*, 1990). The reduced dissipation in strained layer InGaAlP quantum well lasers favors the realization of high-power lasers because the dissipation is no longer a limiting factor. For lasers with tensile strain, the increased band gap due to the higher Ga content leads to a laser wavelength of 660 nm or below, depending on the degree of ordering of the material of the active layer. If moreover quantum wells are used, this shift toward shorter wavelengths is even enhanced.

Because of what has been discussed, it is not surprising that the results reported for the use of tensile strained (quantum well) lasers mainly relate to the shorter wavelength lasers, notably in the wavelength region of 630–640 nm. Threshold current densities down to 1.2 kA/cm^2 are reported (Welch *et al.*, 1991). Compressively strained lasers emitting near 632 nm have also been reported (Valster *et al.*, 1992), with very low threshold current densities (760 A/cm^2), high power operation, and improved 632 nm stable operation at elevated temperatures. Results on L/I- and V/I characteristics are presented in Fig. 1.16 together with the optical emission spectrum. The threshold current at room temperature is slightly below 50 mA even for the gain-guided structure used. When use is made of index-guiding, this value will drop even below 40 mA. Because all these results show that both compressive and tensile strains lead to a reduction of threshold, it is of interest to know what is the general relationship between threshold current-density and strain. In long wavelength lasers, near 1.55 microns, Thijs *et al.* (1991) have shown that this relationship is a W-type curve with minima in both the compressive and in the tensile strain range, the minimum threshold in the tensile region being lower than that of the compressive minimum (see also Sect. 1.4). Results for lasers emitting near 633 nm are presented in Fig. 1.17; it is observed that a similar behavior is present in this wavelength range in the InGaAlP system. It should be noted that the wavelength has been kept constant in these plots and that the quantum well thickness varies with strain. At higher tensile strains, the quantum wells become relatively thick, and the elastic deformation limit mentioned earlier may be exceeded. Another reason for this increase of threshold current at high tensile strains is that as a result of the strain, two of the X minima decrease in energy and at high strain values will approach the Γ-minimum (see Chapter 2 of Volume 1 by Adams and O'Reilly).

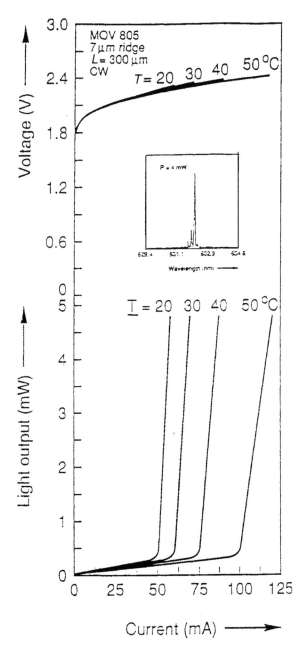

Figure 1.16: L-I characteristics of compressively strained 632 nm lasers, after Valster *et al.* (1992). © 1992 *Electronics Letters.*

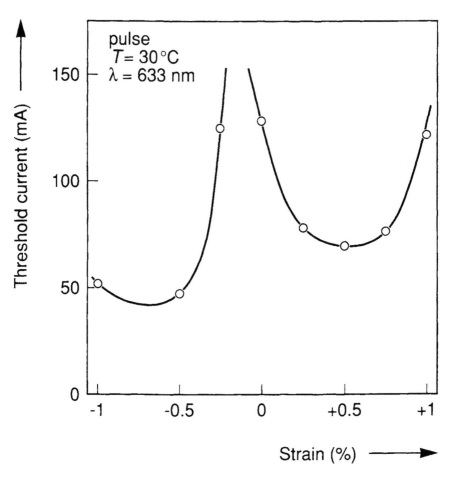

Figure 1.17: Variation of threshold current of 633 nm lasers as a function of strain.

The use of strained layer quantum wells has reduced the threshold current densities to such an extent that lasers with relatively wide stripes can easily be operated; c.w. and high (non–diffraction-limited) powers can be achieved. Powers obtained with broad stripe lasers are around 600 mW c.w. near 670 nm and about 600 mW pulsed near 630 nm (Geels *et al.*, 1992b, and Serreze *et al.*, 1991). Hence it is expected that progress will be made in the area of high-power phase-coupled array lasers with compressive or tensile strained quantum wells.

Recently, the use of strained active layers has reduced the threshold current and hence the dissipation to such an extent that substrate-bonded mounting has become possible. This has led to the realization of dual-spot, separately addressable lasers (Geels *et al.*, 1992a, and Valster and van der Poel, 1994). Junction-up operation of strained layer lasers with InGaAlP secondary AlGaAs cladding and gold heat spreader has also been reported (Unger *et al.*, 1992).

To summarize, the strained layer quantum technology has already proven to be very effective in reducing thermal limitations in red laser diodes and hence is considerably increasing the possibilities for high-power operation in red laser diodes and in reducing the shortest wavelength at which InGaAlP lasers can operate in the c.w. mode down to 600 nm.

1.5.3   Lateral laser structure: gain-guided lasers

In gain-guided lasers, there is usually a stripe-shaped current constriction in the upper part of the *p*-type cladding layer. Thompson (1980) has described many of such AlGaAs structures. Below the stripe, current spreading takes place so that often the width of the optical mode in the lateral direction considerably exceeds the width of the stripe, especially for the narrower ones (approximately 4–5 microns). A significant difference between the AlGaAs and InGaAlP gain-guided lasers is that the current spreading is appreciably less in the latter, being approximately 1 micron for InGaAlP versus about 3 microns for AlGaAs. This difference, which is due to the higher resistivity of the *p*-cladding, leads to a somewhat different behavior of the lasing mode and to a different variation of the threshold current versus stripe width.

The structures for the employed gain-guided lasers can be split up into so-called inner stripe structures and normal stripe structures. The inner stripe structures are realized by a two-step epitaxial process in which, referring to Fig. 1.18, after the growth of the *p*-InGaP intermediate layer an *n*-type GaAs blocking layer is grown. In this *n*-GaAs blocking layer, V-grooves are etched down to the *p*-InGaP intermediary layer. Next, a *p*-GaAs layer is grown on top of the structure on which a broad-area metallized Ohmic contact is made. In such structures, current can only flow in the V-groove regions (Ishikawa *et al.*, 1986 a, b). Laser structures of this type need two (OMVPE) epitaxial steps. The devices generally

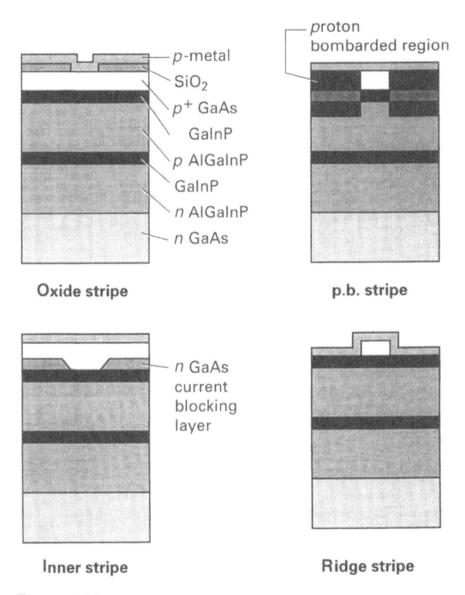

proton
bombarded region

p-metal

SiO$_2$

p$^+$ GaAs

GaInP

p AlGaInP

GaInP

n AlGaInP

n GaAs

Oxide stripe

p.b. stripe

n GaAs
current
blocking
layer

Inner stripe

Ridge stripe

Figure 1.18: Gain-guided 670 nm InGaAlP lasers.

show a pronounced multilongitudinal mode behavior characteristic for gain-guided lasers.

A laser structure that can be obtained with only a single epitaxy step is the proton-isolated structure (Ikeda *et al.*, 1987; see also Fig 1.18.). A proton implantation is used for making the top *p*-GaAs-layer and part of the *p*-type cladding layer on both sides of the stripe highly resistive. Sometimes this structure is made using a tapered stripe that is narrower near the laser mirrors in order to obtain a wider lateral far-field. Another structure grown in a single epitaxial step is the shallow ridge structure (Valster *et al.*, 1990b; see also Fig. 1.18). It also contains a *p*-type InGaP intermediate layer; but its function is somewhat different since it is used as an etch-stop layer. Using photolithography, the top GaAs is selectively etched away on both sides of the stripe using an etch that stops at the InGaP intermediary layer. Then, the top metallization is applied, chosen such that an Ohmic contact is made on the P-GaAs top layer but a blocking contact is obtained on the InGaP layer. Of course, the presence of the InGaP intermediary layer also helps to reduce the required forward voltage. The properties of the structure have been described by Valster *et al.* (1990b). In the following we will denote this structure as the "shallow ridge structure" (SRS). Some of the gain-guided laser data to be discussed later in this chapter are obtained on the SRS, but other structures will yield similar results. Finally, it should be noted the well-known types of AlGaAs gain-guided lasers, like the oxide stripe laser or the planar Zn-diffused stripe laser, have hardly been reported for InGaAlP.

1.5.4   Electro-optical properties of gain-guided lasers

Gain-guided lasers are usually made with a stripe width of around 7 microns. The threshold current as a function of stripe width shows a minimum around 7 microns, as the stripe width decreases the threshold increases steeply. This is related to the fact that for narrower stripes, the lasing mode penetrates on both sides into the less-pumped regions on both sides of the stripe. The transverse (perpendicular) far-field is determined by the active layer thickness and can be fitted to the theory of symmetrical optical waveguides using the refractive index data from literature (Moser *et al.*, 1994). The lateral far-field shows a central maximum with small sidelobes. Such far-field distributions are similar to those of AlGaAs gain-guided lasers, although the sidelobes are much more pronounced in AlGaAs narrow-stripe lasers. The presence of the side lobes in the far-field is related to antiguiding produced by the lowering of the refractive index below the stripe due to the presence of the injected carriers (Thompson, 1980), and they can be associated with leaky waves radiating away from both sides of the stripe. Reliability data are available for both the inner stripe lasers and the SRS. Results show in all cases stable operation up to 60 °C over many thousands of hours (Ishikawa *et al.*, 1989, and Valster *et al.*, 1990b).

1.5.5   Lateral laser structure: index-guided lasers

Usually, the index-guiding of index-guided InGaAlP lasers is attributable to the presence of an absorbing (usually n-GaAs) blocking layer on both sides of the stripe. The distance of the GaAs layer to the thin active layer is about 0.3 microns. This implies that the index-guiding mechanism is basically the same as that of the AlGaAs CSP laser (Aiki *et al.*, 1978). The structure usually takes the form of a buried ridge and is made by a three-step OMVPE epitaxial process. First a normal planar structure is grown (see Fig. 1.19). Then, using photolithography, oxide stripes of several microns are made on the top layer. Next, ridges are etched down to about 0.3 micron from the active layer. In the next step, a layer of n-GaAs is grown on the structure. In this step, the oxide on top of the ridge ensures that the n-GaAs blocking layer grows only on both sides of the ridge and not on top of it. Next, the oxide mask is removed and a final layer of p-GaAs is grown in order to planarize the structure. A schematic of the finalized structure is presented in Fig. 1.19. At room temperature (Ishikawa *et al.*, 1986a) lasers with an InGaP active layer emitting near 670 nm yield threshold currents of 40–50 mA, which is lower than the above-mentioned values for gain-guided lasers (70–80mA). Still, this reduction is relatively moderate, which relates to the fact that a GaAs absorbing blocking layer is present. For InGaAlP, some index-guided laser structures without an absorbing blocking layer have also been reported. The first index-guided laser without absorbing blocking layer reported is a deeply etched ridge guide laser (Chang *et al.*, 1994) analogous to that commonly used in AlGaAs. A ridge-waveguide laser etched by RIE and buried with Al_2O_3 on both sides showed the lowest threshold current ever reported in the AlInGaP system, namely about 5 mA at room temperature. A recent report decribes an overgrown structure resembling the aforementioned SBR-laser but with a nonabsorbing AlInP blocking layer (Kobayashi *et al.*, 1995). Finally, it should be noted that buried-hetero-junction

Fabrication of a selectively buried ridge (SBR) laser-diode

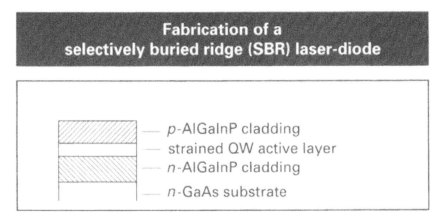

First Epitaxy

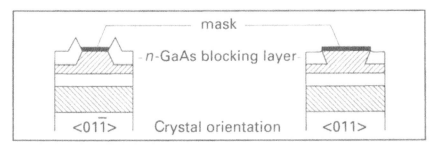

Etching Ridge and Second Epitaxy

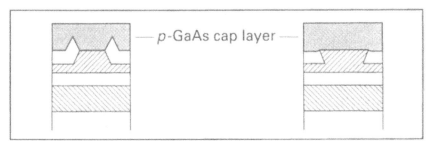

Third Epitaxy

Figure 1.19: Selectively Buried Ridge (SBR) index-guided laser structure.

type lasers, which are well-known for AlGaAs (Tsukada, 1974) and are routinely fabricated in InGaAsP, have not yet been reported for the InGaAlP laser system.

1.5.6   Electro-optical properties of index-guided lasers

Apart from the somewhat lower threshold and operating currents, the advantages of index-guided lasers are twofold. The first advantage is that the stripe width can be narrowed down to 4–5 microns without increased (saturable) absorption losses on either side of the stripe. However, when an absorbing blocking layer is used a (nonsaturable) absorption resulting in a high threshold and a lower differential efficiency will result at stripe widths significantly below 4 microns. At a stripe width of 4 microns the lateral far-field width is around 10 degrees, which is considerably wider than that of most gain-guided lasers. This leads to a smaller asymmetry of the far-field compared to that of most gain-guided lasers. The second advantage is a smaller degree of astigmatism for the index-guided lasers (less than 10 microns compared to about 30 microns in the case of gain-guided lasers).

1.5.7   High-power operation

Generally, for diffraction-limited (narrow-stripe) high-power lasers, the following mechanisms limit the operating power:

1. Catastrophic optical damage (COD) at the laser mirrors.
2. Gradual degradation at the laser mirrors.
3. Kinks in the L/I curve, which are usually accompanied by distortions of the lateral far-field distribution.
4. Thermal limitations, which lead to a thermal rollover of the L/I characteristic.

The experience gained so far shows that the most serious limiting factors for InGaAlP lasers are (1) and (2). It turns out that COD-levels are comparable or even somewhat lower than in AlGaAs. Usually, aluminum oxide sputter coatings are used to protect laser mirrors. Treatments before the deposition of the coatings can have a beneficial effect on COD-levels (Kamiyama *et al.*, 1991). The thermal limitations result from the fact that usually the threshold current is somewhat higher than in AlGaAs, whereas T_0 is lower (*cf.* Eq. 1.2A), and the thermal resistance of InGaAlP is higher than that of AlGaAs.

From the work that has been done on AlGaAs high-power laser diodes, it is known that nonabsorbing regions near the laser mirrors (nonabsorbing mirrors or NAMs) help in reducing gradual mirror degradation. The reason for this is that gradual mirror degradation is generally a consequence of the recombination of electron-hole pairs generated by injection and photon absorption at the laser mirrors. Such NAM regions are obtained, for instance, if the band gap in the regions near the laser mirrors is somewhat higher than in the rest of the active region. As the InGaAlP material can be ordered or disordered, depending on growth and processing conditions (see Sect. 1.3) and the band gap of the disordered material is about 90 meV higher than that of the ordered material, local disordering in otherwise ordered laser material provides a unique way of creating transparent mirror regions. Local disordering can be realized, *e.g.* by locally applying a zinc diffusion, but it is envisaged that local implantation and subsequent annealing may also serve the same purpose. Facet heating of the mirror regions of high-power lasers has also been reduced using local current blocking regions near the facets (Hamada *et al.*, 1991a). Yet, up to now, reliability data on such lasers have been scarce. A disadvantage of the use of locally disordered mirror regions is that the remainder of the active layer should be ordered, which leads to higher internal optical losses (*cf.* Fig. 1.6).

Another way of enhancing the power handling capacity of the laser mirrors, both with respect to COD and to gradual mirror degradation, is by using quantum wells, as has been demonstrated for AlGaAs high-power lasers (Harm, 1990). For the COD level it is found that, expressing the COD in mW per micron stripe width,

$$\text{COD} \propto 1/\Gamma,$$

where Γ is the vertical confinement factor that can be made low by using quantum wells. Of course, the use of quantum well configurations may be combined with local disordering at the laser mirrors. This is even more so since, apart from the disordering of the natural superlattice of the InGa(Al)P, an intermixing of the quantum wells can also be achieved. For these reasons it is evident that quantum-well lasers are interesting candidates for InGaAlP high power lasers, the more so because the use of quantum wells and especially strained quantum wells will reduce the threshold and hence the operating current. Some data will be presented later on high-power strained-layer quantum well index-guided lasers.

In principle, L/I kinks are a problem for InGaAlP high power lasers. However, it was found that it is possible to shift the kink power toward appreciably higher values by optimizing the length of the laser (Schemmann *et al.*, 1995) because the occurrence of the kink in the L/I curve and the associated deformation of the near- and far-fields is associated with a beat between the fundamental- and first-order lateral waveguide modes. Results obtained for high-power SBR-lasers with optimized cavity length and with facet coatings with a mirror reflectivity of 10% for the front mirror and 80% for the rear mirror are shown in Fig. 1.20. The COD power for these lasers is about 90 mW for lasers whose front mirror is provided with standard aluminum oxide magnetron sputtered facet coatings. When the lasers are equipped with optimized coatings (see Sect. 1.5.10), the COD power increases to about 150 mW, and reliable c.w. operation can be obtained at powers as high as 45–50 mW and 50 C for many thousands of hours.

1.5.8   Laser diode operation at shorter wavelengths (630–650 nm)

As was already mentioned in the introduction, the visibility of the laser radiation increases considerably when the wavelength is shortened below 670 nm down to say, 630 nm, the wavelength region of the HeNe lasers. This wavelength reduction can be brought about in several ways:

1. By adding Al to the active layer, *e.g.* using an InGaAlP active layer instead of an InGaP one (Ishikawa *et al.*, 1990, and Kawata *et al.*, 1987).

2. By disordering the material. This gives rise to an increase in band gap both in the active layer and in the cladding layers (Suzuki *et al.*, 1993). The use of a disordered InGaP active layer reduces the wavelength from 675 nm down to 655 nm.

3. By using either unstrained (Dallesasse *et al.*, 1988; Valster *et al.*, 1990c) or strained quantum wells. The quantum wells usually consist of InGaP with excess In or Ga for compressive or tensile strain, respectively. By disordering the InGaP, thicker quantum wells can be used for a given wavelength. Since excess indium reduces the band gap, compressively strained quantum wells will

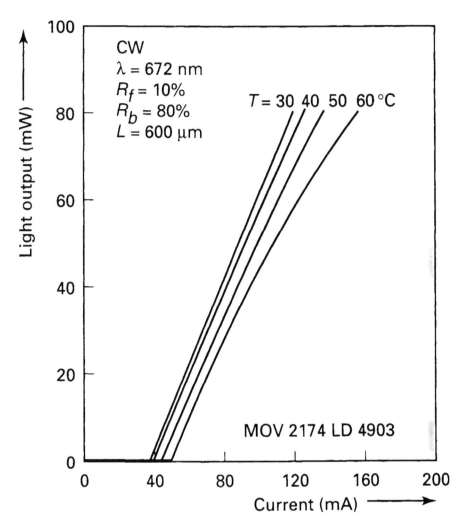

Figure 1.20: L-I characteristics of high-power index-guided 670 nm SBR lasers.

be thinner at the same wavelength compared to unstrained ones, whereas the application of tensile strain will lead to thicker quantum wells. Some typical results for strained material will be presented.

Shortening of the wavelength by adding aluminum to the active layer was reported by Ishikawa *et al.* (1990) and Kawata *et al.*, (1987). As expected, at shorter wavelengths the threshold increases and T_0 decreases due to a combination of increased nonradiative recombination and increased leakage. When the wavelength was reduced to 630 nm, the threshold current increase reported for index-guided lasers was typically 50 mA, while T_0 decreased by 50 K (Hatakoshi *et al.*, 1991). From the results reported it is not always clear whether the InGaAlP active layers were disordered or not.

An example of wavelength reduction effected by using disordered material is shown in Fig. 1.21 for a disordered 655 nm laser as a function of the temperature. The results are compared with those of a 675 nm laser at the same structure but with ordered material. It can be seen that the temperature dependence of the threshold current is similar. It has been found that the disordering process has no negative effect upon the device reliability.

An example of the use of tensile strained InGaP quantum wells for the realization of 633 nm operation up to 50 °C is given in Fig. 1.22. The InGaP quantum wells are disordered. The temperature dependence of the threshold current is generally better for tensile strained quantum wells because thicker quantum wells can be used. It is advantageous to combine quantum wells with tensile strain with slightly compressively strained barriers. It was found (Honda *et al.*, 1994) that by applying this strain compensation, lower thresholds can be obtained at high tensile strain values in the quantum well.

1.5.9   Temperature dependence of the threshold current

The temperature dependence of the threshold current was studied in detail on 670 nm lasers with a bulk InGaP active layer (Hagen *et al.*, 1990; Ishikawa *et al.*, 1991). In both studies it was found that the temperature dependence of the threshold was conveniently described by Eq. (1.3). For the activation energy, values of 0.27 eV (Hagen *et al.*, 1990) and 0.43 eV (Ishikawa, *et al.*, 1991) were reported. In both studies, the results were interpreted as being related to the leakage of electrons to the *p*-type cladding layer. The temperature-dependence of threshold for lasers emitting at shorter wavelengths is steeper. For 655 nm lasers in which the InGaP bulk active layer and the InGaAlP claddings are disordered, the temperature dependence of threshold is still rather similar, as can be seen from Fig. 1.21. Below 650 nm data reported in literature are scarce

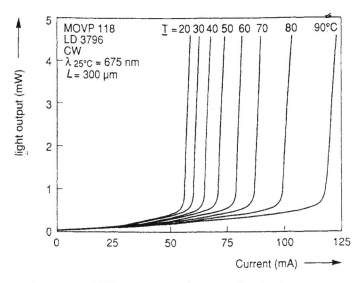

Ridge laser with a Ga$_{0.5}$ In$_{0.5}$ P active layer

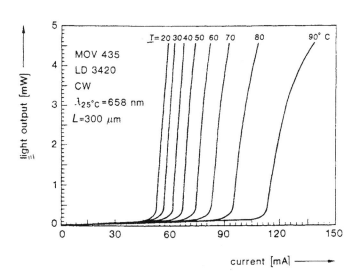

Ridge laser with a disordered Ga$_{0.5}$In$_{0.5}$ P active layer

Figure 1.21: Comparison of 655 nm laser diodes made from disordered InGaAlP/InGaP material with 675 nm lasers made from ordered material.

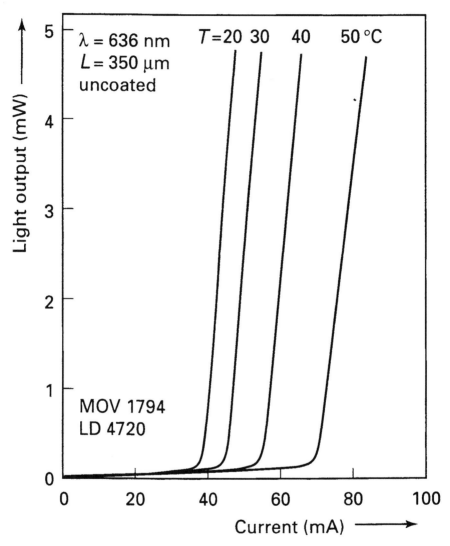

Figure 1.22: L-I characteristics of index-guided 635 nm lasers with tensile-strained layer quantum wells.

and difficult to compare because bulk active layers, unstrained quantum wells, and strained quantum wells have been used. Nevertheless, a general trend can be extracted (Hatakoshi *et al.,* 1991). In eq. (1.2) we observed that the maximum temperature T_{max} for which c.w. operation is possible is an important quantity. Fig 13 of Hatakoshi *et al.* (1991) presents an overview of T_{max} for visible red lasers as a function of wavelength, showing a decrease from about 130 °C near 680 nm wavelength down to about 50 °C near 630 nm, relatively independent of whether bulk active layers or quantum wells have been used. However, it should be noted that this overview did not yet include the more recent results on strained layer quantum well lasers. For quantum well lasers, thermal leakage of electrons from the wells into the barriers may well be the limiting process as it is for AlGaAs quantum well lasers (Blood *et al.*, 1988).

Studies on lasers with InGaP active layer at high pressure (Meney *et al.*, 1995) indicate an increase of the threshold accompanied by a decrease of differential efficiency at high pressures when the Γ minimum approaches the X minima. This indicates, together with the wavelength variation observed as a function of the pressure, that in lasers with bulk active layer and wavelength below 650 nm, loss of carriers in the active layer via the X minima may be an important loss process. Nevertheless, it has been established (Meney *et al.*, 1995) that, near 630 nm, carrier leakage toward the p-type cladding layer is the most important loss mechanism. This conclusion is supported by the observation that the incorporation of an electron-reflecting element such as a multi-quantum barrier (MQB) structure (Iga, *et al.*, 1986; Takagi *et al.*, 1991) into the p-type cladding is claimed to improve the temperature dependence of threshold (Kishino *et al.*, 1990; Rennie *et al.*, 1992) of lasers emitting below 670 nm. It is expected that the use of a multi-quantum barrier reflector together with the use of strained layer structures for the active layer will help to shift the wavelength region of room temperature c.w. operation down to 600 nm, which is really orange-colored. A first indication of this tendency is the reported laser operation near 610 nm in the c.w. mode using a strained MQW structure combined with a MQB (Hamada *et al.*, 1992).

1.5.10 Reliability of InGaAlP lasers

Most data in literature on reliability of InGaAlP lasers refer to 670 nm devices. Experience shows that for these lasers, living more than 10 khrs at 50 °C, facet coatings help in achieving low degradation rates. For the facet passivation, sputtered aluminum oxide is usually used. In fact, most data refer to devices that have been coated that way. From the degradation rates as a function of temperature, activation energies are derived that are believed to be related to the physics of the degradation process like for instance defect motion. In Endo *et al.* (1994), various values for the activation energy between 1.4 and 2.4 eV are reported for InGaAlP lasers; these values are much higher and also show a much larger spread than the values of 0.8–0.9 eV reported for AlGaAs (Ueda, 1988). However, these activation energies, although being useful for lifetime extrapolation from accelerated lifetest results, are not at all representative for the physics of the degradation process. The reason for this is that the degradation rate in InGaAlP lasers depends strongly on the current density. Therefore, more relevant information concerning the temperature dependence of the degradation process is obtained from life tests at elevated temperatures at a constant current and various current densities (Endo *et al.*, 1994). From these data it is inferred that the degradation rate, D, is conveniently described by the following relationship:

$$D = D_0 \exp(J_{op}/J_{ox})\exp(-E_a/kT), \tag{1.10}$$

where D_0 is a constant, J_{op} the operating current density, $J_{ox} = 1.24$ kA/cm^2 and E_a the activation energy for the degradation process. For the activation energy, 0.80 +/− 0.05 eV was found for InGaAlP lasers.

In section 1.5.7, it was already briefly mentioned that the disordering process has no negative effect upon laser reliability. Life tests performed using 655 nm lasers made in our laboratories from disordered material yielded similar results as those obtained on (ordered) 675 nm lasers as described by Valster *et al.* (1990b). Hamada *et al.* (1991b) describe similar results on inner stripe lasers made from disordered material. Life-testing results on lasers made from disordered material are also presented in Fig. 1.23.

Different reliability behavior is observed if the operating temperature approaches the temperature T_{max} (*cf* Eq. 1.2), *i.e.*, if the laser approaches its thermal limits due to a high threshold current and a low T_0. In such a case, any amount of degradation becomes "thermally enhanced" because the higher operating current, the resultant increased heating, and the decreasing T_0 bring the laser even closer to its thermal limits. This has certainly played a role in early life-testing results obtained for 630 nm lasers (Valster *et al.*, 1990b). The recent reduction of threshold brought about by the application of strained layer quantum wells in 630 nm lasers has shifted the operation from the thermal limits and has thus resulted in an increased lifetime.

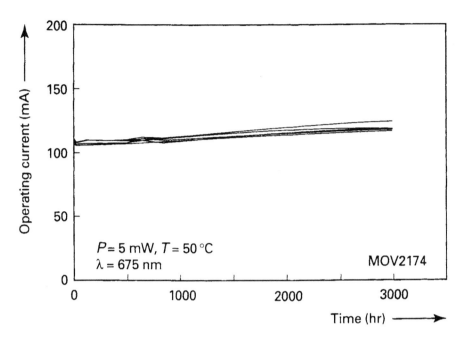

Figure 1.23: Life-test results at 50 °C and 50 mW output power of strained layer 670nm MQW index-guided SBR lasers.

High-power lasers usually operate in the wavelength region of 670 nm. At present, compressively strained quantum wells are used in such lasers. In the past, life tests of index-guided lasers made from such material usually showed stable behavior at power levels of 30–35 mW, but at higher powers, 40 mW or more, degradation took place. It was established that this degradation was entirely due to erosion at the laser mirrors. As mentioned in Sect. 1.5.7, possible remedies are optimization of the facet coating or by the use of a nonabsorbing mirror. Indeed, optimization of facet coating has led to improved laser lifetime at high output powers. The standard coating for high-power AlGaAs lasers is diode-sputtered aluminum oxide. It has been found that diode-sputtered coatings on InGaAlP lasers do not work well, as there is evidence that, due to the diode-sputtering process, the laser facets are damaged, which leads to irreproducible phenomena during life-testing. Therefore, magnetron-sputtered aluminum oxide coatings are generally used, and the use of the latter coatings leads to the aformentioned reliable power of 30–35 mW.

It was already stated in Sect. 1.5.7 that sulfurization of the laser mirrors prior to coating improves the COD level of the laser mirrors. Experiments at the authors' laboratories indicated that sulfurization alone does increase the COD level, but that it does not lead to a significant improvement of the gradual degradation at high c.w. output powers. However, the results still indicate that sulfur has a beneficial effect on the laser mirrors, which probably can be attributed to the fact that, due to the sulfurization, oxygen is removed from the mirror surface. We found that, if the sulfurization is combined with other measures such as the application of a thin layer of an oxygen getter at the facet, the degradation at high c.w. output powers is significantly improved. Life-testing results at 50 mW, 50 °C, of index-guided 670 nm strained layer QW lasers with such coated mirrors are shown in Fig. 1.23. It was found that the results are as good as those obtained at 30 mW on the lasers with the unmodified coatings. Some life-testing results obtained on 5mW 633 nm strained layer quantum well lasers are shown in Fig. 1.24.

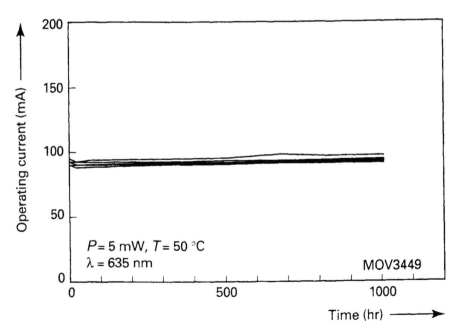

Figure 1.24: Life-test results at 50 °C, 5 mW output power of 635 nm tensile strained layer MQW lasers.

1.6 Short-wavelength visible lasers (blue-green and blue)

1.6.1 II–VI injection lasers

1.6.1.1 Introduction

With the AlGaInP system, the shortest wavelength that can be reached is about 600 nm at room temperatures and about 550 nm at 100 K (Valster *et al.*, 1990a). At shorter wavelengths, the other material systems discussed in Sect. 1.2 should be used. Of those, the II–VIs are by now the most advanced. As was mentioned in Sect. 1.2, the ZnSe/ZnSSe heterojunction structure is not very promising because the carrier and optical confinements are low. In fact, only *e*-beam pumped lasers have been realized with this structure (Bhargava, 1988; Cornelissen *et al.*, 1992) Since then, much improvement has been obtained by using alternative II–VI laser structures, which will be discussed now.

1.6.1.2 The use of CdZnSe strained layer quantum wells and quaternary claddings for blue-green II–VI lasers

Considerable progress was made by inserting a thin (about 65 Å) CdZnSe active layer in the ZnSe layer (Haase *et al.*, 1991). This effectively strained quantum well active layer has a band gap that is smaller than that of the ZnSe, its emission wavelength corresponding to 490 nm at 77K at a Cd-content of about 20%. The conduction band step of the $Cd_xZn_{(1-x)}$ Se/ZnSe quantum wells with a Cd content of about 0.20 is around 230 meV (Olego *et al.*, unpublished) and the valence band step around 80 meV (Jeon *et al.*, 1991).

A schematic representation of the structure is presented in Fig. 1.25a. Narrow-stripe lasers with a width of around 20 microns made on such materials yielded threshold currents around 75 mA near 77 °K. Pulsed lasing was observed up to about 300 °K. The ZnSSe/ZnSe/CdZnSe laser structure of Fig. 1.25 is a separate confinement SQW strained-layer structure. Recently, this structure has been extended toward multi-quantum wells. Furthermore, besides the use of *n*-GaAs substrates, inverted structures on *p*-GaAs were investigated. Using II–VI quaternaries such as ZnMgSSe, one could expect shorter wavelengths, as has been demonstrated (Okuyama *et al.*, 1992). Moreover, significant improvements around 500 nm have been obtained using ZnMgSSe cladding layers, a

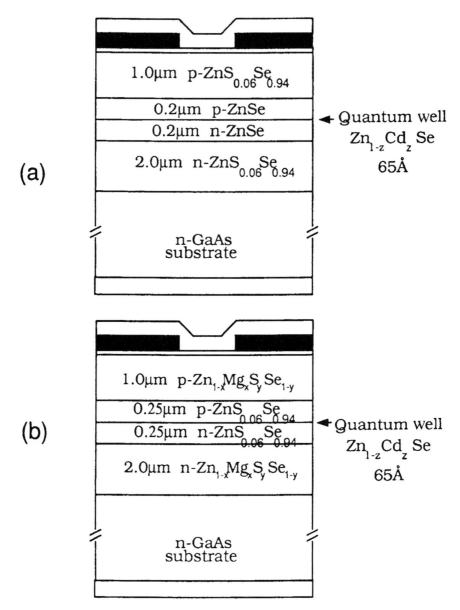

Figure 1.25: (a) Schematic representation of CdZnSe/ZnSSe 490 nm laser structure after Drenten *et al.* 1995. © 1995 *Philips J.of Research.* (b) Schematic representation of 490 nm laser structure made with MgZnSSe cladding after Drenten *et al.* (1995). © 1995 *Philips J.of Research.*

ZnSSe waveguide layer, and a CdZnSe active layer. Low threshold current densities of 500 A/cm^2 at room temperature and lasing up to 394 °K were obtained (Gaines *et al.*, 1993). A schematic of this structure, which is now generally used for II–VI lasers, is shown in Fig. 1.25b. It has been found that the presence of the higher-gap quaternary cladding layers leads to an improved T_0 due to a strong decrease in thermal carrier leakage to the cladding layers. The fact that the difference between the structures with ternary cladding layers and those with quaternary cladding layers is primarily due to the difference in thermal leakage current is illustrated in Fig. 1.26. This figure plots the threshold current as a function of temperature for both types of devices together with some theoretical fits to a model that includes thermal leakage currents (Drenten, 1996) for various values of the difference between the photon energy and the band gap of the cladding layers, which is a measure for the carrier confinement. In addition the structure with quaternary cladding can be made fully pseudomorphic and also has a better optical confinement. T_0-values are around 210 °K for 520 nm lasers and about 120 °K for 490 nm lasers (Ishibashi, 1995). Most of the lasers studied up to now are of the gain-guided type. Recent experiments by Drenten *et al.*, (1995) show that for long current pulses the local heating below the stripe creates a thermal index-guide that considerably reduces the threshold current and leads to a narrowing of the far-field in the junction plane. II–VI blue-green lasers with built-in index guiding made by two-step epitaxy have been reported by Haase *et al.* (1993).

1.6.1.3   Reliability of II–VI lasers

After the introduction of the MgZnSSe cladding layers and recent results on the reduction of forward voltage through the application of improved contacting technologies (such as the use of a *p*-type ZnTe intermediary layer), c.w. operation of MgZnSSe lasers at room temperature has been regularly achieved. CdZnSe compressively strained quantum wells are generally used in the active layer; the corresponding emission wavelength is around 490–530 nm. General experience at the time of writing (1996) is that, when driving the lasers c.w. at room temperature even at low powers (typically 1 mW), degradation occurs usually within several hours. The degradation process is not related to the lasing process itself, but it also occurs in LEDs or in lasers operated below the threshold. Guha *et al.* (1993 and 1994) established that the degradation is related to grown-

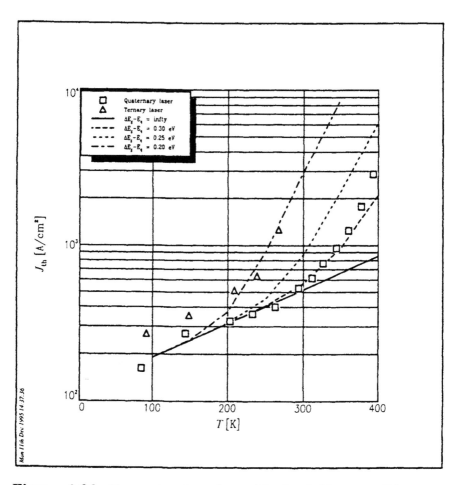

Figure 1.26: Temperature dependence of the threshold current of blue-green II–VI lasers with ternary and quaternary cladding layers, respectively after Drenten (1996). © 1995 Philips Electronics.

in defects. These defects are mainly stacking faults nucleated at the sub-strate–buffer layer interface and subsequently growing out into the remaining part of the layer structure. Normally, densities for this type of defects are rather high, *i.e.* of the order of 10^5–10^7 cm^{-2}. As defect densities decrease, lifetime increases. Guha et al. found that at the positions where the stacking faults intersect the active layer, current injection into the structure leads to the formation of dark line defects extending into the

⟨100⟩-direction, rather like the well-known DLD formation in early AlGaAs lasers. However, a major difference is that in the AlGaAs case, the dark line defect formation took place mainly at threading dislocations extending from the substrate through the layer structure. The subsequent development of GaAs substrates with low dislocation densities since then has led to the virtual elimination of DLDs in AlGaAs lasers, and in that system the effect is now mainly of historical interest. As far as the II–VI lasers are concerned, the situation seems to be fundamentally more difficult. First, the lower growth temperatures required for II–VI compounds and the resulting lower surface mobility of the atoms together with the II–VI materials' known tendency toward polytype formation may lead to the easy formation of high densities of stacking faults. Secondly, part of the problem may be that usually a II–VI buffer layer is grown first, so that the first grown layer in the epitaxial process is already a II–VI layer. For this reason it is expected that future experiments focussing on the reduction of stacking faults will concentrate on the use of GaAs buffer layers. It is believed that reduction of the stacking fault density down to values of the order of one thousand per cm^2 will be required to make the application of II–VI lasers feasible, just as it was the case in the past for the III–V lasers. So far, no evidence of mirror degradation has been found in experiments with high-power c.w. II–VI lasers.

1.6.2   Gallium nitride-based lasers

At the time of writing (1996) gallium-nitride-based lasers are just emerging. However, the present state of gallium nitride technology is such that in the year 1996 GaN/AlGaN or InGaN/AlInGaN lasers are breaking through. Double heterostructures have been grown (Ito *et al.*, 1991) and optically pumped laser action in GaN has been observed (Amano *et al.*, 1990). In addition, *p*-type doping of GaN (Nakamura, 1991) and AlGaN have been achieved (see Sect. 1.3) and subsequently high-brightness blue-emitting LEDs have been reported (Nakamura *et al.*, 1994) having a strained InGaN active layer emitting near 450 nm. One of the surprising aspects of this development is that, although the material was grown on sapphire, which is not a very well-matched substrate to GaN, and the material contained quite high defect densities ($10^8/cm^2$ or even more), practically no degradation of the LEDs was observed. Since then, even better results at different wavelengths were obtained using InGaN strained-layer quantum wells (Nakamura *et al.*, 1995). This means that by mid-1995 all the ingredients for a successful demonstration of injection lasers in the nitrides were available, and indeed first lasers were reported by end of 1995/beginning of 1996 (see the following discussion). A monograph on GaN emitters has recently been published (Nakamura and Fasol, 1997).

More details about the growth of GaN and related compounds are presented in Sect. 1.3. All we need mention here is that in the past the main bottleneck was the impossibility of realizing p-type conductivity. This roadblock was eliminated in 1992, when it was shown (Nakamura, 1991) that Mg-doped GaN grown by MOVPE after a suitable post-growth radiation treatment proved to be p-type. AlGaN, which in the past was always semi-insulating, can now be made both n-type and p-type, provided the aluminum content is not too high (of the order of 20–30%). In the past the choice of a suitable substrate material has always been a problem. By now, successful growths on sapphire as well as on the better-matched and conducting SiC substrates have been demonstrated, and GaN-based blue-emitting high radiance LEDs grown on sapphire as well as on SiC substrates have been produced.

Bandgap electroluminescence of GaN would be around 365 nm, so for blue LEDs thin-strained layer InGaN active layers are used. In general, the growth of InGaN poses some problems, since lower growth temperatures are required to prevent loss of indium, whereas high temperatures are needed to enable the dissociation of the NH_3 for the supply of N. The growth by means of MO-VPE of Mg-doped p-type InGaN has been reported (Yamasaki *et al.*, 1995).

Of particular interest is a paper by Nakamura *et al.* (1995), in which they describe an extension of the wavelengths covered by the InGaN/AlGaN system realized by increasing the In content of strained-layer InGaN quantum well active layers. They thus succeeded in increasing the wavelength of high-brightness LEDs from the UV up into the orange around 600 nm. This means that, at least for LEDs, an overlap between the AlInGaN system and the InGaAlP system has been achieved. This opens up the perspective that, once GaN-based lasers have been developed, an overlap between the InGaAlP and InGaN lasers might be realized. By the year 2000 it may be possible to produce diode lasers at all visible wavelengths. The appearance of quantum well structures brings laser realization even closer. In fact, stimulated emission by electrical pumping of an AlGaN/GaN/InGaN quantum well structure in which the GaN acts as the barrier layers has been reported (Akasaki *et al.*, 1995).

Since then, the first AlInGaN laser operating in a pulsed mode at 410 nm at room temperature has been announced (Nakamura *et al.*, 1996a). A multi-quantum well InGaN active layer consisting of 25 QWs was used. The forward voltages are still rather high, indicating that among others contact resistances need some improvement. These laser structures were grown by OMVPE on sapphire substrates. More recently, Nakamura *et al.* (1996b) reported similar laser structures having comparable laser characteristics in which the laser structures were grown on 111-oriented magnesium aluminate substrates. These substrates have a cubic structure, but by using the 111 direction, the three-fold symmetry is well suited to the growth of the hexagonal wurtzite GaN-based materials. Furthermore, the lattice match between the layer structure and the substrate is somewhat better, which should reduce problems like film cracking.

Although the GaN-based lasers are just emerging, one may easily speculate on what an optimized layer structure could look like. A possible structure, preferably grown on a conductive *n*-type SiC substrate, is shown in Fig. 1.27. On the hexagonal SiC substrate, an AlN-containing buffer layer could be grown followed by the *n*-type AlGaN cladding layer. AlGaInN barrier layers could be used. Compressively strained InGaN

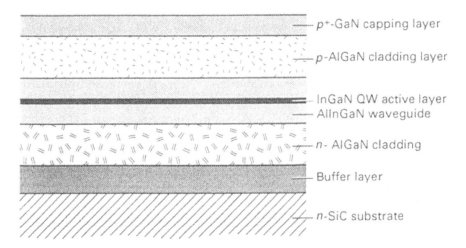

Figure 1.27: Schematic representation of a possible optimized GaInN/AlGaInN laser structure.

quantum wells should constitute the active layer. A heavily p-type GaN capping layer should be grown on top in order to facilitate contacting.

It is also interesting to speculate on what to expect for the lifetimes of lasers at the very high defect densities involved. The reliability of LEDs was found to be surprisingly good in spite of the high defect densities present. However, lasers are inherently more sensitive to degradation. Therefore, although the first AlGaInN lasers are now appearing, it is in our view too early to state that the feasibility of these lasers is already proven.

1.7   Conclusions and perspectives toward future developments

The magnitude of the threshold current and its temperature dependence are vital in limiting the properties of InGaAlP laser diodes and of visible laser diodes in general. In the InGaAlP system, the use of strained layer quantum wells reduces the threshold current considerably for both compressive and tensile strains. The field of InGaAlP lasers is now rapidly maturing.

In blue-green lasers based on the MgZnSSe system, the use of strained active layers combined with the use of quaternary cladding layers has made c.w. operation at room temperature possible. However, lifetime at present is still a problem and much work toward reducing the defect density will be required for bringing about lifetimes in the order of thousands of hours even at elevated temperatures.

On the basis of the present results, some predictions can be made regarding results to be expected within the next years and about the physical limitations.

In red lasers, where today 630 nm lasers are about the shortest wavelength c.w. devices at and above room temperature, the use of strained-layer quantum wells together with index-guided structures with a low internal optical loss is expected to shift the c.w. limit at room temperature down to about 600 nm. Here, the use of multi-quantum barrier reflectors could help in further reducing the temperature dependence of threshold. Around 670 nm, high-power lasers emitting more than 50 mW diffraction limited power c.w. at reasonably high temperatures (say, 50 °C) with long operating lifetimes will be developed within the next two years. These lasers will be the basis for new high-density optical recording systems such as the Digital Video Disk (DVD). In the area of II–VI diode lasers, defect reduction in the growth of the MgZnSSe system is expected to lead to improved lifetimes for c.w. lasing at room temperature and even above for lasers emitting near 500 nm and possibly even at somewhat shorter wavelengths. The recent developments in the AlGaN system, such as the demonstration of growth of AlGaN/GaN heterostructures and the realization of InGaN/AlGaN p–n junction quantum well high radiance LEDs are now leading to the appearance of the first injection lasers emitting near 400 nm. It is expected that this early development will gain strong impetus during the following years and that diode lasers will become available in the whole visible wavelength range. It should be noted, however, that the feasibility of long-lived AlGaInN lasers has not been established yet.

Another interesting recent development is that vertical cavity surface emitting lasers (VCSELs) (Sale, 1995) are now entering the range of visible diode lasers. So far, the best VCSEL results have been obtained to the AlGaAs and InGaAs-AlGaAs near infrared (800–1000 nm) wavelength range. At 690 nm AlGaInP VCSELs have yielded about 2.3 mA threshold current and peak c.w. powers of 2.4 mW, at shorter wavelengths the thermal limitations (which are much more important in VCSELs than in edge-emitting lasers) reduce the output powers to below 1 mW because the wavelength reduction leads to a decrease in T_0 (see Sect. 1.5) (Schneider *et al.*, 1995). Optically pumped II–VI VCSELs have also recently been reported (Floyd *et al.*, 1995). It is evident that special features like the ossible realization of two-dimensional VCSEL arrays will present more advantages for visible lasers than for infrared ones. Hence we foresee that the continuous improvements presently envisaged for normal edge-emitting visible lasers will greatly boost research in the field of visible VCSELs.

Acknowledgments

Numerous colleagues in the Research Group of Philips Optoelectronics Centre and Philips Laboratories, Briarcliff Manor, contributed to the the results on the III–V lasers and II–VI lasers described here. The authors are grateful to P. J. de Waard, J. M. Gaines, and L. Weegels for critical reading of the manuscript and many suggestions for improvement. Part

of the work on InGaAlP high-power lasers was carried out under an EC
contract (ESPRIT 6134 "HIRED").

References

Abernathy, C. R. (1993). *J.Vac. Sci. Technol. A11,* 869.

Adesida, I.; Mahajan, A.; Andideh, E.; Khan, M. A.; Olsen, D. T.; and Kuznia,
J. N. (1993). *Appl. Phys.Lett.* **63**, 2777.

Aiki, K.; Nakamura, M.; Kuroda, T.; Umeda, J.; Ito, R.; Chinone, N.; Maeda, M.
(1978). *I.E.E.E. J. Quantum. Electron. QE-14,* 89.

Akasaki, I.; Amano, H.; Sota, S.; Sakai, H.; Tanaka, T.; and Koike, M. (1995). *Jap.
J. Appl. Phys.* **34**; L 1517.

Amano, H.; Asahi, T.; and Akasaki, I. (1990). *Jap.J.Appl.Phys.* **29**, L 205.

Amano, H.; Tanaka, T.; Kunii, Y.; Kato, K.; Kim, S. T.; and Akasaki, I. (1994).
Appl.Phys.Lett. **64**, 1377.

Ambrosius, H. P. M. M., to be published.

Bartels, W. J. (1983). *J.Vac.Sci.Technol. B1,* 338.

Bastard, G. (1991). "Wave Mechanics Applied to Semiconductor Heterostructures"
Les Editions de Physique, Les Ulis., 248.

Bhargava, R. N. (1988). *NATO Advanced Research Workshop on Growth and Optical Properties of Wide Gap II–VI's and Low Dimensional Semiconductors,
Regensburg.*

Blood, P.; Fletcher, E. D.; Woodbridge, K.; Heasman, K. C.; and Adams, A. R.
(1988). *I.E.E.E. J. Quantum. Electron. QE-25,* 1459.

Boermans, M. J. B.; Hagen, S. H.; Valster, A.; Finke, M. N; and van der Heijden,
J. M. M. (1990). *Electron. Lett.* **26**, 1438.

Born, M., and Wolf, E. (1975). *The Principles of Optics.* New York: Pergamon
Press.

Bour, D. P.; Carlson, N. W.; and Evans, G. A. (1989). *Electron. Lett.* **25**, 1243.

Bouwhuis, G.; Braat, J.; Huiser, A.; Pasman, J.; van Rosmalen, G.; and Schou-
hamer Immink, K. (1985). *Principles of Optical Disc Systems.* Boston: Adam
Hilger Ltd.

Casey, H. C., Jr.; and Panish, M. B. (1978). *Heterostructure Lasers, Part B.* New
York: Academic Press.

Chang, C. V. J. M, and Rijpers, J. C. N. (1994). *J.Vac. Sci. Technol. B12,* 536.

Cook, D. D., and Nash, F. R. (1975). *J.Appl.Phys.* **46**, 1660.

Cornelissen, H. J., Savert, C. J., and Gaines, J. M. (1992). *Philips J. Res.* **46**, 137.

Dallesasse, J. M.; Nam, D. W.; Deppe, D. G.; Holonyak, N.; Fletcher, R. M.; Kuo, C. P.; Osentowski, T. D.; and Craford, M. G. (1988). *Appl.Phys.Lett.* **53**, 1826.

Dawson, M. D., and Duggan, G. (1993). *Phys.Rev.B 47* **12**, 5598.

Drenten, R.; Petruzzello, J.; and Haberern, K. (1995). *Philips J. Res.* **49**, 225.

Drenten, R. R. (1996). *Waveguiding aspects of advanced semiconductor lasers and miniatiure blue-green lasers.* Doctoral thesis, Eindhoven University of Technology.

Endo, K.; Kobayashi, K.; Fujii, H.; and Ueno, Y. (1994). *Appl.Phys. Lett.* **64**, 146.

Floyd, P. D.; Merz, J. L.; Luo, H.; Furdyna, J. K.; Yokogawa, T.; and Yamada, Y. (1995). *Appl.Phys.Lett.* **66**, 2929.

Forstmann, G. G.; Barth, F.; Schweizer, H.; Moser, M.; Geng, C.; Scholz, F.; and O'Reilly, E. P. (1994). *Semicond. Sci. Techn.* **9**, 1268.

Frijlink, P. M. (1988). *J.Cryst.Growth* **93**, 207.

Gaines, J. M. (1995). *Philips J. Res.* **49**, 245.

Gaines, J. M.; Drenten, R. R.; Haberern, K. W.; Marshall, T.; Mensz, P.; and Petruzello, J. (1993). *Appl.Phys.Lett.* **62**, 2462.

Gaines, J. M., and Petruzzello, J., unpublished.

Geels, R. S.; Welch, D. F.; Scifres, D. R.; Bour, D. P.; Treat, D. W.; and Bringans, R. D. (1992a). *Electron. Lett.* **28**, 1460.

Geels, R. S.; Welch, D. F.; Scifres, D. R.; Bour, D. P.; Treat, D. W.; and Bringans, R. D. (1992b). *Electron.Lett.* **28**, 1810.

Glenn, W. E. (1993). Displays, high-definition television and lasers, *Techn Dig. Conf. on Advanced Solid-State Lasers and Compact Blue-Green Lasers,* New Orleans, 293.

Gooch, C. H. (1969). *Gallium Arsenide Lasers.* London: Wiley Interscience, 112–116.

Guha, S.; Cheng, H.; Haase, M. A.; DePuydt, J. M.; Qiu, J.; Wu, J.; and Hofler, G.E. (1994). *Appl.Phys. Lett.* **65**, 801.

Guha, S.; Qiu, J.; Haase, M. A.; DePuydt, J. M.; Cheng, H.; Guha, S.; Baude, P. F.; Hagedorn, M. S.; Hofler, G. E.; and Wu, B. J. (1993). *Appl.Phys.Lett.* **63**, 2315.

Haase, M. A.; Baude, P. F.; Hagedorn, M. S.; Chu, J.; DePuydt, J. M.; Cheng, H.; Guha, S.; Hofler, G. E.; and Wu, B. J. (1993). *Appl.Phys. Lett.* **63**, 2315.

Haase, M. A.; Qiu, J.; DePuydt, J. M.; and Cheng, H. (1991). *Appl.Phys.Lett.* **59**, 1272.

Hagen, S. H.; Valster, A.; Boermans, M. J. B.; and van der Heyden, J. J. M. (1990). *Appl.Phys. Lett.* **57**, 2291.

Hamada, H.; Shono, M.; Honda, S.; Hiroyama, R.; Matsukawa, K.; Yodoshi, K.; and Yamaguchi, T. (1991a). *Electron. Lett.* **27**, 661.

Hamada, H.; Shono, M.; Honda, S.; Hiroyama, R.; Yodoshi, K.; and Yamaguchi, T. (1991b). *I.E.E.E. J. Quantum. Electron.* QE-27, 1483.

Hamada, H.; Tominaga, K.; Shono, M.; Honda, S.; Yodoshi, K.; and Yamaguchi, T. (1992). *Electron.Lett.* **28**, 1834.

Hara, K.; Shinozawa, T.; Yoshino, J.; and Kukimoto, H. (1991). *Jap. J. Appl. Phys.* **30**, L437.

Harm, A. O. (1990). *Philips J. Res.* **45**, 177.

Hatakoshi, G.; Itaya, K.; Ishikawa, M.; Okajima, M.; and Uematsu, Y. (1991). *I.E.E.E. J. Quantum. Electron.* QE-27, 1476.

Heuken, M. (1995). *J.Cryst. Growth,* **146**, 570.

Ho, E.; Fisher, P.A.; House, J. L.; Petrich, G. S.; Kolodziejski, L. A.; Walker, J.; and Johnson, N. M. (1995). *Appl.Phys.Lett.* **66**, 1062.

Honda, S.; Shono, M.; Yodoshi, K.; Yamaguchi, T.; and Niina, T. (1994). *7th LEOS Annual Meeting, Boston.*

Iga, K.; Uenohara, H.; and Koyama, F. (1986). *Electron.Lett.* **22**, 1008.

Ikeda, M.; Sato, H.; Ohata, T.; Nakano, K.; Toda, A.; Kumagai, O.; and Kojima, C. (1987). *Appl. Phys. Lett.* **51**, 1572.

Ishibashi, A. (1995). *I.E.E.E. J. Select. Top. Quantum. Electron.* **1**, 741.

Ishikawa, M.; Ohba, Y.; Nagasaka, H.; Watanabe, Y.; Sugawara, H.; Yamamoto, M.; and Hatakoshi, G. (1986a). *10th I.E.E.E. Int. Semiconduct. Laser Conf., Kanazawa,* Paper A-1.

Ishikawa, M.; Ohba, Y.; Sugawara, H.; Yamamoto, M.; and Nakanisi, T. (1986b). *Appl. Phys. Lett.* **48**, 207.

Ishikawa, M.; Okuda, H.; Itaya, K.; Shiozawa, H.; and Uematsu, Y. (1989). *Jap. J. Appl. Phys.* **28**, 1615.

Ishikawa, M.; Shiozawa, H.; Tsuburai, Y.; and Uematsu, Y. (1990). *Electron.Lett.* **26**, 211.

Ishikawa, M.; Shiozawa, H.; Itaya, K.; Hatakoshi, G.; and Uematsu, Y. (1991). *I.E.E.E. J. Quantum. Electron.* QE-27, 23.

Itoh, K.; Kawamoto, T.; Amano, H.; Hiramatsu, K.; and Akasaki, I. (1991). *Jpn. J. Appl. Phys.* **30**, 1924.

Jeon, H.; Ding, J.; Patterson, W.; Nurmikko, A. V.; Xie, W.; Grillo, D. C.; Kobayashi, M.; and Gunshor, R. L. (1991). *Appl. Phys. Lett.* **59**, 3619.

Jongerius, M. J. (1995). *Philips J. Res.* **49**, 293.

Kamiyama, S.; Mori, Y.; Takahashi, Y.; and Ohnaka, K. (1991). *Appl.Phys.Lett.* **58**, 2595.

Kapon, E.; Katz, J.; and Yariv, A. (1984). *Optics Lett.* **9**, 125.

Katsuyama, T.; Yoshida, I.; Shinkai, J.; Hashimoto, J.; and Hayashi, H. (1990). *Electron. Lett.* **26**, 1375.

Kawata, S.; Fujii, H.; Kobayashi, K.; Gomyo, A.; Hino, I.; and Suzuki, T. (1987). *Electron. Lett.* **23**, 1327.

Kishino, K.; Kikuchi, A.; Kaneko, Y.; and Nomura, I. (1990). *12th I.E.E.E. Int. Semicond. Laser Conf., Davos*, Paper PD-10.

Kobayashi, R.; Hotta, H.; Miyasaka, F.; Hara, K.; and Kobayashi, K. (1995). *I.E.E.E. J. Select. Top. Quantum. Electron.* **1**, 723.

Liedenbaum, C. T. H. F. and Valster, A., unpublished.

Liedenbaum, C. T. H. F.; Valster, A.; Severens, A. L. G. L.; and 't Hooft, G. W. (1990). *Appl. Phys. Lett.* **57**, 2698.

Lin, M. E.; Fan, Z. F.; Ma, Z.; Allen, L. H.; and Morkoc, H. (1994). *Appl.Phys.Lett.* **64**, 887.

Marchant, A. B. (1990). *Optical Recording.* Reading, MA: Addison Wesley.

Meney, A. T.; Prins, A. D.; Phillips, A. F.; Sly, J. L.; O'Reilly, E. P.; Dunstan, D. J.; Adams, A. R.; and Valster, A. (1995). *I.E.E.E. J. Select. Top. Quantum. Electron.* **1**, 697.

Moser, M.; Winterhoff, R.; Geng, C., Queisser, I.; Scholz, F.; and Dornen, A. (1994). *Appl. Phys. Lett.* **64**, 235.

Nakamura, S. and Fasol, G. "The Blue Laser Diode (GaN based light emitters and lasers)" Springer, Berlin, 1997.

Nakamura, S. (1991). *Jpn. J. Appl. Phys. 30*, L 1705.

Nakamura, S.; Harada, Y.; Senoh, M.; (1991a). *Appl. Phys. Lett.* **58**, 2021.

Nakamura, S.; Mukai, T.; and Senoh, M. (1994). *Appl. Phys. Lett.* **64**, 1687.

Nakamura, S.; Senoh, M.; and Mukai, T. (1991b). *Jpn. J. Appl. Phys.* **30**, L 1708.

Nakamura, S.; Senoh, M.; Iwasa, N.; and Nagahama, S. (1995). *Jpn. J. Appl. Phys.* **34**, L797.

Nakamura, S.; Senoh, M.; Nagahama, S.; Iwasa, N.; Yamada, T.; Matsushita, T.; Kiyoku, H.; and Sugimoto, S. (1996a). *Jpn. J. Appl. Phys.* **35**, L74.

Nakamura, S.; Senoh, M.; Nagahama, S.; Iwasa, N.; Yamada, T.; Matsushita, T.; Kiyoku, H.; and Sugimoto, Y. (1996b). *Appl. Phys. Lett.* **68**, 2105.

Neugebauer, J., and van de Walle, C. G. (1996). *Appl. Phys. Lett.* **68**, 1829.

Nurmikko, A. V., and Gunshor, R. L. (1994). *I.E.E.E. J. Quantum. Electron.* QE- *30*, 619.

Okuyama, H.; Miyajima, T.; Morinaga, Y.; Hiei, F.; Ozawa, M.; and Akimoto, K. (1992). *Electron. Lett.* **28**, 1798.

Olego, D.; Petruzello, J.; and Sun, G., unpublished.

Prins, A. D.; Sly, J. L.; Meney, A. T.; Dunstan, D. J.; O'Reilly, E. P.; Adams, A. R.; and Valster, A. (1994a). *Proc. 22nd Intern. Conf. on the Physics of Semiconductors*, Vancouver, August 14–19, 727.

Prins, A. D.; Sly, J. L.; Meney, A. T.; Dunstan, D. J.; O'Reilly, E. P.; Adams, A. R.; and Valster, A. (1994b). *Proc. 22nd Int. Conf. on the Physics of Semiconductors*, Vancouver, August 14–19, 719.

Rennie, J.; Watanabe, M.; Okajima, M.; and Hatakoshi, G. (1992). *Electron. Lett.* **28**, 150.

Sale, T. E. (1995). *Vertical Cavity Surface Emitting Lasers*. New York: John Wiley and Sons.

Schemmann, M. F. C.; van der Poel, C. J.; van Bakel, B. A. H.; Ambrosius, H. P. M. M.; Valster, A.; van den Heijkant, J. A. M.; and Acket, G. A. (1995). *Appl.Phys. Lett.* **66**, 920.

Schneider, R. P., Jr.; Hagerott Crawford, M.; Choquette, K. D.; Lear, K. L.; Kilcoyne, S. P.; and Figiel, J. J. (1995). *Appl.Phys.Lett.* **67**, 329.

Serreze, H. B.; Chen, Y. C.; Waters, R. G.; and Harding, C. M. (1991). *Conference on Lasers and Electro-Optics, Boston,* Paper CPDP-1.

Serreze, H. B., and Harding, C. M. (1992). *Electron. Lett.* **28**, 2115.

Shay, J. L., and Wernick, J. H. (1975). *Ternary Chalcopyrite Semiconductors: Growth, Electronic Properties and Applications*. Oxford, England: Pergamon Press.

Shul, R. J.; Kilcoyne, S. P.; Hagerott Crawford, M.; Parmenter, J. E.; Vartuli, C. B.; Abernathy, C. R.; and Pearton, S. J. (1995). *Appl.Phys.Lett.* **66**, 1761.

Streifer, W.; Osinski, M.; Scifres, D. R.; Welch, D. F.; and Cross, P. S. (1986). *10th I.E.E.E. Int. Semiconductor Laser Conf., Kanazawa,* Paper F-3.

Strite, S.; Lin, M. E.; and Morkoc, H. (1993). *Thin Solid Films* **231**, 197.

Suzuki, T., and Gomyo, A. (1993). In *Semiconductor Interfaces at the Subnanometer Scale* (eds. H. W. Salemink and M. B. Pashley), 11. Dordrecht, Boston, London: Kluwer.

Sun, C. J.; Yang, J. W.; Chen, Q.; Khan, M. A.; George, T.; Chang-Chien, P.; and Mahajan, S. (1996). *Appl.Phys.Lett.* **68**, 1129.

Takagi, T.; Koyama, F.; and Iga, K. (1991). *I.E.E.E. J. Quantum. Electron.* QE-27, 1511.

Tanaka, T., Minagawa, S., Kawano, T., and Kajimura, T. (1989). *Electron.Lett.* **25**, 905.

Thijs, P. J. A.; Binsma, J. J. M.; Tiemeijer, L. F.; and van Dongen, T. (1991). *Proc. 17th European Conference on Optical Communication, Paris,* 31.

Thompson, G. H. B. (1980). *The Physics of Semiconductor Laser Devices.* Chichester, NY: John Wiley and Sons.

Tsukada, T. (1974). *J.Appl.Phys.* **45**, 4899.

Ueda, O. (1988). *J.Electrochem. Soc. Reviews and News,* 12C.

Unger, P.; Roentgen, P.; and Bona, G. L. (1992). *Electron.Lett.* **28**, 1531.

Valster, A.; Finke, M. N.; Boermans, M. J. B.; van der Heijden, J. M. M.; Spreuwenberg, C. J. G.; and Liedenbaum, C. T. H. F. R. (1990a). *12th I.E.E.E. Semiconductor Laser Conf., Davos,* Paper PD-12.

Valster, A.; Liedenbaum, C. T. H. H.; van der Heijden, J. M. M.; Finke, M. N.; Severens, A. G. L.; and Boermans, M. J. B. (1990c). *12th I.E.E.E. Int. Semiconductor Laser Conf., Davos,* Paper C-1.

Valster, A.; Liedenbaum, C. T. H. F.; Finke, M. N.; Severens, A. L. G.; Boermans, M. J. B.; Vandenhoudt, D. E. E.; and Bulle-Lieuwma, C. W. T. (1991). *J.Cryst. Growth* **107**, 403.

Valster, A.; van der Heijden, J.; Boermans, M.; and Finke, M. (1990b). *Philips J.Res.* **45**, 267.

Valster, A., and van der Poel, C. J. (1994). *14th I.E.E.E. Int. Semiconductor Laser Conf., Hawaii,* Paper W 3.1.

Valster, A.; van der Poel, C. J.; Finke, M. N.; and Boermans, M. J. B. (1992). *Electron. Lett.* **28**, 144.

Van der Poel, C. J.; Ambrosius, H. P. M. M.; Linders, R. W. M.; Peeters, R. M.; Acket, G. A.; and Krijn, M. P. C. M. (1993). *Appl.Phys.Lett.* **63**, 2312.

Van der Poel, C. J.; Valster, A.; Finke, M. N.; and Boermans, M. J. B. (1992). *13th I.E.E.E. Int. Semiconductor Laser Conf., Takamatsu,* Paper J2.

Welch, D. F.; Wang, T.; and Scifres, D. R. (1991). *Electron.Lett.* **27**, 693.

Yamamoto, S.; Hayashi, H.; Hayakawa, T.; Miyauchi, N.; Yano, S.; and Hijikata, T. (1983). *I.E.E.E. J. Quantum.Electron.* QE-19, 1009.

Yamasaki, S.; Asami, S.; Shibata, N.; Koike, M.; Manabe, K.; Tanaka, T.; Amano, H.; and Akasaki, I. (1995). *Appl.Phys.Lett.* **66**, 1112.

Chapter 2

Long Wavelength (λ > 2 μm) Semiconductor Lasers

P. G. Eliseev

P. N. Lebedev Physics Institute, Moscow, Russia; present address: Center for High Technology Materials, University of New Mexico, Albuquerque, NM. E-mail: eliseev@chtm.unm.edu

Abstract

The state of the art of semiconductor lasers emitting in the wavelength range $\lambda > 2$ μm is presented with some historic excursions. Laser types based on both interband and intraband optical transitions are described. Material systems of III–V, II–VI, and IV–VI semiconductor compounds and alloys for laser heterostructures are presented. The most important actual problems are the laser action at room temperature, the longest wavelength limit, the lowest threshold for laser oscillations, and the increase of the mode tuning range in tunable lasers. The applications of long-wavelength semiconductor lasers are reviewed briefly.

2.1 Introduction

Semiconductor lasers have been successfully introduced into everyday practice. However, they still remain the focus of intense research and development. One direction is a study of new materials and structures

71

for lasers in the middle and far IR range. The scientific issues relevant to IR semiconductor lasers are numerous and may be classified according to the transitions used to obtain the stimulated emission. There are the following mechanisms:

1. *Interband (bipolar)* mechanism based on the interband optical transitions; the emission wavelength λ is closely associated with an energy band-gap E_g.

2. *Unipolar* intraband mechanism of transitions between sublevels formed in the high magnetic field, including the cyclotron resonance levels and levels of shallow impurity (hot-hole lasers, etc.).

3. *Unipolar* intraband mechanism of transitions between levels in quantum wells with a tunneling pumping (quantum-cascade lasers, etc.).

4. *Optical conversion* mechanisms by a parametric interaction of electromagnetic waves in semiconductor media (difference-frequency generation, laser action based on nonlinear scattering).

Interband-transition lasers were introduced in 1962, and they are most widely known as the conventional laser diodes (LDs). There are currently commercial productions and large-scale applications of these lasers. Scientific publications on these lasers predominate the semiconductor laser literature. Consequently, we devote the main part of this chapter to interband semiconductor lasers. Other types are in a laboratory stage, although they can be considered as quite competitive devices to the interband ones. The hot-hole lasers are most suitable for the far-IR region, whereas tunneling-based quantum-cascade lasers (QCLs) are promising for high-temperature operation in the mid-IR region.

There are numerous versions of the conventional laser structures, based mostly on semiconductor alloys. Some of them (mainly lead-salt–based lasers) have found applications in the high-resolution IR spectroscopy of atoms and molecules where the tunability of semiconductor lasers plays an important role. Spectroscopic sensors and high-sensitivity heterodyne-type IR receivers are also demonstrated using the narrow-line IR semiconductor lasers. However, most of these lasers are capable of operating below room temperature (RT) and need cryogenic accessories. RT-operating IR devices would be very desirable to increase the scale of application of these laser diodes.

Several approaches of the development aimed for RT-operating laser diodes: One is an improvement in the design and technology of conventional laser diodes, and another is the construction of quantum-size tunneling structures with unipolar laser mechanism. In this review chapter, we consider some results for lasers of both types operating in the range $\lambda > 2\ \mu$m with attention to material and structure problems. All compound types suitable for laser application in the LDs as well as in other laser devices are presented here (Sect. 2–7). Most important materials relate to families of III–V, II–VI, and IV–VI compounds and alloys (Sect. 4–6). A comparative discussion and actual issues of conventional LDs are given in Sect. 7. Sections 8–10 relate to some nonconventional laser devices, and we give a short survey of applications of mid-IR semiconductor lasers and conclusions in Sect. 11 and 12, respectively.

2.2 Wavelength ranges of various laser materials

The accessible spectral coverage of semiconductor lasers based on various compounds and alloys is shown in Figs. 2.1–2.4. There are a number of

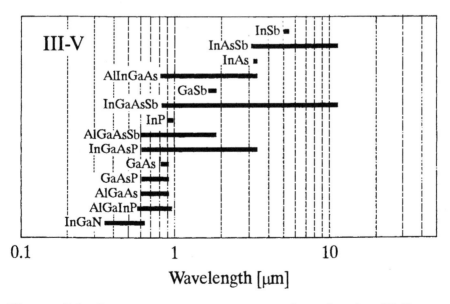

Figure 2.1: Spectral coverage of semiconductor lasers based on III–V compounds and alloys.

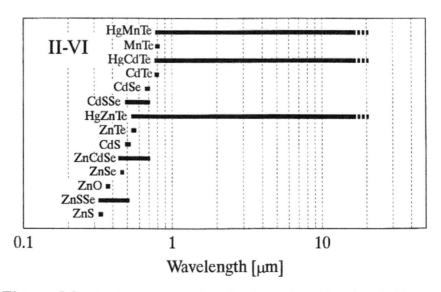

Figure 2.2: Spectral coverage of semiconductor lasers based on II–VI compounds and alloys.

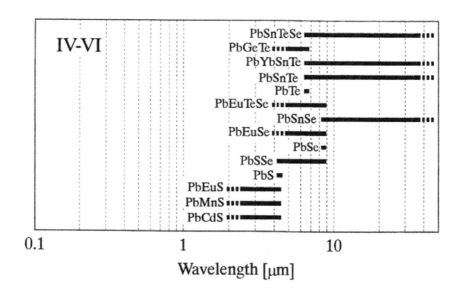

Figure 2.3: Spectral coverage of semiconductor lasers based on IV–VI compounds and alloys.

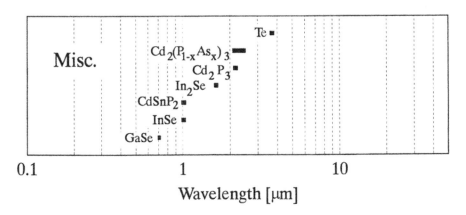

Figure 2.4: Spectral coverage of semiconductor lasers based on compounds and alloys from miscellenous groups.

groups of compounds, and for long-wave emission some of them are of most importance, namely, groups of III–V, II–VI, and IV–VI types. A list of narrow band gap active compounds (with no alloys) for the laser applications in the long-wavelength is shown in Table 2.1. Binary active materials are GaSb, InAs, InSb (III–V type); and PbS, PbSe, and PbTe (IV–VI type). Representatives of other groups have been tested for the laser action under optical (Cd_3P_2) or electron-beam excitation (only elementary semiconductor Te). Incidentally, tellurium was predicted as a probable laser material by Watanabe and Nishizawa (1957) in their very early patent application on the "semiconductor optical maser." The reason to choose tellurium was that the expected emission wavelength would be rather long, and it might be relatively easier to use maser experience for making the cavity to provide a stimulated emission. Tellurium appeared indeed to be a laser-active medium under electron beam pumping, with an emission near the wavelength of 3.7 μm (see Table 2.1), but there have been no further laser-related publications on this material since 1965.

A typical requirement for interband-transition laser materials is quite obvious: The energy band gap E_g has to correspond to the desirable wavelength according to the relationship

$$E_g = 1.239/\lambda, \qquad (2.1)$$

where E_g is in electron-Volts and λ is in micrometers. The requirements for the quantum-cascade laser materials follow from the consideration of

Material	Crystal Structure	Lattice Constant, Å	Band Gap, meV		Laser Applications	References on Early Laser Applications
			LT	RT		
GaSb	zincblende	6.0959	813	720	DH (RT)	Dolginov et al. (1976b)
InAs	zincblende	6.0584	425	360	HJ (LT), DH (RT)	Melngailis (1963) Kobayashi and Horikoshi (1980)
InSb	zincblende	6.4794	236	180	HJ (LT)	Bernard et al. (1963) Phelan et al. (1963)
PbS	rock salt	5.936	290	410	DH (LT)	Ishida et al. (1989)
PbSe	rock salt	6.124	165	290	DH (near RT)	Spanger et al. (1988a)
PbTe	rock salt	6.462	190	320	DH (LT)	Feit et al. (1991b)
Cd_3P_2	tetragonal	$a = 8.75$ $c = 12.12$	550	—	OP (LT), $\lambda = 2.12 \, \mu m$	Bishop et al. (1969)
Te	trigonal	$a = 4.48$ $b = 3.74$ $c = 5.912$	320	380	EP (LT), $\lambda = 3.7 \, \mu m$	A la Guillaume and Debever (1965a)

DH = double heterostructure; HJ = homojunction structure; OP = optical pumping; EP = electron-beam pumping; RT = room temperature; LT = low (cryogenic) temperature.

Table 2.1: Narrow-band semiconductors used in mid-IR lasers

the energy difference between the quantum-confined states as the transitions occur between such states in one (conduction) band. The contemporary experience in this field is limited to utilization of quantum-well and superlattice structures based on GaAlAs/GaAs and InGaAlAs/InP. For unipolar lasers operating in the far-IR ($\lambda > 100$ μm), only p-Ge has been used so far as an active material.

2.3 Growth techniques

The bulk semiconductor materials are used in lasers mainly as substrates for multilayer structures. An exception is germanium, used in hot-hole lasers with a relatively large active volume. Also, bulk active material can be utilized in semiconductor lasers with electron-beam pumping and photopumping and in semiconductor spin-flip lasers. Fabrication techniques of crystalline ingots of Ge, GaAs, InSb and other semiconductors are well known. High-quality GaAs and InP substrate wafers 2 inches in diameter are commercially available. The opportunity to prepare ingots of alloys is very limited. Only a few alloy types are available as bulk material: $In_xGa_{1-x}As$ (with small x), HgCdTe, PbSnTe, and some others. Practically, a large-scale production of laser structures has to be based on the utilization of binary substrates (GaAs, InP, GaSb InAs, etc.).

In order to prepare the p–n structure in one material (homostructure), both diffusion and epitaxial technique are used. However, all modern laser structures are heterostructures, *i.e.*, they include chemically different layers. There are also diffusion and epitaxial approaches in the fabrication of heterostructures. The former is applicable to very simple structures containing only heterojunctions (single heterostructures). The technique of compositional interdiffusion (CID) was used to prepare single heterostructures in PbSSe and PbSnSe (Linden *et al.*, 1977). This technique allows change in the chemical content of the material via substituting by diffusion of some component. In IV–VI compounds, the equilibrium vapor pressure of the narrower-gap binary component is higher than that of wider-gap component. By anneal of PbSnSe wafer in a sealed ampoule with a sample of another composition (less or no tin content), one can obtain a surface layer with a wider gap. The depth of the composition change is dependent on the time and the temperature of the anneal. Typically, a CID in PbSnSe alloy system of 6 h at 600 °C results in a depth of 5 μm. The migration exchange process is based on self-diffusion, which is normally very slow. Therefore CID technique is applicable to few material systems.

More successful and productive techniques for a preparation of heterostructures are the epitaxial growth procedures using the starting phase of liquid (liquid phase epitaxy, LPE), vapor (vapor phase epitaxy, VPE) or molecular flux in vacuum (molecular beam epitaxy, MBE). LPE appeared to be very fruitful in the past but appeared to be not competitive with modern variants of VPE. The LPE technique is relatively simple and cheap. An attractive property is the gettering of rapid diffusing contaminant impurities during the growth process and their strong segregation to a liquid phase. It allows preparation of high-purity structures using starting materials in a wide range of purity.

A disadvantage of LPE is a limited composition and thickness control during the process. A consequence is poor thickness reproducibility in the growth of ultrathin layers. Because of this, many technologists consider LPE to be unsuitable for the quantum well and superlattice structures. Another consequence is a compositional change during the process. The component segregation produces a deficiency of some of the component in the liquid phase. As a result, one has to limit the layer thickness if composition uniformity is desired. The limited number of layers that can be grown in one LPE run is also a shortcoming.

As mentioned before, in some alloys for a device application, the composition miscibility presents a difficulty. The instability of the solid phase leads to a nonequilibrium state at the growth front. The LPE is a near-equilibrium technique; therefore its applicability to the preparation of a such mixture is strictly limited. This difficulty restricts a fabrication of InGaSbAs alloys, particularly at composition lattice matched to GaSb with In content exceeding 20–25% (emission wavelength more than 2.4–2.5 μm at RT). A near-equilibrium technique is also not efficient in the growth of mixtures with strongly segregating components. This is the case of alloys like InGaAlAs or InAlSbAs, where non-nearest neighbors of the periodic table are mixed.

VPE and MBE are growth techniques allowing larger deviation of the growth conditions from the equilibrium. A hot-wall epitaxial (HWE) method is also a technique of this sort, similar to MBE. These techniques seem to be more fruitful for the above-mentioned problematic alloy systems. The HWE is suitable for preparing high-quality layers, including ultrathin ones, and is not expensive. It appears to be useful for lead-salt structures.

The MBE is a highly flexible technique with precise flux control for growing a wide range of materials and structures. It became a frontier technique in the semiconductor quantum well and other microstructure preparations. The VPE and its advanced metalorganic version (MOVPE or metalorganic chemical vapor deposition, MOCVD) are quite competitive in this field and give excellent commercial results in shorter wavelength laser fabrication.

A sizable miscibility is typical for many mixtures, particularly of InAs–InSb, InAs–GaAs, InSb–GaSb, PbSe–SnSe, etc. Inside the miscibility ranges, there are numerous alloy compositions covering a wide range of E_g. The immiscibility phenomenon was found in mixtures GaAs–GaSb and in quaternaries of InGaAsSb and InGaAlSb. The alloy is thermodynamically unstable in some composition ranges and cannot be prepared by the near-equilibrium growth technique (like the liquid phase epitaxy, LPE). This does not mean that the material does not exist at all: The unstable alloy can be prepared by a nonequilibrium technique like molecular beam epitaxy (MBE). However, the usage of unstable material may be limited due to the diffusion-assisted segregation in the solid phase. In any case, devices based on the unstable materials have to be tested for their reliability under long-time operation and storage.

Growth of metastable compositions of InGaSbAs alloys inside the miscibility gap using MOVPE has been reported by Cherng et al. (1986). The material had been prepared with the band gap energy of 0.45 eV (In content $x = 0.32$) and 0.37 eV ($x = 0.74$) being lattice-matched to GaSb substrate and also with $E_g = 0.74$ eV ($x = 0.29, y = 0.76$) lattice-matched to InP substrate. The LT photoluminescence of these metastable alloys consists of a single spectral peak attributed to the interband emission. The half-widths of the peaks are broader than those for the metastable ternary alloy GaAsSb, indicating that a significant amount of composition clustering occurs during the growth. Another report on the metastable alloy was given by Chuiu et al. (1987), who performed the MBE growth of the alloy with $x = 0.26$ lattice-matched to GaSb substrate.

Most alloys can be prepared only as a thin epitaxial layer. Therefore the substrate material has to be optimized for each alloy composition. The optimization means the lattice matching and the available substrate materials are few: Only binaries can be grown in the form of high-quality bulk ingots. Therefore, the lattice-matched epitaxy is limited by alloys having the same lattice parameters as one of the binary substrate materials. Ternaries are typically mismatched to any of the composing binaries. The preparation of the high-quality epilayers is much easier in the case of quaternary alloys that have two chemical degrees of freedom: A quaternary alloy may be composed in such a manner that both degrees of freedom compensate each other in the influence on the lattice parameter. As a result, families of quaternaries (or multinaries) lattice-matched to a binary compound could be prepared. Quaternary alloys of InGaSbAs and InSbAsP are of that sort that can be lattice-matched with respect to either InAs or GaSb.

More generally, heterostructures can be classified according to crystalline quality into the following groups:

1. Lattice-matched (isomorphic interface, no misfit dislocations, no stress).

2. Strained (with a pseudomorphic interface and no misfit dislocations, but the stress produces an elastic deformation that modifies the crystal symmetry).

3. Poorly matched (the misfit stress is relaxed by a dislocation formation, causing incoherent interface; the density of the misfit dislocation network increases as the relative misfit grows). At a large misfit, the epilayer can become polycrystalline and therefore useless for device applications.

In fact, the device heterostructures may be related to the first group if the relative misfit is less that about 10^{-3}, and absolute ideal matching is not possible over all temperature ranges, as the temperature expansion coefficient cannot be well adjusted simultaneously with the lattice parameter matching. A second group consists of device heterostructures with intentionally introduced strain to modify the active medium characteristics in favor of a desirable emission property. In this group, there are only the quantum-well structures because the opportunity to avoid the dislocation formation exists only for very thin epilayers. The critical thickness appears to be dependent on the relative lattice misfit $\Delta a/a$ and on mechanical properties of materials. Most widely investigated lattice-matched heterostructures are listed in Table 2.2 as they had been introduced in pioneering publications. All of them were tested as active materials in laser diodes.

Heterojunction	Lattice-Matching Condition	References on Early Experiments
$Al_xGa_{1-x}Sb_{1-y}As_y/GaSb$	$y = 0.09x/(1 + 0.06x)$	Dolginov *et al.* (1976a)
$In_xGa_{1-x}Sb_{1-y}As_y/GaSb$	$y = 0.87x/(1 - 0.05x)$	Dolginov *et al.* (1978a)
$In_{1-x}Ga_xSb_yAs_{1-y}/InAs$	$y = 0.88x/(1 - 0.05x)$	Aidaraliev *et al.* (1989)
$InSb_xAs_{1-x-y}P_y/InAs$	$y \sim 2.2x$	Kobayashi and Horikoshi (1980)
$PbS/Pb_{1-x-y}Cd_xSr_yS$	$y \sim 0.57x$	Koguchi and Takahashi (1991)
$PbS/Pb_{1-x}Cd_xSe_yS_{1-y}$	$y \sim 2.19x$	Koguchi *et al.* (1989)
$PbSe/Pb_{1-x}Sn_xTe_ySe_{1-y}$	$y \sim 2.4x$	Davarashvili *et al.* (1977)
$Pb_{1-x}Sn_xTe/PbTe_{1-y}Se_y$	$x \sim -2.5y + a$	Kasemset *et al.* (1980)
$Pb_{1-x}Eu_xTe_{1-y}Se_y/PbTe$	$y \sim 0.4x$	Partin (1984a)

Table 2.2: Some lattice-matched heterojunctions applicable to diode laser structure.

2.4   Compounds and alloys of III–V type

2.4.1   Brief review of III–V materials for long-wavelength laser applications

Binary compounds GaSb, InAs, and InSb are similar to GaAs in some respects, having the same zinc-blend crystal structure and direct gaps suitable for laser applications. They are chemically and mechanically stable. Cleavage along (110) planes is the routine technique for preparation of high-quality mirror facets of the optical resonator. It is worthwhile to mention that with a decrease of the band gap in the III–V family of compounds, some regular trends are observable: decrease of the effective masses of carriers, decrease of heat conductivity, and decrease of the mechanical hardness and strength. As to the latter property, the indium antimonide InSb is significantly more sensitive to the mechanical treatments than GaAs: Even a usual lapping procedure may produce rather deep damage—microcracks, dislocation loops, etc. The cleavage technique needs more delicate elaboration in this case.

The interrelation between E_g and the lattice parameter is shown in Fig. 2.5. Ternary compositions are presented by lines, whereas quaternaries may be found in the areas confined by these lines.

The alloy of InGaSbAs has been widely studied since 1978, and some commercial production exists for IR emitting diodes and photodiodes based on this material. The dependence of the energy gap on the composition x at RT along the line for the alloy of $In_xGa_{1-x}Sb_{1-y}As_y$ lattice-matched to GaSb is given by Dolginov et al. (1983) as

$$E_g(x) = 730 - 990x + 560x^2 \text{ (meV)} \tag{2.2}$$

The heterojunctions formed by these materials are of two basic types, I and II. In the case of the type-I heterojunction, the band edge offsets are of signs such that the carrier energy at the band edge is lower in the narrow band gap side. This provides a confinement effect for both conduction electrons and holes in the narrow band gap side. This side represents the potential well for both types of carriers. This heterojunction type is found in GaAs/AlGaAs and in numerous other combinations. In contrast to this, the type-II heterojunction is characterized by band-edge offsets providing a separation of carriers: One side is a potential well for electrons, whereas the other side is the potential well for holes. It appears that the heterojunctions GaSb/InAs and some GaSb/InGaSbAs and InAs/

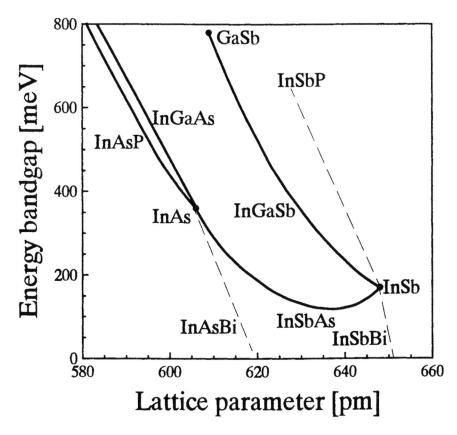

Figure 2.5: Interrelation between the lattice parameter and energy gap in long-wavelength III–V materials.

InGaSbAs are of type II. The difference between these heterojunction types is shown in Fig. 2.6.

Another relative alloy is $InSb_{1-y}As_y$, which is a part of the quaternary system ($x = 0$). The $E_g(y)$ passes through a minimum near $y = 0.64$, where its value is about 0.1 eV at RT (therefore it is less than E_g in both InAs and InSb). According to Fang *et al.* (1990b), at 10 K the value of $E_g(y)$ is given by

$$E_g(y) = 235(1 - y) + 415y - 672y(1 - y) \text{ (meV)} \qquad (2.3)$$

Some III–V alloys have even narrower band gap, but the technology of their preparation is not well developed. For example, there are Bi-

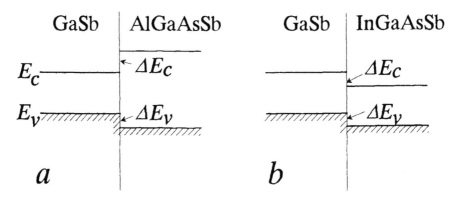

Figure 2.6: Examples of type I (a) and type II (b) heterojunctions.

containing compositions $InP_{1-x}Bi_x$, $InAs_{1-x}Bi_x$, and $InSb_{1-x}Bi_x$, preliminarily analyzed by Berding *et al.* (1988). It was concluded that to obtain E_g = 100 meV the composition has to be x = 0.44 for InPBi, x = 0.14 for InAsBi, and x = 0.05 for InSbBi. A zinc-blend–type binary compound of InBi is a semimetal, and the transition semiconductor–semimetal occurs probably at $x \sim 0.11$ in $InSb_{1-x}Bi_x$. Fang *et al.* (1990a) investigated InAsBi alloy at $x < 2.3\%$ and found that the energy band gap decreases by 55 meV per molar percent of Bi, whereas introducing Bi into the InSbAs alloy gives a shift of the low-temperature luminescence peak at a rate of -46 meV per molar percent of Bi.

A possible laser material of GaN_xAs_{1-x} (cubic modification) can be mentioned for long wavelength applications. Due to the a strong mixture effect by substitution of N and As atoms in the V-group sublattice of the alloy, the energy band gap is calculated to go to a zero value at relatively low N content (about 10%). This alloy is a direct-gap material. Thus, by growing GaN_xAs_{1-x} in the composition range of 0.05–0.1, a long-wavelength laser structure may be projected.

III–V alloy systems for long-wavelength lasers are presented in Table 2.3 with selected references on corresponding laser experiments.

2.4.2.   Lasers based on III–V materials

The laser action at LT in III–V compound materials has been known since the 1960s. In the long-wavelength range the pioneering demonstrations

Material	λ, μm	Comments	References
InSbAs	3.2	LD (LT)	Basov et al. (1966)
	3.4	LD (CW, LT)	Choi et al. (1996)
	3.9	OP (LT)	Van der Ziel et al. (1986b)
InAsP	2.0	LD (LT)	Quinn and Manley (1966)
InSbAsP	3.1–3.7	EP (LT)	Dolginov et al. (1978d)
	2.6–3.0	LD (LT)	Kobayashi and Horikoshi (1980)
	3.9	LD (LT)	Zotova et al. (1983)
	3.0–4.0	OP (LT)	Van der Ziel et al. (1985)
	4.2	IRED(RT)	Krier (1990)
InGaSbAs	1.86–2.4	EP (RT)	Dolginov et al. (1976a)
	1.9	IRED (LT)	Dolginov et al. (1978c)
	2	LD (LT)	Kano and Sugiyama (1980)
	1.8–2.3	LD (RT)	Bochkarev et al. (1985)
	2.4	LD (RT)	Drakin et al. (1987)
	2.38	LD (CW, RT)	Bochkarev et al. (1988a)
	2.0	LD (CW, RT)	Baranov et al. (1988a)
	2.48–2.52	LD (RT)	Baranov et al. (1988b)
	2.51	LD (RT)	Tournie et al. (1990)
	2.78	LD (LT)	Garbuzov et al. (1995)
	3.95	LD (LT)	Aidaraliev et al. (1989)

LD = laser diode; OP = optical pumping; EP = electron-beam pumping.; IRED = IR emitting diode; CW = continuous-wave operation; LT = low temperature; RT = room temperature.

Table 2.3: III–V alloy systems used in mid-IR semiconductor lasers.

of the laser action were given by Melngailis (1963) in an InAs diode laser ($\lambda = 3.2$ μm) and by Phelan *et al.* (1963) in an InSb diode laser ($\lambda = 5.2$ μm). The latter remains as the longer emission wavelength III–V laser, but operating only below 77 K. Quinn and Manley (1965) demonstrated a laser action in InAsP alloy at about 2 μm. Basov *et al.* (1966) obtained laser action at 3.2 μm in InSbAs homojunction diodes. A detailed study of electrical and optical characteristics of InAs homojunction lasers was given by Patel and Yariv (1970). They had measured the threshold current density near 400 A/cm^2 and determined the internal optical loss coefficient to be 40 cm^{-1} at 20 K. All these lasers operated at low temperatures; therefore their practical application was restricted due to the necessity of using cryogenic equipment. In order to increase the operating temperature it was necessary to elaborate new types of emitting structures.

   Important steps were the development of heterostructures at the end of the 1960s and the introduction of lattice-matched quaternary heterostructures in the 1970s. The development of the long-wavelength III–V-type heterostructure lasers had begun by demonstration of quaternary AlGaSbAs/GaSb lasers operating at 1.5–1.8 μm at room temperature by Dolginov *et al.* (1976a, 1976b). The interest was shifted then to the heterostructures of InGaSbAs/GaSb, which were recognized as a base for emitters in the range $\lambda > 1.5$ μm. Pioneering works were reported in the 1970s on the laser action in InGaSbAs bulk epilayers by the electron-beam pumping by Dolginov *et al.* (1976a) and also Bondar *et al.* (1976). The double-heterostructure laser InGaSbAs/GaSb grown by liquid-phase epitaxy (LPE) was reported to operate at 1.9 μm at low temperature by Dolginov *et al.* (1978a, 1978b). It became clear that deeper studies on the LPE process and phase equilibrium were necessary for improved growth technique of these materials to obtain RT operation and to extend the wavelength range. Kano and Sugiyama (1980) achieved the CW operation in the InGaSbAs/GaSb double-heterostructure laser, also at low temperature (80 K). A difficulty with RT operation might be related to a poor electronic confinement and also to optical antiguiding in the double heterostructure of this type. Therefore, a cladding material other than GaSb would be desirable.

   Next step was introducing the quaternary AlGaSbAs as the wide band gap side of the heterojunction. Kobayashi *et al.* (1980a, 1980b) worked out an improved version of the laser with the lattice-matched alloys of AlGaSbAs/InGaSbAs/AlGaSbAs forming the double heterostructure on the GaSb substrate. The RT laser action was obtained at wavelength of 1.8 μm.

The coherent emission from InGaSbAs/GaSb structures and wave-guiding properties have been studied by Aleksandrov et al. (1984), who realized the distributed-feedback (DFB) corrugation and observed the laser action in DFB-mode at low temperature.

Later Dutta et al. (1985) obtained RT laser action in this type of heterostructure at 2.0 μm under optical pumping. Also by such a pumping, Van der Ziel et al. (1986a, 1986b) observed laser emission at 2.07 μm in similar structures prepared by the molecular beam epitaxy.

Caneau et al. (1985, 1986) have made improvements in the hetero-structure by increasing the aluminum content in claddings up to 34% and by optimizing the active layer thickness. They reported the threshold current density of 3.5 kA/cm^2 at RT in diodes with 0.5-μm-thick active layer. The wavelength was 2.2 μm.

Baranov et al. (1986b) obtained the RT laser action in such a structure at 2.06 μm under current density 3.7–6.0 kA/cm^2. Srivastava et al. (1986) reported a decrease of the threshold current density and achievement of the CW operation at 220–235 K. Drakin et al. (1986) described the extension of the laser wavelength to 2.4 μm achieved by an increase of the In and As molar fractions in the lattice-matched alloy just near the boundary of the miscibility gap. Details of these developments were given also by Dolginov et al. (1983), Bochkarev et al. (1985), Bochkarev et al. (1986), and Drakin et al. (1987). Akimova et al. (1988) reported a study of the laser performance of InGaSbAs-based DH lasers operating in the wavelength range 2.0–2.4 μm at RT. The threshold current density was measured to be 5.4 kA/cm^2 at $\lambda = 2$ μm and 7.6 kA/cm^2 at $\lambda = 2.4$ μm (active layer thickness 1.0 and 1.6 μm, respectively). At low temperature (77 K) a threshold current density of 0.1 kA/cm^2 was observed at $\lambda = 2.1$ μm, and CW operation was obtained.

Further decrease of the threshold current density to 1.7 kA/cm^2 in 2.226-μm laser diode has been reported by Caneau et al. (1987). The active layer thickness was reduced to ~0.5 μm. The nominal current density (related to the 1-μm thickness) was 3.4 kA/cm$^2\mu$m. In principle, such a low threshold could be enough to obtain the CW operation at RT.

The CW-RT laser action was obtained at first in the low-mesa stripe diodes for the wavelength range 2.2–2.4 μm by Bochkarev et al. (1988a). The LPE wafer preparation is described also by Bochkarev et al. (1988b) and Baranov *et al.* (1988a). A strong influence of the lattice mismatch on the morphology of the heterostructures was found. The wafers with $\Delta a/a < 10^{-3}$ appeared to be suitable for laser action at RT in the pulse regime of operation. However, only a few of the lasers show the sufficiently low threshold for CW-RT lasing as it was noticed by Bochkarev *et al.* (1988a). It was found that diodes suitable for CW-RT operation have to have a pulsed threshold current as low as 75–100 mA or less. The stripe width was about 10 μm, and apparent current confinement was observed in those samples where the cap layer was entirely etched outside the stripe and the thickness of the upper cladding layer was reduced to about 1 μm. Otherwise the current spreading was large causing the undesirable increase of the lasing area and the increase of the threshold current. Improvement of the laser characteristics was found when the aluminum content in cladding layers was increased to 0.54. The CW-RT laser action was obtained also in diodes with a "terraced-substrates" stripe geometry providing the internal "self-aligned" current confinement due to a formation of the stripe window in the blocking layer at the step of the substrate terraces (Bochkarev *et al.*, 1988a, 1991). The lowest RT threshold current in these diodes was 45 mA.

About the same time, Baranov *et al.* (1988a) reported CW at RT in the stripe-geometry laser diodes emitting at 2.0 μm with much higher threshold current (200–300 mA at RT). A high quality of the heterostructures has been confirmed by very low threshold current density of about 50 A/cm^2, as observed at 77 K. Baranov *et al.* (1988b) reported an observation of the pulsed laser emission at a longest wavelength of 2.52 μm for this type of heterostructure.

A short review of the studies concerning laser diodes in the 2.0–2.5 μm wavelength range is presented by Baranov *et al.* (1989). They indicated the observation of an output optical power as large as 1 W in the above mentioned spectral range, the CW operation in the range 2.0–2.2 μm, and pulsed laser operation up to 2.5 μm at RT. The miscibility gap was indicated in the $In_xGa_{1-x}Sb_{1-y}As_y$ alloy in the composition range $0.25 < x < 0.75$. It was stated to be difficult to obtain the epilayers with a content close to the miscibility gap, so the longest wavelength of this type of lasers was concluded to be near 2.4–2.5 μm. Three different classes of heterostructures were tested for laser applications:

1. *n*-AlGaSbAs/*n*-InGaSbAs/*p*-AlGaSbAs, the conventional double heterostructure.

2. n-AlGaSbAs/n-InGaSbAs/pGaSb/p-AlGaSbAs, type-II heterojunction double heterostructure.

3. n-AlGaSbAs/n-GaSb/n-InGaSbAs/p-GaSb/p-AlGaSbAs, separate-confinement type-II heterojunction double heterostructure.

With respect to the type-II heterojunction structures, a new version of the laser action has been proposed by Baranov *et al.* (1986a), who suggested the quantum well formation in the depletion region of the II-type heterojunction formed by InGaSbAs/GaSb. These two layers were inserted in the double heterostructure confined by AlGaSbAs layers. The "non-injection" (but rather tunneling) mechanism was postulated for this structure. We shall discuss this type of pumping in Sect. 2.4.4.

Channeled-type buried stripe laser diodes with a crescent-shaped active region were fabricated for CW operation at RT in the 2 μm wavelength range. The emitting region was formed by p-GaSb layer and n-InGaSbAs layer ($x = 0.1$, Te-doped) with a donor concentration in the range $(1–3) \times 10^{17}$ cm^{-3} and with a total thickness 0.6 μm. Both layers are confined by cladding layers of GaAlSbAs ($x = 0.34$) 2–3-μm thick. The structure was grown by LPE on the (001) oriented n-GaSb substrate (Te-doped, donor concentration 9×10^{17} cm^{-3}) with 10-μm-wide groove channels made by wet process using photolithography. Current blocking outside the channel stripe was achieved by introducing p-GaSb barrier layer. In order to obtain the low-resistance contacts to the p-side, p^+-GaSb cap layer was grown (Ge-doped, acceptor concentration 5×10^{19} cm^{-3}). Ohmic contacts to this cap layer were prepared by alloying the eutectic Au-Ge mixture, and to the n-GaSb side by alloying of Au-Te. The specific series resistance of contacts was not more than 2×10^{-4} Ohm.cm^2. Diode chips were 200–500 μm long as prepared by cleaving. They were mounted by soldering on a standard copper holder, junction down. It was observed in the spontaneous emission spectrum that a single band dominates in the 4.2–300 K range. The threshold current was, in some samples, as low as 6–15 mA at 4.2 K, and it increases up to 14–30 mA at 77 K and 240–500 mA at 300 K. The temperature constant T_0 was 89 K. The CW operation was observed at a maximal temperature of 20 °C. Laser emission was linearly polarized (polarization degree of 0.97) with TM mode predominance. The spontaneous emission was also polarized but to a lower degree (polarization degree of 0.23–0.3, TM-mode). At a current exceeding the threshold by 1.5–2 times, only the laser mode remains in the spectrum. The main lobe of the laser beam was 30° wide in vertical plane with some side lobes; the divergence in the horizontal plane was 25°, also with a subsidiary structure. The central part of the far-field pattern was nearly circular.

Baranov *et al.* (1988c) reported the preparation of the longest wavelength laser by LPE growth of InGaSbAs alloy with x in the range 0.24–0.28. The Al content in the cladding layers was increased to 0.52. The growth process was performed on (111)B GaSb substrates. The diodes of the mesa-stripe type and also three-side cleaved (high-Q-factor cavity) type were prepared. It was observed that the laser emission wavelength is dependent on the resonator length and increases when the length grows. The wavelength appears to be maximal in the all-cleaved-side devices. This was used to demonstrate a laser action at the longest wavelength of 2.52 μm at RT in a DH-structure with In content x = 0.29 and As content y = 0.24.

Baranov (1991) reported a study concerning the linewidth enhancement factor (α factor) for GaInSbAs/GaSb double-heterostructure lasers emitting near λ = 2 μm at room temperature (1.8 μm at 82 K). The structures were grown by LPE on n-GaSb substrates with an active medium composed of two layers of n-In$_x$Ga$_{1-x}$Sb$_{1-y}$As$_y$ ($x \sim 0.1$) and n-GaSb, each 500-nm thick, sandwiched between the wider band gap claddings of Al$_x$Ga$_{1-x}$Sb$_{1-y}$As$_y$ (x = 0.34) to provide optical and electron confinements. All the epitaxial layers were lattice-matched to the substrate. The lasers had a substrate-channel buried structure with a channel width of about 10 μm. The measurements were performed at 82 K and the threshold current amounted to about 20 mA at this temperature (200–400 mA at room temperature). The laser emission was TE-polarized, and the gain spectra were measured for this polarization. The α-factor was measured to be 3.1 at the lasing wavelength and varied from $\sim$3 to 8 in the wavelength range from 1.77 to 1.81 μm. These measurements related to the routine "type-I" heterojunction formed at the n-GaSb/p-AlGaSbAs junction. In addition to this, if the junction was formed at the p-GaSb/n-GaInSbAs interface, the type-II heterojunction would be involved in the pumping mechanism. Baranov (1991) stated the observation of the appearance of modes corresponding to the type-II heterojunction laser action with α-factor being at least two times smaller, which is suggested to be attributed to the formation of a quantum-well structure at the heteroboundary under lasing conditions.

DFB laser diodes made of InGaSbAs/GaSb were reported by Vasil'ev *et al.* (1990). The samples were prepared by the two-step LPE procedure.

At the first step the p-GaSb layer and the quaternary layer were grown progressively on the p-GaSb substrate at 630 °C. Then the surface corrugation was produced by the holographic lithography. The corrugation period was 468 nm to provide a laser action at a wavelength controlled by the DFB grating. At second LPE step the surface of the quaternary layer was overgrown by the n-GaSb layer. Laser action was observed at 77 K in the DFB-mode near 1.78 μm. The temperature shift of the laser wavelength was 0.12 nm/K; the refractive index of quaternary alloy emitting at that wavelength was estimated to be 3.82.

Choi and Eglash (1991a) reported the emission properties of MBE-grown stripe geometry gain-guided DH laser operating at 2.2 μm. The laser structure consists of the following layers: 0.8 μm thick n-GaSb buffer (at GaSb substrate); 2 μm thick n-$Al_{0.75}Ga_{0.25}Sb_{0.94}As_{0.06}$ cladding; 0.4 μm thick undoped p-$Ga_{0.84}In_{0.16}As_{0.14}Sb_{0.86}$ active layer; 2 μm thick p-$Al_{0.75}Ga_{0.25}Sb_{0.94}As_{0.06}$ cladding; and 0.05 μm thick p^+-GaSb cap layer. All layers are closely lattice matched.

MBE growth technique allowed preparation of high Al content (75%) cladding layers, which had not been used earlier in LPE grown diodes, as the latter epitaxy technique is very difficult for lattice-matched quaternary material with Al content higher than 0.4. But wide band gap cladding is desirable to prevent the excess carrier leakage and to provide stronger optical confinement. Low-threshold 30-μm-wide stripe diodes were prepared by etching both cap and p-cladding layers to a depth of 0.2 μm through a photoresist mask and by metallizing of entire surface with Ti/Au to form an Ohmic contact to cap layer stripe and Schottky contact to the cladding layer. Along with the stripe-geometry diode, the broad-area (300 μm wide) laser diodes were also made with cavity length L from 300 to 1000 μm. As L increases, the threshold current density decreases monotonically from 1.3 kA/cm^2 to 940 A/ cm^2 at RT. The latter value gives normalized threshold (J_{th}/d) as low as 2.35 kA/cm$^2\mu$m. The emission spectrum was centered near 2.19 μm. In narrow stripe devices, spectral peak was observed at 2.17 μm, and the shift was a result of greater band filling, which occurs because J_{th} increases with decreasing stripe width.

Output powers as high as 10.5 mW, 4.6 mW, and 0.5 mW per facet were measured at heat sink temperatures of 5 °C, 20 °C, and 30 °C, respectively. The CW threshold current increases from 117 mA at 5 °C to 250 mA at 30 °C. These values are 12 and 77 mA higher than the respective pulsed values because the junction temperature is higher under CW operation. Values of T_0 were 49 K between 5 and 35 °C and 44 K between 35 and 55 °C. These measurements demonstrated the decrease in the absolute and normalized threshold current density and the increase in the CW output power in MBE-grown laser diodes as compared to the LPE-grown ones described earlier. The maximal CW operation temperature was elevated to 303 K.

In another report Choi and Eglash (1991b) described pulsed (broad stripe, 300 μm wide) lasers operating at 2.27 μm prepared by MBE technique on n-GaSb (100) Te-doped substrates. The active layer was 0.4 μm thick. The relative lattice matching was within $\pm 1.5 \times 10^{-3}$. Differential quantum efficiency was 50%, optical loss coefficient was about 43 cm^{-1}, and the threshold current density was 1.5 kA/cm^2 at the cavity length 700 μm. A peak optical power of 900 mW/facet was measured in 500-ns-long pulses. The parameter T_0 was found to be 75 K between -120 and -50 °C and 50 K near RT. The internal quantum efficiency was estimated to be close to 100%.

The first quantum-well structures of InGaSbAs/AlGaSbAs were fabricated by Choi and Eglash (1992) by MBE on GaSb substrates. The active region consisted of five 10-nm-thick wells of In$_{0.16}$Ga$_{0.84}$Sb$_{0.86}$As$_{0.14}$ separated by 20-nm-thick AlGaSbAs barriers. Some compressive strain was measured in the wells by X-ray diffraction in spite of the fact that compositions chosen to obtain the lattice-matching. The broad-stripe laser diodes emitting at $\sim$ 2.1 μm exhibited pulsed operation up to 423 K with T_0 parameter of 113 K at RT and 45 K at highest temperature. Lowest RT threshold current density was 260 A/cm^2 in 1-mm-long diodes, and the differential quantum efficiency was $\sim$0.7 in 300-μm-long diodes. The internal quantum efficiency was estimated to be 0.87 and optical loss coefficient was $\sim$10 cm^{-1}. CW operation was obtained up to 293 K with output power of 190 mW/facet.

Garbuzov et al. (1995) reported the laser action at 2.7-μm wavelength in InGaSbAs/AlGaSbAs MQW structure grown by MBE. Four wells of the active region were formed by 10-nm-thick layers of In$_{0.24}$Ga$_{0.76}$Sb$_{0.84}$As$_{0.16}$. Stripe diodes were fabricated with a stripe width of 22 μm. It was noticed that diodes had a tendency to operate in a dominant single mode over well-defined temperature and current intervals. CW operation was obtained at highest temperature of 234 K in a 700-μm long diode.

Data on laser diodes based on InGaSbAs/GaSb are given in Table 2.4. The laser performance has significantly improved since 1978, and this material remains the most promising candidate for the wavelength range 1.9–2.7 μm.

Active Medium, Structure	Wavelength, μm	T, K	Threshold Current Density, A/cm²	References
InGaAsSb, DH	1.9	90	900	Dolginov et al. (1978b)
InGaAsSb, DH	2.2	300	1700	Caneau et al. (1987)
InGaAsSb, DH	2.2	300	5400	Akimova et al. (1988)
	2.4	300	7600	
InGaAsSb, DH	~2	300	940	Choi and Eglash (1991a)
$In_{0.16}Ga_{0.84}As_{0.14}Sb_{0.86}$, MQW	2.1	300	260	Choi and Eglash (1992)
$In_{0.25}Ga_{0.75}Sb/InAs$, QSL	3.28	80	100–400	Chow et al. (1995)
InGaAsSb, DH	2.8	300	—	Lee et al. (1995)
$In_{0.24}Ga_{0.76}As_{0.16}Sb_{0.84}$, MQW	2.7	234 CW	—	Garbuzov et al. (1995)

DH = double heterostructure; MQW = multi-quantum well; QSL = quantum superlattice.

Table 2.4: Selected data on InGaAsSb/GaSb and InGaAsSb/AlGaAsSb injection lasers (GaSb as a substrates).

2.4.3   Type-II heterojunction InGaSbAs/GaSb

According to a traditional classification (see, for example, Kroemer and Griffiths, 1983; Esaki, 1985), heterojunctions with the same sign of both band-edge offsets at the interface are termed "type II". InGaSbAs/GaSb is just of this type, whereas many more known heterojunctions like GaAlAs/GaAs and InGaAsP/InP are of type I, with opposite signs of the offsets. The interface may serve to confine both excess electron and hole carriers at one side of the heterojunction in type I, but to separate electrons and holes in type II. The former is more favorable for electron and hole confinement in DH laser structures.

In the case of GaSb/InGaSbAs/GaSb DH structures, the leakage of holes may take place, because a narrow band gap middle layer provides a potential well for electrons but a barrier for holes. This leakage can be modified if edge bending provides a carrier localization in the middle layer. By adding Al into cladding layers, the problem may be avoided, because the heterojunction of AlGaSbAs/InGaSbAs becomes type I at a sufficiently large amount of Al. According to Shen *et al.* (1995), the transition from type I to type II may occur for the composition of $x < 0.7$ in $In_{1-x}Ga_xSb_{0.96}As_{0.04} / Al_{0.22}Ga_{0.78}Sb_{0.98}As_{0.02}$ system (for light holes). The conduction-band offset ratio $\Delta E_c/\Delta E_g$ was estimated as 0.66 ± 0.01 for the type I heterojunction range.

On the other hand, a type II heterojunction provides some new features, unusual in ordinary laser diodes. A bending of the condition band edge near the heterojunction interface can provide a potential well for electrons. It suggested that in this well the quantum-confined states may appear and participate in the laser action in a manner similar to that in quantum-well lasers. The localization of carriers near the type II heterojunction gives rise to a specific interface recombination process due to tunneling of carriers from localized states. The process is known as "photon-assisted tunneling" (PAT). Kroemer and Griffiths (1983) proposed a tunable light source based on this mechanism in the type II heterojunction. The tunability is predicted due to the influence of the applied forward bias on the energy of the photon that is assisting the tunneling.

The laser action in InGaSbAs/GaSb DH structure is attributed by Baranov *et al.* (1988b) to the nonconventional mechanism specific to the type II heterojunction. It may be connected with the above-mentioned photon-assisted tunneling (with no injection) identified as one sort of interface radiative recombination. The model of noninjection laser diode proposed by Baranov *et al.* (1988b) suggests the quantum-size confinement of carriers, accumulated near the interface but at opposite sides. Therefore the model includes a consideration of quantum-well properties of such laser emission similar to that in normal injection QW-lasers, but with no preparation of ultrathin layers.

Baranov *et al.* (1988c) reported a study of the photoluminescence in the type II p-GaSb–p-InGaSbAs heterojunction obtained by the LPE-growth of a quaternary layer ($x = 0.2$) on the (111)B GaSb substrate of the same conductivity type in both the layer and substrate. Under the optical excitation at wavelength 1.52 μm, several lines were observed in the spectrum of the emission. One of them, denoted A_0-line, was explained as due to the interband transitions in the bulk of the narrow band gap layer, whereas others having longer wavelength (I-lines) were explained as a result of the recombination of excess electrons, localized in the potential well at the heterojunction (an interface emission). It was stated that these electrons can recombine with holes at both sides of the well. In some cases, the interface emission is produced by the recombination of electrons with holes in the bulk of the narrow band gap side (line I_1); in other cases the emission is produced by recombination with holes, localized at the opposite side of the interface (line I_2).

Titkov *et al.* (1990a, 1990b) investigated the properties of the interface radiative recombination in the n-GaSb-n–InGaSbAs–p-GaSb heterostructure containing n–n and p–n heterojunctions of type II. Measurements were performed at a temperature of 2 K under optical and current excitation. Four spectral bands were found in the range 2.0–2.8 μm. They observed a bias-tunable spectral band of the interface emission (sometimes a doublet) in the wavelength range 2.2–2.6 μm denoted as I-line, and bulk emission was also observable. Corresponding lines are denoted as A-line for radiative transitions of electrons from conduction band to the acceptor level and A_0-line for interband transitions. The shift of spectral lines was measured in a magnetic field at 2 K. The 2-D character of the electronic states in the wells was demonstrated by the vanishing of the shift of the I-line in the magnetic field as the latter was oriented parallel to the interface plane. It was noticed that the I-line emission is linearly polarized at a TM mode manner, whereas A and A_0 lines are not polarized. The time-resolved behavior of I-line emission is found to be much slower than that of bulk lines, and this is to be expected for the PAT mechanism. It was stated also that, at high bias, the lines A and A_0 become more intense than the interface recombination I-line, namely when the applied bias V becomes sufficiently large, *i.e.*, $V > E_g/e$, where E_g is the band gap of the narrow band gap side.

The TM-polarization predominance is explained as a result of the light hole tunneling through the interface barriers. As to the doublet structure of the I-line, it was suggested that two quantum-confined states of electrons participate in the radiative transitions.

Consider briefly the experimental evidence for the noninjection model of laser action. First, the laser photon energy does not follow the temperature dependence of the band gap, as otherwise expected, but deviates due to the influence of the quantum confinement in the heterojunction wells. But, due to the band-filling effect, the photon energy of laser emission is not fixed with respect to the band gap and may vary to some temperature-dependent extent for a given diode. Therefore, lasing photon energy does not follow the band gap generally, and the deviations from the band gap temperature dependence may have other causes than the quantum confinement. A second argument is that the laser emission is TM polarized, in contrast to the one in normal AlGaAs/GaAs lasers where TE polarization predominates. A polarization of the laser emission seems to be dependent on the residual strain in the active region. One of the usual causes of the strain is the lattice mismatch, and another one is the mounting stress. The lattice mismatch in the quaternary lasers may be of different sign due to more degrees of freedom in composition variations, in contrast to GaAlAs/GaAs laser. If the active region is subjected to biaxial tension, the TM mode of polarization appears to be dominant. Therefore, the predominance of TM emission may have an origin other than the tunneling mechanism suggested in the noninjection model. A third argument is that nonclassical I–V characteristics were observed in these diodes. This relates to low-current range where accurate dc measurements are possible. In this range, the carrier transport through the junction is nonclassical due to a contribution of different tunneling mechanisms. The question remains whether such tunneling mechanism appears to be responsible for the laser action at higher currents.

In the injection model, the excess carriers (with a density of $\sim 10^{18}$ cm^{-3} for RT laser action) of both signs are accumulated in the same volume producing a quasi-neutrality. They provide the optical gain in proportion to the bulk recombination rate in the active volume. In the noninjection model, the same optical gain has to be supplied by recombination of carriers spatially separated by the potential barrier in the heterojunction. A smaller wave function spatial overlapping is characteristic for this case as compared with the bulk process. We would overestimate the recombination rate if we assumed the same recombination probability for the tunneling-assisted process as in the bulk case. But such assumption allows us to estimate the upper level of the current density supplied by the tunneling. An accumulation of carriers increases the electric charge at both sides of the interface. The only reason for preventing them from diffusing toward each other is the interface barrier. It is normally of a height close to a band gap E_g. Accumulated charge reduces the barrier height, and this leads to the thermal injection. When a charge Q and $-Q$ of carriers on both sides of the junction barrier causes a potential depression comparable to the diffusion potential, the thermal diffusion would dominate. A rough estimation for an upper limit of the radiative tunneling current is

$$J_{max} < B(\varepsilon \, \varepsilon_0 E_g)^2/e^3 d^3, \tag{2.4}$$

where B is the coefficient of interband radiative recombination, ε and ε_0 are relative and vacuum dielectric permittivity, and d is a thickness of the barrier (Eliseev, 1991). The right side of Eq. (2.4) corresponds to a condition that the accumulated charge Q causes a barrier depression of E_g/e in the plane dielectric capacitor with a gap of d. This value is one crucial parameter of the problem. It influences the current estimation very strongly. Assuming $d = 20$ nm, $E_g = 0.5$ eV, and $\varepsilon = 15$, one obtains $J_{max} < 35$ A/cm^2. This value seems too low to provide the laser action at RT. A higher upper limit of current can be reached at smaller d, but this is hardly the case for LPE-grown heterojunction. Thus the noninjection pumping is not a promising mechanism for the RT lasers. Published studies of MBE-grown lasers deal with DH and QW heterostructures involving injection pumping. Those are characterized by reasonable and low-threshold current density at LT. The LT threshold current appears to be of the same order of magnitude in both conventional lasers and in those that are suggested to operate by a noninjection mechanism.

2.4.4 InSbAsP/InAs and some other lasers

Dolginov *et al.* (1978d) demonstrated the electron-beam-pumped laser using LPE-grown InSbAsP in the wavelength range 3.1–3.7 μm. Employing the same alloy, Esina *et al.* (1983) fabricated IR-emitting diodes with a wavelength of 4.6 μm at RT. The optical pumping of quaternary alloy InSbAsP at wavelength up to 3.9 μm (at 125 K) was demonstrated by van der Ziel *et al.* (1985).

The low-temperature (77 K) laser action in InSbAsP/InAs double heterostructure diodes was reported by Zotova *et al.* (1986) in the range of 3.6–3.9 μm and in a series of publications by Aidaraliev *et al.* (1987a, 1987b, 1988) in the range 3.0–3.3 μm. Similarly, laser action was obtained in InSbAsP/InSbAs (77 K) by Mani *et al.* (1988) at 3.3 μm, in InSbAsP/InSbAsP (15–55 K) by Akiba *et al.* (1988) at 2.5–2.7 μm, in InGaAs/InAsP (210 K) by Martinelli *et al.* (1989) at 1.58–2.45 μm, and in InGaAs/InSbAsP (77 K) by Aidaraliev *et al.* (1989) at 2.44–2.52 μm (80–190 K) and at 2.90–3.55 μm. A very low threshold current density of 39 A/cm^2 at LT was stated in the latter report for diodes capable of CW laser operation.

Zotova *et al.* (1986) obtained a laser action in DH diodes with InAs active medium at LT. Further results were published by Aidaraliev *et al.* (1987a, 1987b). The DH structures n-InSbAsP–n-InAs–p-InSbAsP were grown on InAs substrate and studied at 4–110 K. In four-side-cleaved diodes, the laser emission had been observed at 3.05 μm (77 K) with the threshold current density of about 3 kA/cm^2. Aidaraliev *et al.* (1987a) reported a laser action in the range 3.0–3.3 μm (77K) in similar heterostructures. The CW operation was also reported at 3.1 μm with a threshold current density of 240 A/cm^2. Lower thresholds were reported by Aidaraliev *et al.* (1988) for the same type of heterostructures with Sb content of 0.05–0.062 and P content of 0.10–0.19 in the confining layers. The band gap of confining layers was 430–510 meV at 77 K. The refractive index step was estimated as 0.02–0.03. Active layer thickness was 1.5–3 μm. The pulsed operation threshold current density was measured to be 60 A/cm^2 (77 K) at the wavelength 3.05 μm. For CW operation, the density was 70 A/cm^2. The maximal operation temperature of pulsed laser action was 140 K. Temperature parameter T_0 was as small as 17 K, probably due to a strong increase of nonradiative Auger process at higher temperatures. It was noticed that the misfit dislocation network does not seem to limit a laser operation at LT. In a strained DH structure with n-InSb$_{0.07}$As$_{0.93}$ active region operated at 3.55 μm (77 K), the threshold was about 86 A/cm^2 at relative mismatch $\Delta a/a = 1.8 \times 10^{-3}$ (Baranov *et al.*, 1990).

Martinelli and Zamerovski (1990) have studied InGaAs/InSbAsP DH laser structures grown by a hydride VPE technique on the InP substrates. The active layer composition was chosen as In$_{0.82}$Ga$_{0.18}$As (for laser emission in the range about 2.52 μm), not lattice-matched to the substrate. A graded 20-μm-thick InGaAs layer was introduced between the substrate and the DH structure with In content changing from 0.53 to 0.82. The cladding layer composition was $InSb_{0.1}As_{0.14}P_{0.76}$ (providing a conduction-band offset of 0.11 eV and a valence-band offset of 0.3 eV). Active layer thickness was in the range of 0.5–0.7 μm with a calculated optical confinement parameter of 0.5 or more. The threshold current density J_{th} was 0.41 ± 0.11 kA/cm^2 at 80 K as averaged for a number of 40-μm-wide, 200-μm-long stripe laser diodes. The temperature dependence of J_{th} was characterized by parameter $T_0 = 41$ K below 180 K and $T_0 = 29$ K above 180 K. The highest pulse-operating temperature was 234 K at wavelength 2.54 μm. It was noticed that the misfit dislocation network seems not to limit a laser operation at LT.

Choi and Turner (1995) prepared by MBE the strained quantum well InAsSb/InAlAs laser structure on GaSb substrate. The active layer of InAsSb was compressively strained. The laser emission power of 540 mW was obtained under pulsed conditions at the wavelength of ~4.5 μm (95 K). The laser threshold current density was as low as 350 A/cm^2 at 50 K, and the highest operation temperature was 85 K using electrical pulses and 144 K using pulsed optical pumping.

Fabrication and study of strained InAs/In$_{0.53}$Ga$_{0.47}$As single quantum-well heterostructure lasers grown by MBE on InP substrates was reported by Tournie et al. (1994). The active layer thickness were varied from 2 to 23 monolayers of InAs. An extremely high biaxial compressive strain was applied to the InAs active well (~3.2%). The emission wavelength was found to be shifted to 2.38 μm at RT. The LT laser action was tuned to 1.84 μm with the threshold current density of 475 A/cm^2 (80 K). The LT CW operation was achieved with narrow stripe devices but at shorter wavelengths (from 1.74 to ~1.58 μm) due to increased lateral losses.

Superlattice structures of InGaSb/InAs and InAs/AlSb for laser diodes operating in the 3–5 μm wavelength range were grown by Chow et al. (1995). For example, 5- or 6-period superlattice of In$_{0.25}$Ga$_{0.75}$Sb quantum wells/InAs barriers was sandwiched by InGaSbAs alloy claddings, all grown on the GaSb substrate. The laser wavelength ranged from 3.28 μm (maximum operation temperature of 170 K) to 3.9 μm (maximum operation temperature of 84 K). Hasenberg et al. (1995) reported lasing at a 3.5 μm wavelength in a broken-gap type superlattice laser diode structure of In$_x$Ga$_{1-x}$Sb/InAs.

Compressively strained quantum-well InSbAs/AlInSbAs lasers are reported by Choi and Turner (1995) and Turner et al. (1995). The laser action was obtained at $\lambda \sim 3.9$ μm. InSbAs-based MQW laser diodes for the 3.5–3.6 μm range are reported by Kurtz *et al.* (1996). The active medium was a strained (biaxially compressed) alloy of $InSb_{0.06}As_{0.94}$ and the barrier material was InAs. The thickness of each of 10 quantum wells was 9 nm. Pulsed laser action was observed below 135 K with $T_0 \sim 33$ K. It was proposed that to increase the T_0-parameter, the barrier material has to be chosen such that it provides larger valence band-edge offset and larger heavy–light hole splitting in the QW by higher strain. Choi *et al.* (1996) fabricated InSbAs/AlInSbAs strained MQW laser structures for LT operation at 3.4 μm by MBE. Ten compressively strained 15-nm-thick active layers of $InSb_{0.065}As_{0.935}$ were separated by 30-nm-thick barrier layers of $Al_{0.15}In_{0.85}Sb_{0.1}As_{0.9}$. The structure was grown on InAs substrate. Broad-stripe laser diodes were tested for pulse and CW operation. The threshold current density was as low as 44 A/cm^2 at 80 K in a 1-mm-long laser diode with $T_0 \sim 50$ K at low temperature and ~ 25 K at 200 K. Maximum operation temperature was 220 K for pulse mode and 160 K for CW mode. Output power of 215 mW/facet was measured in the CW mode of operation at 80 K and 35 mW/facet at 150 K. The diode voltage of 4 V in CW operation regime was measured at 80 K and suggested to affect the power limitation by the Joule heat generation in the diode. The series resistance has to be reduced for higher power performance. It was found also that internal optical losses can be described by a loss coefficient of 9 cm^{-1} and internal quantum efficiency in the laser regime was estimated to be 63%.

In Table 2.5 we summarize results related to laser diode structures grown on InAs or InP substrates. These data cover the wavelength range of 2.4–4 μm. All represented laser diodes operate under cryogenic conditions.

2.5   Compounds and alloys of II–VI type

2.5.1   Brief review of II–VI materials for long-wavelength laser applications

Alloys of HgCdTe, HgMgTe, and HgZnTe seem to be suitable for long-wavelength laser applications. All of them are direct-gap semiconductors emitting over a wide wavelength range. CdTe and ZnTe are well-known laser semiconductors. MgTe in zinc blend modification has a band gap of 3.4 eV at LT and lattice parameter of 0.634 nm. However, HgTe is a semimetal and has an inverted band structure with a negative band gap

Active Medium, Structure	Wavelength, μm	T, K	Threshold Current Density, A/cm^2	References
InAs, HJ	3.1	77	5000	Melngailis and Rediker (1966)
InAs$_{0.98}$Sb$_{0.02}$, HJ	3.19	77	1000	Basov et al. (1966)
InAs, HJ	3.23	77	4500	Esina et al. (1967)
InAs, HJ	3.1	77	2000	Brown and Porteous (1967)
InAs, HJ	3.13	12	500	Anisimova et al. (1970)
InAs$_{0.94}$P$_{0.04}$Sb$_{0.02}$, DH	3.0	77	3000	Kobayashi and Horikoshi (1980)
InAs$_{0.95}$Sb$_{0.05}$, DH	3.2	77	4500	Mani et al. (1988)
In$_{0.82}$Ga$_{0.18}$As, DH	2.44	80	400	Martinelli and Zamerowski (1990)
(InP substrate)	2.52	190	—	
InAs$_{0.87}$Sb$_{0.13}$, HJ	3.90	4.2	200	Aidaraliev et al. (1993)
InAs$_{0.92}$Sb$_{0.08}$, SH	3.60	77	4000	
InAs, DH	3.05	77	64	
In$_{0.99}$Ga$_{0.01}$As, DH	3.04	77	100	
In$_{0.93}$Ga$_{0.07}$As$_{0.935}$Sb$_{0.065}$, DH	3.23	77	39	
In$_{0.935}$Ga$_{0.065}$As$_{0.935}$Sb$_{0.065}$, DH	3.29	77	80	
InAs$_{0.93}$Sb$_{0.07}$, DH	3.55	77	87	
InAs$_{0.94}$Sb$_{0.06}$, MQW	3.55	77	250	Kurtz et al. (1996)
InAs$_{0.935}$Sb$_{0.065}$, MQW	3.4	80	44	Choi et al. (1996)

HJ = homojunction; DH = double heterostructure; SH = single heterostructure; MQW = multi-quantum well.

Table 2.5: Injection lasers emitting in the 3.0–4.0 μm range (InAs as a substrate). Pulsed laser operation.

energy. The band gap in the alloys passes a zero value, providing a choice of compositions for far IR applications. In this family of semiconductor materials, HgCdTe is a most widely investigated system that covers a very wide IR range. The lattice parameter change is only 0.3% between endpoints in this ternary system (from 0.648 nm in CdTe to 0.646 nm in HgTe). This weak dependence of the lattice parameter on the alloy content makes it possible to grow the low-mismatch layers and heterostructures on CdTe substrates. The direct band gap can be adjusted from 1.529 eV (0.8 μm) for $x = 1$ (CdTe) to 0 for $x \sim 0.1$. The HgTe endpoint has a band gap of -0.115 eV. For laser applications in the $\lambda > 2.0$ μm range, the composition $x > 0.5$ is desirable (lattice mismatch is about 0.15%).

The composition dependence of the energy band gap of $Hg_{1-x}Cd_xTe$ was given by Hansen *et al.* (1982) as follows:

$$E_g(x, T) = -302 + 1930x + 0.535T(1 - 2x) - 810x^2 + 832x^3 \,(\text{meV}) \qquad (2.5)$$

The wavelength range for these emitting devices is estimated to be from 1 to 14 μm. The band gap of $Hg_{1-x}Zn_xTe$ is

$$\begin{aligned} E_g(x) &= -270 + 1300x + 1270x^2 \,(\text{meV}) \,(\text{at 77 K}) \qquad (2.6) \\ &= -170 + 1100x + 1270x^2 \,(\text{meV}) \,(\text{at 300 K}) \end{aligned}$$

as given by Dziuba *et al.* (1968).

2.5.2   II–VI-based IR semiconductor laser diodes

There are several reports on laser action in these materials under optical pumping. In HgCdTe, the wavelength of stimulated emission was measured for various alloy compositions at 3.8 and 4.1 μm (Melngailis and Strauss, 1966). Harman (1979) also demonstrated the optically pumped IR laser action in $Hg_{1-x}Cd_xTe$. The quantum-well structures grown by the photoassisted MBE technique were shown to yield stimulated emission at 2.8 μm wavelength under optical pumping by Giles *et al.* (1989). Laser emission at 5.4 μm was reported by Ravid *et al.* (1991) in optically pumped $Hg_{0.77}Zn_{0.23}Te$ at low temperatures.

There are several successful realizations of II–VI-based laser diodes. A well-known one is visible strained quantum-well injection lasers on ZnSe by Haase *et al.* (1991). Other versions are IR injection lasers based on II–VI, reported by Becla (1988), Zandian *et al.* (1991), Zucca *et al.* (1992a), and Zucca *et al.* (1992b).

The stimulated emission at 5.33 μm was observed by Becla (1988) from the VPE-grown heterostructure of HgCdMnTe/HgMnTe (active material was the ternary alloy $Hg_{1-x}Mn_xTe$). The threshold current density was above 1.2 kA/cm^2 at 77 K.

The IR electroluminescence from HgCdTe homojunction diodes was described by Verie and Granger (1965), Tarry (1986), Zucca *et al.* (1988), and Bouchut *et al.* (1991). The refractive index of $Hg_{1-x}Cd_xTe$ was studied by Kucera (1987). Haug (1989) theoretically considered the suitability of this alloy for the lasers taking into account the Auger processes as one of the most important sources of the nonradiative losses. He concluded that the alloy may appear to be a more efficient laser material than III–V alloys at $\lambda > 1.5$ μm because a calculated Auger coefficient decreases in HgCdTe strongly as the emission wavelength increases, from 10^{-28} cm^6/s at $E_g = 1$ eV to 10^{-34} cm^6/s at $E_g = 0.5$ eV. Jiang *et al.* (1991) have analyzed interband Auger and radiative recombination lifetimes in bulk and in quantum-well (QW) HgCdTe active media with an emission wavelength in the range of 2–5 μm. It was estimated that Auger recombination process becomes dominant in bulk double-heterostructure material above ~60 K, and it can cause an increase in the laser threshold current at higher temperatures. On the other hand, a decrease in the threshold in the case of QW is found due to the decrease of the Auger coefficient. At RT the calculated threshold was ~10 kA/cm^2 in the bulk double-heterostructure laser with an active-layer thickness of 0.3 μm, whereas it was 2 kA/cm^2 in a multi-QW structure with 15 active wells of HgCdTe spaced by CdTe barriers, each 10 nm thick.

The modeling of HgCdTe lasers was reported later by Singh and Zucca (1992) and Zucca *et al.* (1992a,b). It included the numerical calculation of the laser threshold for a separate-confinement double heterostructure with a series of active region thicknesses ranging from 10 nm to 1.5 μm. An internal optical loss of 30 cm^{-1} was assumed. Lacking direct experimental values of the Auger coefficient, C, for a 3.5-μm-emitting HgCdTe medium, the theoretical expression was used from the paper of Zucca *et al.* (1991), which was well confirmed for the 10-μm emitting material. Taking $C = 10^{-26}$ cm^6/s at 77 K, it was found that the Auger current is dominant only for extremely narrow quantum wells (< 10 nm), and it is about 30% of the radiative current at $d = 20$ nm, then it decreases rapidly as the well size is increased. The calculated total threshold current density at 77 K was 425 A/cm^2, whereas experimental values of 419 and

521 A/cm^2 were reported for 3.4-μm emitting laser diodes. This good agreement gives evidence that the Auger recombination is not expected to play a significant role in the lasers at the above-mentioned wavelength range, at least at low temperature.

The IR diode lasers based on HgCdTe were realized by Zandian *et al.* (1991). These are also reported by Zucca *et al.* (1992a) and Arias *et al.* (1993). Double heterostructures were grown by MBE at 433–463 K on $Cd_{0.96}Zn_{0.04}Te$ semi-insulating substrates. The substrate composition was chosen to minimize the dislocation density in the epilayers by a closer lattice matching. Effusive sources of Hg, CdTe, and Te$_2$ were used to grow the double heterostructures. Active layer thickness was from 0.9 to 1.4 μm. The composition grading layers ($\sim$150 nm) were intentionally inserted between the claddings and the active layer to reduce the mismatch effect. Confinement layers of *p*- and *n*-types were doped up to 10^{18} cm^{-3} with arsenic and indium, respectively. There are some difficulties with *p*-doping, as the arsenic is an amphoteric impurity. The doping was realized by the growth of CdTe:As/HgTe superlattice with consequent interdiffusion of As during the anneal step (typically, at 633 K, 1 h). The latter converts the superlattice into p^+-HgCdTe alloy. Simultaneously, the annealing under excess Hg pressure provides the annihilation of Hg vacancies formed in the alloy during the MBE growth. The edge wavelength of confining layers was 2.2–2.6 μm. A cap layer (200 nm) of HgTe was grown on top of the structure to lower the *p*-contact resistance.

Stripe-geometry laser diodes were prepared using a mesa etching (in Br-HBr solution) to *n*-layer providing both contacts at the same structure side. A 40-μm-wide stripe window was opened in the SiO$_2$ isolation layer at the mesa top. Good facets were formed by a cleavage on (110) planes to obtain a cavity length from 300 to 800 μm (substrate orientation was [211B]).

The devices were operated under pulsed currents at temperatures between 40 and 90 K. The 77 K stimulated emission wavelengths for these lasers were from 2.9 to 4.4 μm. Operation at 5.3 μm was demonstrated at 60 K. The lowest 77 K threshold current density was 419 A/cm^2 (active layer thickness of 1.3 μm), which is very close to the prediction of a numerical calculation. The characteristic temperature parameter T_0 was observed to be 22 K in diodes emitting at 2.9 μm (in the 77–90 K range). The HgCdTe laser diodes in that study were found to be well-behaved, stable devices that operated without failure while being tested.

2.6 Compounds and alloys of IV–VI type

2.6.1 Brief review of IV–VI materials for long-wavelength laser applications

The family of IV–VI laser compounds and alloys (chalcogenides) is known also as lead-salt-based material, as Pb is a necessary component. These binaries are PbS, PbSe, and PbTe. Alloys consist of two subfamilies: true IV–VI alloys (PbSSe, PbSnSe, PbSnTe, PbSnSeTe, PbGeTe) and heterovalent alloys (PbCdS, PbSrS, PbEuSe, etc.). Some important alloy systems are represented in Table 2.6. The crystalline structure of these materials corresponds to the rock-salt (NaCl) cubic type. Not only binaries, but some alloys (for example, PbSnTe) can be grown in a form of bulk ingots from the melt, and most of the alloys are known as epilayer materials. In many cases IV–VI-type heterostructures are not well lattice-matched, and, in spite of this, the laser action was observed at low temperatures. In narrow-band-gap materials like PbSe or PbTe the dominating mechanism of the nonradiative loss seems to be of intrinsic nature (the Auger process, etc.) rather than defect-related. Possibly, due to this, the misfit defects in some cases are not identified to be a factor in performance limitation. However, RT operation of all of IV–VI-type lasers remains an open problem.

Some characteristic properties of this material family may be itemized as follows:

1. Crystals are fragile and are not compatible with a rough mechanical treatments. Cleavage along {001} planes can be easily used for the cavity fabrication.

2. Stoichiometry of most materials is dependent on history of samples. Stoichiometric defects are electrically active, providing a background carrier concentration. As-grown bulk crystals are usually far from stoichiometry and need a long time annealing for decreasing the free carrier density.

3. Different epitaxial methods for the layered structure preparation are available in the device technology. Satisfactory crystal quality can be obtained by growing on heterogeneous substrates like KCl, BaF_2, etc.

4. There is a possibility of obtaining sufficient carrier injection not only in *p-n* junction but also in metal-semiconductor contact, as

Material	λ, μm	Comments	References
PbSeS	5–6	EP (LT)	Kurbatov et al. (1968)
	4–7.2	LD (LT)	Chashchin et al. (1970b)
	4.66	IRED (RT)	Lo and Swets (1980)
	4.0–5.0	LD (LT)	Preier et al. (1976)
PbCdS	3.5	HJ LD (LT)	Nill et al. (1973)
PbSnSe	10	LD (LT)	Shotov et al. (1980)
PbMnS	3.86	LD (LT)	Kowalczyk et al. (1981)
PbMnSe	7–8	LD (LT)	Kowalczyk and Szczerbakow (1988)
PbSnTe	8–10	SH LD (LT)	Walpole et al. (1973a)
		DH (LT)	Groves et al. (1974)
		LMH (LT)	Kasemset and Fonstad (1979)
PbGeTe	5–6	HJ LD (LT)	Anticliffe et al. (1972)
PbSnTeSe	12–16	LMH LD (LT)	Horikoshi et al. (1981)
	11	LMH LD (LT)	Horikoshi et al. (1982a)
	46	D (LT)	Kurbatov et al. (1983)
HgZnTe	5.4	OP (LT)	Ravid et al. (1991)
HgCdTe	4	IRED (LT)	Verie and Granger (1965)
	3.8	OP (LT)	Melngailis and Strauss (1966)
	1.25–2.97	OP (LT)	Harman (1979)
	2.9	OP (LT)	Ravid et al. (1989)
	2.8	OP (LT)	Giles et al. (1989)
HgMnTe	5.33	LD (LT)	Becla (1988)
$Cd_3(P_{1-x}As_x)_2$	2.14–2.45	OP (LT)	Arushanov et al. (1982)

LD = laser diode; OP = optical pumping; EP = electron-beam pumping; IRED = IR emitting diode; CW = continuous-wave operation; LT = low temperature; RT = room temperature; HJ = homojunction structure; LMH = lattice-matched heterostructure; SH = single heterostructure.

Table 2.6: IV–VI and other alloy systems used in mid-IR semiconductor lasers.

described by Nill *et al.* (1970) with PbTe and PbSnTe-based laser diodes provided with Pb, Zn, or In contacts.

5. All lead-salt materials have a high dielectric constant and a high refractive index. The extraction of the spontaneous emission is limited by internal reflections (the reflectivity of natural facets is 40–55% for normally incident light).

6. As all materials containing heavy atoms (Pb, Sn), the lead salts have rather low heat conductivity, and this makes heat sinking difficult in devices operating at high levels of dissipated power.

7. Lead-salt materials are chemically stable and available for the conventional material processing in the device technologies.

All the above-mentioned binaries and most of alloy compositions are direct-gap materials. Exclusions of this rule are some compositions with a zero band gap; examples are $Pb_{1-x}Sn_xSe$ at $x = 0.56$ and $Pb_{1-x}Sn_xTe$ at $x = 0.27$, both at RT. It is interesting to note that band inversion is not an obstacle to obtaining a laser emission in the case of PbSnSe, as was shown by Harman *et al.* (1969). The laser diodes operating at 12 K were prepared with a composition at both sides of the zero-band-gap crossover point. In the $Pb_{1-x}Ge_xTe$ alloys, the temperature coefficient of the energy band gap appears to be composition-dependent and continuously passes the zero point near $x \sim 0.03$, where the laser wavelength seems to be temperature-independent in the range of 5–77 K, as reported by Anticliffe *et al.* (1972).

Quaternary alloys of IV–VI type are important as materials useful for preparation of well-lattice-matched heterojunctions. Existence of solid solutions PbSnTeSe, PbSnSeS, and PbTeSeS was shown by Abrikosov *et al.* (1969). Quaternary double heterostructures were proposed by Davarashvili *et al.* (1977), based on PbSnTeSe and PbTeSeS alloys. The possibility of obtaining the entirely lattice-matched structures on PbSe and PbTeSe substrates was indicated.

Energy band gap data for IV–VI laser materials are given in Fig. 2.7. Some detailed data can be found in review papers by Horikoshi (1985), Partin (1988), and Katzir *et al.* (1989).

2.6.2.   IV–VI-based IR semiconductor laser diodes; comments on history

The IR emission from PbS was reported by Galkin and Korolyov (1953) and also by Garlick and Dumbleton (1954). The luminescence was identified to

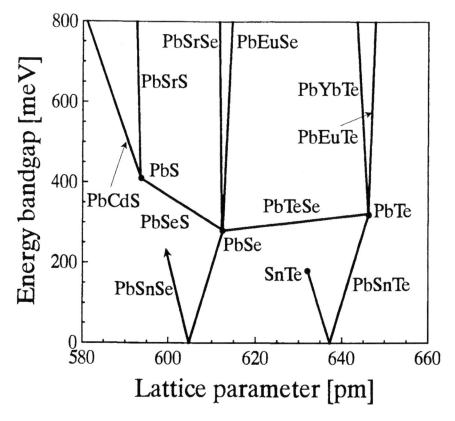

Figure 2.7: Interrelation between the lattice parameter and energy gap in long-wavelength IV–VI materials.

be due to interband radiative recombination. Mackintosh (1956) and later Baryshev (1961) investigated radiative recombination in PbS, PbSe, and PbTe.

After demonstration of the laser action in GaAs (1962) many efforts were directed to search other laser semiconductors in order to cover a wide spectral range. Butler *et al.* (1964a, 1964b) reported laser emission from homojunctions in PbSe and PbTe. Junga *et al.* (1964) reported laser action in PbTe and discussed the significance of the band structure comparing lead-salt semiconductors and III–V compounds. They concluded that lead salts may have an advantage in the long-wavelength region, pointing out the presence of four equivalent extrema in both conduction and valance bands (in the *L*-point of the Brilluoin band) instead of one in InSb. This provides a high density of states near the band edges and favors the radiative recombination to overcome a nonradiative one. The PbS-based diode lasers were introduced by Butler *et al.* (1965). Next step was the utilization of alloys. Dimmock *et al.* (1966), Butler *et al.* (1966), and Butler and Harman (1967) reported laser action in ternary IV–VI alloys. This gave a wide choice of laser material to cover the IR range. Calawa *et al.* (1969) extended the laser wavelength of lead-salt-based diodes to 34 μm. In the lead-salt type of lasers the single-heterostructures of PbSnTe/PbTe were reported by Walpole *et al.* (1973a). They observed a very low threshold density of 45 A/cm^2 at 4.2 K in diodes emitting at $\sim$ 10 μm. These structures were not lattice-matched and contained a high density of misfit dislocations. In spite of this, such laser diodes, reported by Yoshikawa *et al.* (1977), could operate CW quite reliably during 1500 h at 77 K. This led to a conclusion that the misfit defects are not a serious obstacle to a stable laser operation at low temperature. When the metallurgical problems of stable contacts to the lead salts were solved, as reported by Lo (1981), these heterostructure lasers became quite reliable devices.

Further developments were associated with an introduction of the lattice-matched heterostructures and the quantum-well structures. As it was mentioned, Davarashvili *et al.* (1977) proposed some quaternary alloys of IV–VI type for improved lattice matching. Following this proposal, Gureev *et al.* (1978) fabricated the lattice-matched PbSnTeSe/PbSe heterojunctions and observed a laser action under the optical pumping at low temperature.

Kasemset and Fonstad (1979) obtained laser action in a PbTeSe/PbSnTe double heterostructure at the highest temperature of 166 K with a threshold current density $\sim$500 A/cm^2 at 100 K. Horikoshi *et al.* (1981) fabricated low-threshold (350 A/cm^2 at 77 K) PbSnTeSe/PbTeSe double heterostructure lasers covering the ranges of 7–16 μm at 4.2 K and 6–12 μm at 77 K. Kurbatov *et al.* (1983) used PbSnTeSe/PbTeSe double heterostructures for the longest wavelength laser diode with a wavelength of 46.2 μm at 10 K.

New heterovalent quaternary alloys have been considered by Partin and Trush (1984), PbYbSnTe; by Shani *et al.* (1986), PbEuSnSe; and by Partin (1988), PbEuTeSe. Other quaternaries were prepared by mixture of the lead salts with chalcogenides of Cd and Sr. These are PbCdSeS as reported by Koguchi *et al.* (1989) and PbCdSrS by Koguchi and Takahashi (1991).

Very low threshold current density was reported for the lattice-matched heterostructure laser diodes. For example, Horikoshi *et al.* (1982b) obtained a threshold as low as 13 A/cm^2 at 4.2 K in PbSnTeSe/PbTeSe double heterostructure laser. The low value was explained as resulting from excitation of internal reflection modes with very low optical losses. More typical values for Fabry-Perot modes in these devices were ~90 A/cm^2 with a wavelength near 11 μm at 4.2 K (Horikoshi *et al.*, 1982a). An even lower threshold, 3.9 A/cm^2 at 15 K, was reported by Feit *et al.* (1989) from studies of PbEuTeSe/PbTe double heterostructure lasers with an active layer thickness of 0.75 μm. Stripe diodes based on this heterostructure have operated with a threshold current of 1.9 mA at 100 K. The CW operation was observed up to 195–203 K (Feit *et al.*, 1989, 1990b). A single quantum-well (SQW) PbEuTeSe/PbTe laser structure was reported by Partin (1984b) with a maximal CW operation temperature of 174 K. Quantum-confined states were identified from a study of emission spectra and of the temperature dependence of the threshold current. Shinohara *et al.* (1985) reported PbSnTe multiple quantum-well (MQW) lasers at λ ≈ 6 μm.

The properties of the quantum-well emission were described by Shotov and Selivanov (1990), who reported a SQW laser of PbS/PbSeS/PbSnSe/PbSeS/PbS with a separate-confinement structure. These devices operated near 10.6 μm at 77 K, and the highest operation temperature was 218 K. The progress in the technology of the IV–VI laser devices allowed an increase in the operation temperature close to RT. Particularly, Spanger *et al.* (1988a) reported the laser action at 290 K with PbSrSe/PbSe double heterostructure grown by MBE. The laser wavelength was 4.4 μm in a pulse regime.

2.6.3   Homojunction lasers diodes of IV–VI type

PbS laser diodes were studied since the 1960s when a homostructure with a diffused *p-n* junction was prepared by Butler *et al.* (1965). The diodes operated at deep cooling at a 4.3 μm wavelength. Chashchin *et al.* (1969a, 1969b, 1970a) also reported the laser action in PbS homojunction at 77 K and studied the dependence of laser threshold on the cavity length. They found an optical loss coefficient of 2.7 cm^{-1} at 77 K and a differential gain coefficient of 1.76×10^{-3} cm/A. The wavelength at 77 K was 3.95–3.97 μm with a peak output power as high as 0.7 W at the 35-A current. The beam divergence angles were 66° in the vertical plane and 8° in the horizontal plane. The coherent spot width in the near-field pattern was estimated to be about 30 μm for a 4-μm-thick active region. The internal quantum yield was measured to be not more than 28%, and the external efficiency was about 8%. The threshold current density at 77 K was 20 kA/cm^2.

Nill *et al.* (1973) added 2% of cadmium to PbS and observed the laser action in $Pb_{0.98}Cd_{0.02}S$ diodes at 3.5 μm. The stripe-geometry PbS lasers were investigated by Ralston *et al.* (1974). A peak power of 0.2 W/facet (0.35 W from both facets) was measured at 15 K. The diode emitted at 4.3-μm wavelength with a differential quantum efficiency of 24%. The single longitudinal mode output power was as high as 50 mW.

Homojunction lasers of PbSeS were prepared by Chashchin *et al.* (1970b) to cover the range of wavelength between 4 and 7.2 μm at 77 K. At a Se mole fraction of 0.4 and 0.7, the wavelength of laser emission was 4.74 and 5.52 μm, respectively. Linden *et al.* (1977) measured output power of 70 mW at 4.5-μm wavelength from a PbSeS homojunction laser.

Karczewski *et al.* (1981) reported laser action in PbMnS homojunction diodes at ~15 K. The composition of the active region was $Pb_{0.986}Mn_{0.014}S$, and the small amount of manganese shifts the laser wavelength from 4.31 μm (pure PbS) to 3.86 μm. The threshold current density was about 2 kA/cm^2. The emission was magnetically tunable between 3.848 and 3.866 μm at the field up to 14 kG.

LT lasing in PbSe homojunction diodes was described by Butler *et al.* (1965). The threshold current density was about 2 kA/cm^2 at 12 K. CW laser diode made of PbSe was reported by Zasavitski *et al.* (1975). Further studies of a tunable PbSe CW laser are also described by Zasavitski *et al.* (1978).

Nill *et al.* (1970) obtained laser emission using metal-semiconductor barriers on *p*-PbTe and *p*-PbSnTe; the wavelength was 6.4 μm from a PbTe alloy and 28.9 μm from a $Pb_{0.72}Sn_{0.28}Te$ alloy. This is the only case in laser diodes where sufficient injection pumping was achieved using a nonlinear contact to a metal.

Anticliffe *et al.* (1972) obtained CW operation in PbGeTe diodes grown by VPE with mole fraction of germanium telluride varying from 1.5% to 3%. Diodes were prepared by diffusion of Sb at 700 °C to a depth of 40 μm. Etched mesastripe structures were prepared with sizes of 250 to 380

μm. At a liquid helium temperature, the CW operation was observed at a threshold current density ~0.1 kA/cm^2.

PbGeTe diodes prepared by Sb diffusion were also studied by Kurbatov *et al.* (1981) with a GeTe mole fraction of 2.5%, 5%, and 7% (at higher contents of GeTe the precipitation of a second phase occurred). At 10 K CW operation was obtained with J_{th} in the range of 50–100 A/cm^2 ($\lambda = $ 6.2 μm at $x = 5\%$). The threshold at 77 K was not more than 1 kA/cm^2 ($\lambda = 5.6$ μm at $x = 5\%$).

Donnelly *et al.* (1972) used the proton bombardment in the technology of PbSnTe diode lasers. Stripe-geometry PbSnTe lasers were investigated also by Ralston *et al.* (1973). By polishing the facet mirrors of PbSnTe, Walpole *et al.* (1973b) improved the slope quantum efficiency to 0.16. They obtained a single mode output power of 6 mW in the long wavelength range.

2.6.4 Heterojunction laser diodes of IV–VI type

In addition to the lattice-matched heterostructures shown in Table 2.2, some other heterostructures used in laser diodes are listed in the Table 2.7. At the beginning, the structures containing one heterojunction (single heterostructures, SH) were studied. Linden *et al.* (1977) investigated SH lasers prepared by the composition interdiffusion (CID) applied to the PbSSe and PbSnSe materials. Stripe geometry *p-n* junctions were subsequently fabricated by diffusion through a mask producing 50-μm-wide stripes.

SH lasers showed better performance characteristics than those in homostructure lasers (Walpole *et al.*, 1973a; Britov *et al.*, 1976b). CW operation was obtained for compositions in the wavelength range from 4.1 to 12.7 μm. Single-ended CW output power as high as 70 mW with an external efficiency of about 44% was observed in PbSSe SH devices (Linden *et al.*, 1977).

The next step of the IV–VI laser development was the introduction of double heterostructures (DH) with a configuration providing the electron confinement, *i.e.*, the limitation of diffusion spreading of excess carriers, and optical confinement (the waveguide effect). The active layer consisted of a narrower band gap component of the structure playing the role of the potential well for excess carriers. The DH with an ultrathin active layer appears to manifest quantum size-dependent properties.

Type	Materials	$\Delta a/a$	Range	References
SH	$Pb_{1-x}Sn_xTe/PbTe$	0.021	$x = 1$	Walpole et al. (1973a)
	$Pb_{1-x}Sn_xSe/PbSe$	0.026x	$x < 0.4$	Britov et al. (1976)
DH	$PbS/PbSe_xS_{1-x}/PbS$	0.032	$x = 1$	Preier et al. (1976)
	$PbS/PbSe/PbS$	−0.031	—	Preier et al. (1977)
	$PbTe/Pb_{1-x}Sn_xTe/PbTe$	0.021	$x = 1$	Walpole et al. (1977)
	$Pb_{1-x}Sn_xTe/Pb_{1-y}Sn_yTe$	$0.02(y-x)$	—	Yoshikawa et al. (1977)
	$Pb_{1-y}Sr_xSe/PbSe/Pb_{1-x}Sr_xSe$	0.0173	$x \ll 1$	Moy and Reboul (1982)
	$Pb_{1-x}Eu_xTe/PbTe/Pb_{1-x}Eu_xTe$	0.078x	$x \ll 1$	Spanger et al. (1988a)
	$Pb_{1-x}Sr_xS/PbS/Pb_{1-x}Sr_xS$	0.034x	$x < 0.04$	Nishijima (1989)
	$Pb_{1-x}Eu_xSe/PbSe/Pb_{1-x}Eu_xSe$	0.011		Ishida et al. (1990)
			$x = 1$	Schlereth et al. (1990)
SCH-SQW	$PbS/PbSe_xS_{1-x}/Pb_{1-y}Sn_ySe//$	0.032x,	—	Selivanov and Shotov (1987),
	$PbSe_xS_{1-x}/PbS$	$0.032{-}0.021y$	—	Shotov and Selivanov (1990)

Table 2.7: Some strained (lattice-mismatched) laser heterostructures.

Ishida *et al.* (1989) reported a PbSrS/PbS DH laser prepared by the hot-wall VPE epitaxy with a lattice mismatch of 10^{-3}. The laser emission was obtained at 2.97 μm with maximal operation temperature of 245 K for pulsed conditions and of 174 K for CW conditions.

Koguchi and Takahashi (1991) reported a study of PbCdSrS/PbS DH lasers with LT threshold currents reduced more than 10 times as compared to PbSrS/PbS DH lasers. Quaternary composition of confinement layers is chosen to get more precisely matched heterojunctions, taking into account that Sr and Cd additions to PbS give lattice parameter changes of opposite sign, but both additions give an increase of E_g. As a result, the band gap offset ΔE was obtained to be 0.4 eV at a rather small lattice mismatch.

An example of low-threshold stripe-geometry laser diodes was given by Feit *et al.* (1990a), who fabricated narrow-stripe buried heterostructure (BH) devices operating in CW mode at currents in the milliampere range. Some data on these laser diodes are represented in Table 2.8. The active region consisted of PbTe or a quaternary alloy of PbEuTeSe. Output power of 1 mW was measured at temperatures above 100 K. Low-threshold PbEuSeTe/PbTe SCH laser diodes with a buried active region were reported by Feit *et al.* (1996). The highest CW-operation temperature was 223 K. An active medium of PbTe emits at that temperature near the wavelength of 4.3 μm.

2.6.5   Quantum-well IV–VI lasers

The next step of the technical improvement of the IV–VI lasers was the development of ultrathin active layer structures. This became possible due to advances in epitaxial growth techniques, such as vacuum deposition (MBE and HWE). These growth techniques provide a precise thickness control and allow the obtaining of atomically flat interfaces. Due to the small effective mass of carriers in lead-salt semiconductors, the quantum confinement states may be observed at active layer thickness below 0.1 μm. A single quantum-well (SQW) laser based on PbEuSeTe/PbTe structures was introduced by Partin (1984b). The lattice-matched structure was grown by MBE; dopants were Bi (donor) and Te (acceptor); the active layer was made of 30-nm-thick PbTe, and the composition of the barrier layers was $Pb_{1-x}Eu_xTe_{1-y}Se_y$ with $x = 0.018$ and with y adjusted for minimal lattice misfit. The difference of band gaps was $\Delta E = 0.1$ eV at 80 K. The concentration of Eu was increased to 0.04 outside the active and barrier regions to form a separate confinement heterostructure (SCH), as

Composition of Active Layer, at.%		Wavelength Range, μm	Threshold Current, mA		Maximal Temperature of CW Operation, K	Maximal Output, mW
Eu	**Se**		**80 K**	**140 K**		
0	0	4.35–6.52	3.2	36.2	195	1.22
24	34	4.3–5.96	2.9	23.4	183	1.22

Table 2.8: Characteristics of PbEuTeSe/PbTe BH lasers (Feit *et al.*, 1990a).

the refractive index of the quaternary alloy decreases with increasing Eu concentration. Stripe diodes were prepared by anodic oxidation with stripe widths from 16 to 22 μm and by cleaving to form cavity lengths from 325 to 450 μm.

The evidence for quantum-confined states of carriers was found in a spectral behavior of the emission near T = 120 K. Specifically, in a laser diode operating in a single mode from a laser threshold of 85 mA up to about 0.35 A, the mode structure changed at 0.47 A and showed a new mode with a much higher (about 10 meV) photon energy. This was interpreted as a result of a switching from optical transitions between ground levels of quantum-confined states of electrons and holes (*1e–1h*) to those between second levels (*2e–2h*). At T = 140 K, the transition *2e–2h* became the dominant one.

Using the spectral positions of the quantum-confined states, Partin (1988) discussed the band offsets ΔE_c and ΔE_v, as their sum ΔE_g is known (103 meV). He concluded that ΔE_g is roughly evenly divided between ΔE_c and ΔE_v.

Multiple quantum well (MQW) PbEuSeTe/PbTe lasers were also grown and studied by Partin (1985). In these lasers the transition *1e–1h* appeared to be dominating with no observation of *2e–2h* transition. This may be expected taking into account the increase of gain at the *1e–1h* transition due to summation of the contribution from every active layer in the multilayer structure. Optical absorption studies of lattice-mismatched PbEuTe/PbTe superlattices also indicated that $\Delta E_c \simeq \Delta E_v$, as noticed by Ishida *et al.* (1987).

PbSnTe/PbSeTe lattice-matched MQW lasers were prepared and studied by Shinohara *et al.* (1985) and also by Ishida *et al.* (1986). An anomalous temperature dependence of the threshold current was found suggesting that the heterojunction PbSnTe/PbSeTe appears to be of type-II; namely, the conduction band edge in the PbSnTe active region is higher than that in the cladding PbSeTe layer. These laser diodes operated up to 130 K within CW mode and up to 204 K in pulsed mode at 6 μm.

The single quantum well separate confinement heterostructure (SQW SCH) laser with active region of PbSnSe (Sn content x = 0.05) was reported by Shotov and Selivanov (1990). The structure was prepared by HWE on PbSe substrate of n-type doped by Bi. Cladding layers consisted of PbS (3 μm thick) and waveguide layers consisted of $PbS_{0.4}Se_{0.6}$ (1 μm thick). The thickness of active layer d could be reduced to 40–100 nm. It was observed that the photon energy of emission from laser diode with active layer thickness greater than 100 nm corresponds to the band gap E_g = 109 meV of the active region. At the thickness of 40 or 50 nm the photon energy appeared to be greater than E_g. When d = 40 nm and T = 77 K, two emission lines were identified: $hv(1e–1h)$ = 127 meV (λ = 9.8 μm), which is 9.7 meV larger that E_g, and another one, $hv(2e–2h)$ = 155 meV (λ = 8 μm), which is 38 meV larger than E_g. To fit the photon energy to definite transitions between quantum-confined levels in the wells, the values ΔE_c = ΔE_v, m_e = 0.035m_0, and m_h = 0.31 m_0 were proposed. These assumptions allow obtaining a good fit to the experimental values of photon energy. It is interesting that both laser emission lines (8 μm and 9.8 μm at 77 K) were observed simultaneously in the range of 60–120 K, but with different threshold currents. This means that the capture time to ground levels appeared to be comparable with recombination time of the carriers.

Schlereth *et al.* (1990) reported SQW lasers of PbSe/PbSrSe in the 4-μm wavelength range with a binary PbSe as an active material. Some advantage of structures with wider well attributed to the influence of the gain saturation was pointed out. Gain saturation is more effective in a smaller active volume, and so it was more suitable to use a larger active volume. It could be achieved also by introducing a multiple quantum-well structure. PbSe/PbSrSe MQW laser diodes were grown by Shi *et al.* (1995) using the MBE technology on p-PbSe substrates. The active region consisted of seven periods of alternating $Pb_{0.9785}Sr_{0.0215}Se$/PbSe layers. The total width of active PbSe was 0.35 μm. A pulsed laser action was obtained at the highest temperature of 282 K, rather close to RT. The emission wavelength was 4.2 μm. The HWE was used by Ishida *et al.* (1990) to fabricate PbS/PbSrS quantum wells. Structures of this type demonstrated a pulsed laser action at 225 K. In spite of expectation, the quantum-well IV–VI lasers did not yet surpass the characteristics of the IV–VI DH lasers by maximal operation temperature but rather appear to approach the best laser performance.

2.7 Resume of interband laser properties

2.7.1 Threshold and power characteristics

Some selected data on threshold current density in long wavelength laser diodes are presented in Table 2.9. The very low values shown there may

Structure	λ, μm	T, K	J_{th}, A/cm^2	References
InGaSbAs/GaSb	2.0	77	50	Baranov et al. (1988a)
InAsSb/InAlAsSb	3.4	80	44	Choi et al. (1996)
InGaSbAs/InSbAsP	3.95	77	39	Aidaraliev et al. (1989)
PbSnTe/PbTe	10.8	4.2	45	Walpole et al. (1973a)
PbSnTeSe/PbSeTe	~11	77	350	Horikoshi et al. (1982a)
		4.2	65	
		4.2	13*	
PbTe/PbEuTeSe	6.5	15	3.9**	Horikoshi et al. (1982b) Feit et al. (1989)
InAlAs/InP QCL	~5	10	1500–2000	Faist et al. (1996)

*Internal reflection mode.
**One sample.

Table 2.9: Low-temperature threshold current density in some long-wavelength laser diodes (minimal values measured in devices of various sizes).

be interpreted in terms of traditional threshold considerations, developed by Lasher and Stern (1964) and adequate for bulk active region lasers operating at LT. The threshold current density J_{th} can be expressed as

$$J_{th} = 8\pi e n^2 E_g^2 \Delta E d\alpha / c^2 h^3 \eta \tag{2.7}$$

where ΔE is the spontaneous emission bandwidth, d is the active layer thickness, α is the optical loss coefficient (including internal and external losses), and η is the internal quantum yield of radiative recombination. Using parameters of the GaAs active layer ($n = 3.6$; $E_g = 1.4$ eV, $\Delta E = 5$ meV, $d = 1$ μm, $\eta = 1$, $\alpha = 100$ cm^{-1}), one can obtain the estimate J_{th} ~80 A/cm^2, which is close to LT oscillation threshold in GaAs DH diodes. A corresponding estimate for 2.2 μm wavelength DH laser of InGaSbAs ($n = 3.8$; $E_g = 0.56$ eV, $\Delta E = 5$ meV, $d = 1$ μm, $\eta = 1$, $\alpha = 100$ cm^{-1}) gives 14 A/cm^2, which is several time less than reported values of 39–50 A/cm^2. This "classical" estimate continues to decrease as one goes to longer wavelength lasers. At $\lambda = 6.5$ μm in PbTe DH laser ($n = 6.3$, $E_g = 0.19$ eV, $\Delta E = 5$ meV, $d = 0.75$ μm, $\eta = 1$, $\alpha = 100$ cm^{-1}), one obtains a nominal density $J_{th} = 3.2$ A/cm^2, which is in an agreement with a lowest value of 3.9 A/cm^2 claimed in a paper by Feit *et al.* (1989) for a PbEuTeSe/PbTe DH laser diode with the lattice-matched heterojunctions, operating at 15 K. In these satisfactory estimations it was assumed that $\eta = 1$; therefore the calculated current is entirely the radiative one. This may be the case for well-confined DH structures at LT limit, but not in other circumstances. Considering LT threshold currents, one has to notice that there are some sources of nonradiative losses, increasing the actual threshold. The analysis of the threshold for long-wavelength laser diodes was given by Tomasetta and Fonstad (1975), Anderson (1977), Rosman and Katzir (1982), Bychkova *et al.* (1982), Horikoshi (1985), and some other authors. The sources of losses are: (i) the optical absorption, (ii) the nonradiative recombination, and (iii) the noninjection tunneling. As to absorption effects, they are included in the total optical losses α in the expression for J_{th}. The contribution of nonradiative and noninjection losses may be taken into account by the decrease of the radiative quantum yield η.

The gain calculation by Anderson (1977) for PbSnTe DH lasers predicted the threshold density of 25 A/cm^2 in LT limit at $\lambda = 50$ μm and $\alpha = 100$ cm^{-1}. This estimate increases for shorter wavelengths (100 A/cm^2 at 12.4 μm and 200 A/cm^2 at 6.2 μm) and for higher temperatures (300 A/cm^2 at 100 K and 10 kA/cm^2 at 300 K, $\lambda = 6.2$ μm). It was found also that the free-carrier absorption coefficient α_{fc} increases strongly as wavelength and temperature increase. The coefficient may be expressed in a classical model with additive contributions of electron and holes that are proportional to their concentrations N_e and N_h. The expression is

$$\alpha_{fc} = (\lambda^2 e^3 / 4\pi^2 c^3 n \varepsilon_0)[(N_e / m_e^2 \mu_c) + (N_h / m_h^2 \mu_h)] \qquad (2.8)$$

where ε_0 is the vacuum dielectric constant, m_e and m_h are effective masses of electron, and holes, and μ_e and μ_h are their mobilities, respectively. The influence of the wavelength is seen from this expression. The temperature dependence appears due to temperature changes of mobilities and also to changes of N_e, N_h, and λ. The calculations were made for some laser materials taking into account effective mass anisotropy and detailed mobility properties of the laser medium. Anderson (1977) found α_{fc} at 100 K to be about 8 cm^{-1} at 6.2 μm, 60 cm^{-1} at 12.4 μm, 350 cm^{-1} at 25 μm, and more than 10^3 cm^{-1} at 50 μm. We see two opposite tendencies of the wavelength influence on the laser threshold. First is the threshold decrease due to more efficient use of emitted photons (less optical modes over the emission bandwidth as λ grows up). It is presented in the expression (2.7) as it was derived from fundamental principles. The second tendency is a threshold increase due to increase of the wavelength-dependent optical absorption.

Other sources of power losses may also contribute as the wavelength increases, but its magnitude can change nonmonotonically and needs a separate consideration. We shall return to the wavelength influence in the discussion on the highest operation temperature.

The interface recombination via defects at the lattice-mismatched heterojunctions was measured by Kasemset and Fonstad (1979), who showed that in $Pb_{0.86}Sn_{0.14}Te/PbTe$ heterojunctions (used in the DH lasers of the 10–12 μm range) the surface nonradiative recombination velocity may be as large as 10^5 cm/s at 5 K. It leads to reducing the radiation quantum yield η to 80% when the bulk carrier lifetime is 4.3 ns and active layer thickness is 2 μm. The decrease is more pronounced in DH lasers with thinner active layers. The surface recombination velocity S was expressed in proportion with lattice misfit $\Delta a/a$ as $2.9 \times 10^7 \, \Delta a/a$ (cm/s). In well-lattice-matched heterojunctions it can be significantly reduced. It becomes probably negligible in junctions with a coherent lattice interface (free of dangling bonds).

Another important mechanism of losses is the Auger recombination process, which seems to be more significant as the energy band gap approaches the energy difference for any efficient intraband and inter-subband transitions. These processes were extensively studied in connection with the emission properties of long wavelength materials by Takeshima (1972), Emtage (1976), Sugimura (1980, 1982), and Haug (1983). Early theoretical estimations were rather pessimistic for GaSb and InAs, as reported by Sugimura (1980). The Auger recombination coefficient C determines the nonradiative component, which can grow as CN^3 (where N is the excess carrier concentration). Therefore the component may become dominant over the radiative one at a sufficiently large N.

The Auger recombination contribution increases with temperature, so it imposes a limitation on the high-temperature performance of IR semiconductor lasers. In a study of the temperature dependence of the laser threshold in InGaAsSb-based laser diodes, Andaspaeva *et al.* (1991) concluded that above 200 K the carrier lifetime is controled by the Auger processes (transitions of the third carrier to the spin-orbit [SO] split valence subband, 200–300 K, and into the conduction band, above 300 K). The factor providing the Auger process involving the split valence subband is that the SO splitting energy Δ is close to the band gap in materials like GaSb, InAs, and their alloys. Depending on the alloy content of InGaAsSb, the following situations are possible: $E_g \geq \Delta$ or $E_g < \Delta$. In the first case the Auger excitation of free holes into an SO-split valence subband is possible, but the probability is larger when both energy parameters are close to each to other. In the second case, the process is excluded. Sometimes, the effect of strain in the active region can be explained by the change of the valid unequality. The agreement of calculated and experimental dependencies of the threshold current density was excellent (Andaspaeva *et al.*, 1991) for laser diodes operating near a wavelength of 2 μm. Three components of the current were involved in the theoretical calculation. The power temperature dependences were obtained for each of them as T^γ. The exponent was found to be $\gamma = 1.5$ for the radiative component, dominating below 200 K, $\gamma = 4.5$ for the Auger component with SO-split subband involved (dominating in the range 250–280 K), and $\gamma = 6\text{–}8$ for Auger process with an excitation of the conduction electron in a high-temperature range (above 300 K).

Theoretical limits of the laser performance in a number of mid-IR semiconductor structures (λ in a range of 2.1–4.1 μm) are considered by Flatté *et al.* (1995). The active region was assumed to be composed of the following structures:

1. InAs/InGaSb broken-gap superlattices, specifically,
   1a) 2-nm-InAs/3.2-nm-$In_{0.25}Ga_{0.75}Sb$ for 4.1-μm-emission.
   1b) 1.67-nm-InAs/3.5-nm- $In_{0.25}Ga_{0.75}Sb$ for 3.5-μm-emission.
   1c) 0.9-nm-InAs/3.5-nm-GaSb for 2.1-μm-emission.

2. InGaAsSb quantum well, namely, 10-nm-
   $In_{0.16}Ga_{0.84}As_{0.14}Sb_{0.86}$ /20-nm-$Al_{0.2}Ga_{0.8}As_{0.02}Sb_{0.98}$ for 2.1-μm-emission.

3. $InAs_{0.91}Sb_{0.09}$ bulk medium for 4.1-μm-emission.

4. HgCdTe   superlattice,   namely,   1.6-nm-HgTe/1.56-nm
   $Hg_{0.15}Cd_{0.85}Te$ for 3.5-μm-emission.

Calculations included optical gain and radiative and nonradiative recombination rates at 77 K and 150 K. Most promising candidates are found to be InGaAsSb quantum wells and InAsInGaSb superlattices rather than other versions. The calculated threshold at 150 K was as low as 30 A/cm^2 in structure (1a), 36 A/cm^2 in (1b), 99 A/cm^2 in (1c), 107 A/cm^2 in (2), 1558 A/cm^2 in (3), and 191 A/cm^2 in (4).

Comparison with experiments indicated that the threshold current density of InGaAsSb-based devices is about three times greater than those calculated for 25 cm^{-1} gain. In the InAs/InGaSb superlattices, it is ~100 times above the theoretical limit.

There are some interesting comparative data on IR semiconductor lasers under electron-beam (EB) pumping. We show these data in Table 2.10, including compilations from some papers on EB-pumped lasers and the original results of the paper by Gubarev *et al.* (1991). Only figures related to laser action at λ > 2 μm are extracted. The threshold current density of the electron beam is much lower than the current density in laser diodes because of the carrier multiplication under action of the incident high-energy electrons. The minimal equivalent current density at the laser threshold in the InAs laser is about 5 kA/cm^2, which is much larger than can be achieved in the laser diodes at LT (~0.1 kA/cm^2). An important reason for this difference is the larger active volume, which is pumped by high-energy electrons. An increased total output power resulted by using a larger active volume. The peak power reported by Gubarev *et al.* (1991) was 55 mW in InAs laser and 600 mW in InSb laser. In the same paper, a tendency was noticed for EB-pumped III–V lasers: Both the threshold and slope efficiency decrease monotonically with an increase of the emission wavelength from GaAs to InSb. The decrease of

Material	T, K	λ, μm	J_{th}, A/cm^2	Slope Efficiency, %	References
Cd$_3$P$_2$	90	2.17	0.5	—	Kurbatov et al. (1976)
InAs	20	3.0	—	—	A la Guillaume and Debever (1964)
	77	3.0	0.3–1.2	0.6	Gubarev et al. (1991)
	80	2.7–3.0	0.1–0.5	1–2	Kryukova et al. (1979)
InAsSbP	80	3.1–3.7	0.1–0.3	8	Dolginov et al. (1978d)
Te	4.2	3.71	—	—	A la Guillaume and Debever (1965a)
PbS	90	3.85	1.0	—	Kurbatov et al. (1967)
InSb	20*	4.96	—	—	A la Guillaume and Debever (1965b).
	30**	5.0	0.1	1.5–3.5	
	77***	—	~2.0	< 0.1	Gubarev et al. (1991)

*Cavity length was 0.3 mm.
**Electron energy was 75 keV.
***Maximum operation temperature.

Table 2.10: Laser parameters of IR semiconductor lasers with electron-beam pumping. The transverse geometry is used; typically, the energy of electrons is from 20 to 50 keV, the cavity length is 0.4–0.6 mm, beam diameter is about the same size (unless indicated otherwise). Data are taken from Gubarev et al. (1991).

the threshold is a quite expected phenomenon, as the effective masses of carriers are decreasing simultaneously with the decrease of the band gap. On the other hand, the nonradiative Auger recombination is also expected to become more important, and, probably, this effect is responsible for a rapid increase of the threshold of IR lasers with temperature. In InSb lasers this increase was most pronounced, and maximum operation temperature was about 77 K (see Table 2.10). The change of the slope efficiency can be explained by intraband reabsorption of the emission, particularly by inter-valence-subband absorption in GaSb and InAs. Free-carrier absorption also has a tendency to grow with wavelength; therefore, it can contribute to the high threshold in InSb lasers.

Selected data on power performance of mid-IR semiconductor laser diodes are given in Table 2.11. Optical output does not exceed a level of 1 W, and a tendency is seen for a decrease of obtainable laser emission power with the increase of the wavelength. It can be explained by several causes; one is an increase of the free-carrier absorption at longer wavelengths; another is a decrease of the heat conductivity in longer-wavelength materials, which limits the pumping current.

2.7.2   Mode control and tunability of the laser emission

The availability of mixed materials suitable for laser applications is an important advantage of semiconductor lasers, almost continuously covering a wide spectral range. This offers the "chemical tunability," *i.e.*, the opportunity to find the alloy composition for any wavelength (in contrast to gas lasers covering some discrete bands of IR). Another advantage is tunability of all semiconductor lasers using the dependence of the band gap on physical parameters. The penalty for physical tunability is a drift of the laser wavelength in unstabilized ambient and the necessity of using special stabilizing schemes to improve the spectral stability.

As lasers provide narrow emission lines, their tunability is very favorable for high-resolution spectroscopy, and the spectral quality of the emission spectrum becomes an important characteristic. Special spectroscopic requirements include: (i) a narrow-line laser emission, as a small linewidth determines the high spectral resolution; (ii) an absence of subsidiary spectral lines of the laser emission to eliminate parasitic spectral identifications; and (iii) a reduced continuous background level of emission due to spontaneous and superluminescent (amplified spontaneous) emission

λ, μm	Output Power, mW	Mode of Operation	T, K	Structure	References
2.1	190	CW	300	InGaAsSb QW	Choi and Eglash (1992)
	60	CW, SM			
3	10	P	83–130	PbS/PbSrS DH	Ishida *et al.* (1990)
3.23	700	P	77	InAs/InAsSbP DH	Aidaraliev *et al.* (1993)
	15	CW			
3.4	215	CW	80	InAsSb/InAlAsSb MQW	Choi et al. (1996)
	35	CW	150		
3.95	700	P	77	PbS HJ	Chashchin (1970a)
4.3	200	P	15	PbS HJ	Ralston *et al.* (1974)
4.3	1.22	CW, SM	180	PbTe/PbEuSeTe DH	Feit *et al.* (1990a)
7.8	0.6	CW, SM	25–75	PbSnSe/PbEuSnSe DBR-DH	Shani et al. (1988)
5.2	200	P	300	InGaAs/AlInAs/InP QCL	Faist et al. (1996)
	100	P	320		
	25	CW	10		
	15	CW	120		

Table 2.11: Selected data on output power of mid-IR laser diodes.

CW = continuous-wave (mode of operation); P = pulse mode of operation; SM = single mode; HJ = homojunction; DH = double heterostructure; MQW = multi-quantum-well; DBR = distributed Bragg reflector; QCL = quantum-cascade laser.

to provide more accurate spectral measurements. The mode structure of the laser emission is of importance in this respect. Uncontrollable laser oscillations in broad area lasers usually have a multimode nature. To improve the mode content, the experience of mode selection in shorter wavelength lasers could be effectively used. Some methods of mode selection are extendable to the long wavelength lasers. These are presumably the methods compatible with cryogenic accessories, particularly ones that are based on monolithically integrated devices. It appears that a transverse-stabilized stripe-geometry laser can sometimes emit a single-frequency radiation with no additional mode control devices. This occurs because a selection of fundamental spatial modes is favorable for selection of one longitudinal mode. To provide a single mode operation, the stripe width must be about 5 μm or less. Buried heterostructure lasers are found to be capable of emitting a single-longitudinal mode radiation over a large current span, as mentioned by Preier (1990).

Feit *et al.* (1991a) also reported a wide range of the single-longitudinal-mode operation (up to 30 times the threshold current) in low-threshold stripe lasers of PbEuTeSe/PbTe. The mode tuning range in a Fabry-Perot-cavity lasers is typically to ~1 cm^{-1}. Distributed feedback (DFB) or distributed Bragg reflector (DBR) lasers, as well as composite-cavity lasers (for instance the cleaved-coupled cavity, so called C^3-lasers), supply more reproducible single-spectral-mode control. The tuning of laser emission is achievable by the change of optical properties of the active region using a control of the ambient temperature, a stationary heating by the current through the diode, a nonstationary heating by pulse current, an external pressure variation (hydrostatic or uniaxial), and an external magnetic field variation. In all these cases, both real and imaginary parts of the dielectric constant are influenced. Therefore, changes of the refractive index and shifting of the emission spectral band occur simultaneously. As a result, the spectral peak of the gain band and the mode wavelength shift provide a coarse tuning (following the gain peak) and fine tuning (following the mode position). The shifts occur in the same direction but with different rates, namely, mode shift is several times slower. It leads to mode hopping during the coarse tuning over a wider range than can be attained by continuous fine tuning of the individual mode.

Walpole *et al.* (1976, 1977) reported DFB PbSnTe/PbTe lasers, operating in the wavelength range of 12–14 μm at LT. The surface corrugation period was 1.1 μm, corresponding to the first order of the Bragg diffraction. DFB-mode oscillation was observed in a temperature range between 38 and 62.5 K, where the mode tuning rate was about 3.2 cm^{-1}/ K and the tuning range for the strong mode was 6.5 cm^{-1}. Kapon and Katzir (1985) had demonstrated operation of DFB PbSnTe/PbSeTe lasers. Studies of tuning properties in DBR PbEuSnSe/PbSe laser were given by Shani *et al.* (1986, 1988). The DBR containing a grating with corrugation period 0.8 μm and depth 0.5 μm was coupled monolithically to the active section of the diode. The CW operating DBR laser at 77 K was tuned continuously by a change of the current in the wavelength range 7.85-7.89 μm with a single-frequency output. The fine tuning range was about 20 cm^{-1}.

2.7.3 High-temperature operation

Studies of luminescence properties of some Pb-salt semiconductors gave evidence that there are no basic restrictions (including those due to Auger processes and free carrier absorption) for laser action at RT conditions, at least in PbSe and PbSeS. Galeski *et al.* (1977) stated an observation for the first time of the RT stimulated emission from PbSe polycrystalline films under pulsed optical pumping by Nd:YAG laser. The samples were prepared by vacuum deposition at 250 °C and by consequent annealing. This "activating" anneal in air is known as a technique to obtain higher photosensitivity of some lead-salt films. Along with this, the luminescence properties are also improved, probably due to recrystallization and extraction of impurity and defects from the bulk. The threshold of the stimulated emission was observed to be lowered by two to three orders of magnitude after the annealing. The most impressive result of this study was that stimulated emission could be obtained in a wide temperature range, up to 300 K, and photoluminescence up to 330 K, in the annealed samples. The radiation was detected as emitted along the normal direction to the film (a "surface-emitting" geometry), and the identification of stimulated emission was performed by the spectral narrowing to a bandwidth less than kT.

The photon energy of the stimulated emission from the polycrystalline samples differs considerably from the band gap energy of PbSe. This difference has not been explained. Similarly, RT stimulated emission was detected also from some PbSeS films by Yunovich *et al.* (1979). By contrast, with the same pumping technique the stimulated emission from the single-crystal film of PbSe was observed at temperatures not higher than 220 K. Approximately the same result was obtained for PbTe and PbSnTe films, but spontaneous emission spectra from PbSnTe films was followed even up to 430 K (Yunovich, 1988).

A plurality of results concerning the highest operation temperature of diode lasers is summarized in Figs. 2.8 and 2.9. In the pulsed regime (Fig. 2.8), the spectral range at liquid nitrogen cooling is extended up to 25 μm. At RT, the long wavelength limit is near 2.5 μm for lasers of the III–V group, whereas in group IV–VI, only some lasers approach this temperature. These are the PbSrSe/PbSe DH diode, operating at 290 K (Spanger et al., 1988b); the PbEuTeSe/PbTe SQW laser at about 280 K (Partin, 1985); and the PbSrS/PbS DH laser at 245 K (Ishida et al., 1989). In Fig. 2.8 a theoretical line (*3*) is shown according to Britov *et al.* (1976a) and Britov (1983), for the "ultimate limit," which corresponds to the entire compensation of the gain by the free carrier and lattice absorption.

Some CW laser diodes (Fig. 2.9) operate near 200 K, and the range covered by these lasers has an upper wavelength limit near 20 μm. Curves

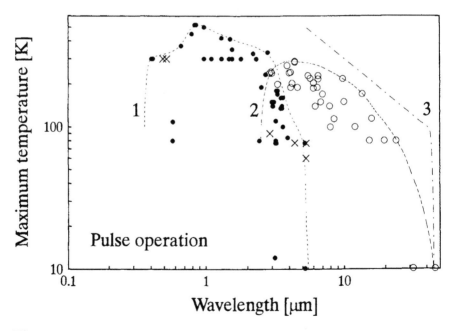

Figure 2.8: Maximum operation temperature versus laser emission wavelength for pulsed diode lasers. Empirical limits are shown by curves (1) for III–V and (2) for IV–VI materials, whereas line (3) is the "ultimate" theoretical limit (see text).

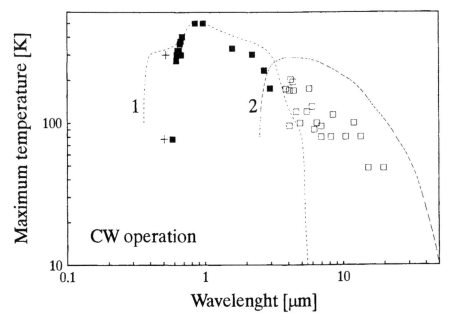

Figure 2.9: Maximum operation temperature versus laser emission wavelength for CW diode lasers. The same empirical limits are shown by curves (1) for III–V and (2) for IV–VI materials.

in this plot are the same as in Fig. 2.8 for empirical limits for III–V (*1*) and IV–VI (*2*) materials.

2.7.4 Longest wavelength of interband semiconductor lasers

We discussed before that the use of narrow bandgap semiconductors in lasers gives rise to a number of principal physical problems, like the tendency to increase the losses by nonradiative recombination and tunneling, by free-carrier absorption, by plasmon and phonon absorptions, etc.

In addition to free-carrier absorption, which strongly increases with increasing wavelength, there are long-wavelength optical losses due to plasma absorption. The resonant peak of this absorption shifts to shorter wavelength with an increase of the free carrier density. On the long wavelength side of the plasma peak, the laser oscillation appears to be impossible to achieve as the plasma becomes nontransparent. Therefore, in order to obtain longer wavelength laser emission, one has to use a low free-carrier concentration medium. The longest wavelength of laser diodes (37 μm in PbSnSe and 46 μm in PbSnSeTe diodes) was probably limited by the plasma absorption.

It is known, however, that there is a range of transparency (magnetoplasma window) for a linearly polarized radiation in magnetic field B beyond the plasma resonance peak. The influence of magnetic field on the long-wavelength absorption at $T < 5$ K in $Pb_{0.8}Sn_{0.2}Te_{0.2}Se_{0.8}$ laser diodes was studied by Kurbatov *et al.* (1986). It was found that at $B >$ 1.5 kG the oscillation threshold decreases by about a factor of two, and the laser spectrum splits onto two bands; at $B > 2$ kG only the longer wavelength band remained, and longest wavelength laser emission was registered at 49.1 μm. At $B > 5$ kG the laser threshold increases rapidly. This behavior was interpreted as the evidence of the magnetoplasma window, and laser transitions were identified as ones between Landau levels in the conduction and valence bands. The band gap was estimated from these experiments to be 21 ± 0.5 meV at $B = 0$. The longest wavelength emission is close to plasma resonance at $h\upsilon = 27$ meV; at optimal B the laser spectrum shifts into the magnetoplasma window. If this interpretation is correct, there is in principle a possibility of obtaining laser action in almost the entire far IR region. Semiconductor materials for these lasers may be found among III–V alloys containing Bi, as discussed above; among II–VI alloys in the system HgCdTe passing the SM-SC transition at a Cd content of about 14%, and among IV–VI alloys in the PbSnSe or PbSnTe systems, also passing the band inversion at some Sn contents. These transition points may also be slowly approached by the application of hydrostatic pressure. In all these cases reducing the band gap energy to zero provides a possibility of a very wide tuning of the laser emission to a longer wavelength.

2.8 Hot-hole (unipolar) semiconductor lasers

The IR stimulated emission is obtainable under strong electric field and transverse magnetic field at low temperature using direct optical transitions between light and heavy subbands, as suggested by Andronov (1979). The stimulated emission appears due to the population inversion resulting from a redistribution of holes between the subbands assisted by inelastic phonon scattering. Laser action was demonstrated with p-Ge crystals as reported by Andronov *et al.* (1984). The laser covers a wavelength range between 70 and 250 μm with output power up to 10 W in some cases. Typical experimental conditions are:

- Carrier (hole) concentration is near 10^{14} cm^{-3}.
- Sample size is near 50×5×5 mm.
- Pulse voltage magnitude is near 1 kV.
- Pulse duration time is near 1 μs.
- Pulse repetition rate is near 5 Hz.
- Magnetic field strength is near 1 T.
- Ambient temperature is 4.2 K.

A similar technique of crossed electric and magnetic fields was used to obtain laser action based on the transitions between sublevels of cyclotron resonance of light holes. Lasing in p-Ge was found at wavelength ranges of 100–150 μm and 200–350 μm. Output power was measured up to 10 mW. With a negative mass, cyclotron resonance maser action in p-Ge was observed in the wavelength range of 0.8–4 mm.

These devices are termed hot-hole lasers (masers). Review of their mechanism is given by Andronov *et al.* (1986). The principle of action in these devices does not give hope of elevating the operation temperature to RT.

It is closer to traditional laser techniques to use the optical transitions between discrete energy levels of impurities incorporated into the crystalline medium. Among claims of laser action based on the intra-center transition in semiconductors, some are obviously identified. One example is IR laser-type emission obtained by Klein *et al.* (1983) from InP:Fe at liquid helium temperature under optical excitation. Another example is the stimulated emission in the wavelength range of 155–200 μm observed by Murav'ev *et al.* (1990) with p-Ge at 4.2 K. The pumping conditions were the same as in the hot-hole lasers (see above), namely, crossed electric and magnetic fields. Some spectral bands in the spectral range were explained to be resulting from transitions between excited and ground states of acceptor centers (Ga, Al, B, or unknown residual impurity in Ge).

2.9   Unipolar tunneling-based semiconductor lasers

2.9.1   Theoretical schemes of the tunneling pumping

There is a promising version of semiconductor lasers in which the photon-assisted tunneling for laser action in the IR region is used. Kazarinov and Suris (1971) first proposed the tunneling mechanisms in a semiconductor superlattice for tunable laser operation. It was pointed out that inverted population between minibands in the superlattice may be obtained as electrons would tunnel from a ground state in one well into an excited state in adjacent one due to tilting of the energy band edge in presence of the electric field. This situation is shown in Fig. 2.10a. The resonant tunneling may be an efficient pump for the emission mechanism of optical transitions from level 2 to level 1 inside the well (sequential photon-assisted tunneling). Sufficient gain can be reached by dense filling of the optical mode volume by emitting wells. Such a laser is based on the optical transitions of electrons passing the structure and emitting photons. Thus, the laser mechanism is associated with unipolar transport through the quantum-confined   states. The unipolar mechanism allows monolithic stacking of the series of optimally designed wells providing the cascade principle of the device. Because of this principle, these lasers are termed quantum cascade lasers (QCL).

The tunneling current appears to be bias-dependent in a reso-nancelike manner, as the electron transfer is easier if level 1 in the left well (see Fig. 2.10a) is at the same energy as level 2 in the right well in every pair of the superlattice. Sweeny and Xu (1989) noticed that direct photon-assisted (nonresonant) tunneling between two wells (see Fig. 2.10b) may occur in addition to the resonant one and may contribute to the stimulated emission. The wavelength of emitted radiation appears also to be tunable by changing of the external bias. This is the case of diagonal optical transitions in the barrier layers.

Yuh and Wang (1987) proposed the IR semiconductor superlattice laser with the use of intersubband optical transitions in a so-called band-aligned structure. In the structure, the miniband discontinuity is created by tailoring the quantum barrier profiles; the discontinuity plays a role of a band offset as in conventional heterostructures, and the population inversion is achieved by current injection. An example of calculation was given for a laser structure operating at $\lambda \sim 10$ μm.

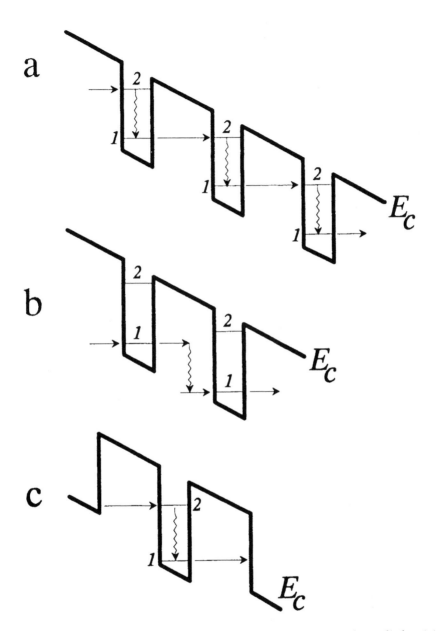

Figure 2.10: Simplified schemes of tunneling-type unipolar laser diodes: (a) superlattice and continuous consequence of quantum wells and barriers; (b) "diagonal" transitions between quantum wells; (c) "vertical" transitions in a quantum well between two barriers.

Belenov *et al.* (1988) analyzed the optical gain at resonant tunneling of electrons in the double-barrier quantum-well structure, shown in Fig. 2.10c. The inversion is proposed to be produced by tunneling from bulk semiconductor layer into a level *2* of quantum confined state in the well and by emptying a level *1* by tunneling from it to the bulk layer at the right side. In this case, the *vertical* optical transitions occur in the wells. It was calculated that using this operation scheme, a sufficient gain for laser action $(10-100 \text{ cm}^{-1})$ may be obtained in the IR range. The structure is cascadable for a proper filling of the volume of the optical mode.

A main problem in tunneling unipolar lasers is very low radiative quantum yield (less than 0.1%). This is due to rapid intraband nonradiative phonon-assisted relaxation of the electrons on the upper level (characteristic lifetime for LO phonon emission is ~ 1 ps), which leads to the increase of the threshold current density for QCL even at LT. The cascade principle allows each electron to emit photons many times when it passes the active region. The scheme of unipolar tunneling-based mid-IR laser was elaborated by Faist *et al.* (1994a, 1994b) and realized using III–V materials (therefore, providing a great advantage of the structure as based on well-developed technologies).

Interband cascade-type lasers were reported by Chow *et al.* (1995). Projects for mid-IR *interband* cascade lasers (ICL) were reported by Meyer *et al.* (1995) and Meyer *et al.* (1996). Two schemes of structures have been proposed, and a feature of each is the tunneling mechanism of the pumping the quantum-well active regions. The working transitions occur between quantum-confined electron levels in InAs to the quantum-confined hole levels in $Ga_{0.7}In_{0.3}Sb$ side of the tunneling type-II heterojunction. Another scheme is based on the type-I heterojunction between InAlAs and InAsSb with a tunneling of electrons through the InAlAs barrier into the active region. A high radiative efficiency was expected for these laser schemes since the phonon processes that dominate relaxation in the intersubband quantum-cascade lasers are eliminated. The Auger losses are assumed to be suppressed owing to a light in-plane hole effective mass and the removal of resonance between the energy gap and intervalence transitions.

2.9.2   Characteristics of quantum cascade lasers

The realization of the unipolar tunneling laser mechanism was reported by Faist *et al.* (1994a). A review paper on QCL was given by Faist *et al.*

(1994b). The typical unipolar laser structure consists of a large number of quantum-size layers (many hundreds). In spite of this, these lasers demonstrated a rapid progress in recent years and surpass other types of mid-IR semiconductor lasers in terms of high-temperature and high-power performance. The practical advantages of the QCL were found to be larger T_0 parameter and higher operating temperature than conventional interband semiconductor lasers.

Faist *et al.* (1996) reported the QCLs operating above RT and also capable of CW operation at LT, based on vertical optical transitions. The whole structure was fabricated in a lattice-matched manner with InAlAs/InGaAs heterojunctions grown by MBE on the InP substrate. These diode structures contain an active region of 25 periods with three different QWs and a chirped superlattice injection/relaxation region ("funnel" injector). The quantum-confined electronic states in active QWs were engineered so that at the applied voltage the lower working level would be emptied by resonant phonon emission. The superlattice region acts as a distributed Bragg reflector for electron waves, preventing the leakage of electrons from an upper working level. The 3-mm-long diodes were tested under pulse pumping (50 ns) near RT. The threshold current density was 8–10 kA/cm^2, and T_0 was 114 K in the temperature range from 160 to 320 K. The emission wavelength was slowly changeable: 5.015 μm at 25 K, 5.088 μm at 134 K, and 5.2 μm at 300 K. Pulse peak optical power was measured at 200 mW at 300 K (average power of 6 mW) and 100 mW at 320 K. Differential efficiency was almost constant, ~250 mW/A at LT, but decreases to 106 mW/A at 300 K and to 78 mW/A at 320 K. The optical loss coefficient in the active waveguide was assumed to be of 10 cm^{-1}. CW operation was obtained at LT, and the maximum operation temperature was 140 K. The optical power was up to 25 mW at 10 K, 15 mW at 120 K, and ~2 mW at 140 K.

2.10   Nonlinear optical conversion in semiconductors

We shall mention here some other mechanisms in semiconductors suitable for generation of coherent IR emission. A well-known technique is nonlinear frequency conversion of laser emission or parametric oscillation in some crystalline material. For example, nonlinear mixing of emission from CO and CO_2 lasers may be used to obtain coherent emission tunable in the range of 2.5–17 μm as reported by Kildal and Mikkelsen (1974), who used a nonlinear $CdGeAs_2$. There are many other works on this technique. Another approach was used by Pidgeon *et al.* (1971) to obtain tunable coherent emission in the 5 μm wavelength region by the Raman stimulated process. Nishizawa and Suto (1980) described a semiconductor Raman laser and obtained IR radiation (near 25 μm) from GaP or GaAs with peak output power of 3 W. The threshold for optical pumping was 5 MW/cm^2 at RT, as reported by Suto and Nishizawa (1983). Later, the threshold in epitaxial Raman laser on GaP was found to be lowered to 0.8 MW/cm^2 as it was reported by Suto and Nishizawa (1986).

Tunable spin-flip lasers were demonstrated by Patel and Shaw (1970) on InSb of n-type with pumping by CO_2 or CO gas laser in a transverse magnetic field (48–100 kG) at 25–30 K. Peak output power of this spin-flip laser was high as a kilowatt for first Stokes line and near 30 W at anti-Stokes line. The tuning range was 9–14.5 μm for InSb, with CO_2 laser pumping and 5.2–6 μm with CO laser pumping. Spin-flip lasers were also demonstrated in other semiconductors (InAs, HgCdTe, etc.). Summarized data on long-wavelength semiconductor spin-flip lasers are given in Table 2.12.

2.11   Brief review of applications

Practical purposes that can be achieved by using the mid-IR and far-IR laser diodes may be summarized as follows:

- Ultrahigh-resolution spectroscopic studies.
- Spectral-selective atmospheric pollution monitoring.
- Process monitoring and diagnostic devices in industries for humidity and water-content control and detection of methane and other gas impurities.
- Spectroscopic sensing in various areas of medicine.
- Communication and lidar technologies: emission sources for atmospheric window transmission and for the next generations of fiber-optics communication based on fluoride, sulfide, and other new glasses, highly transparent in the middle IR region.

There are many (more than five hundred) publications concerning spectroscopic applications of laser diodes. Only some of them can be men-

Pumping Source-laser	Pumping Wavelength, μm	Semiconductor Material	Tuning Range, μm	Maximum Output Power, W	References
CO_2	10.6	InSb	9–14.6	1000*	Aggarwal et al. (1971)
				10**	
				30***	
CO_2/SHG	5.3	InSb	5.2–6	1000*	Wood et al. (1973)
CO (CW)	5.2–6.5	InSb	5.2–6.5	1*	Bruek and Mooradian (1971)
HF (TEA)	2.9	InAs	2.98–3.0	> 200	Eng et al. (1974)
CO_2 (CW)	9.52	HgCdTe	9.7–10.2	1*	Sattler et al. (1974)

*1st Stokes component.
**2nd Stokes component.
***Anti-Stokes component.

Table 2.12: IR spin-flip tunable lasers (Mooradian, 1976)

tioned here, particularly, selected works and review papers on various aspects in this area. Earlier considerations of the topic were given by Hinkley (1970), Zasavitski *et al.* (1971), Butler (1976), and Hinkley *et al.* (1976). The application of the tunable laser diode to heterodyne receiving was reported by Hinkley *et al.* (1968) using a PbSnTe laser operating at 10.6 μm. The principle was successfully used in the heterodyne spectroscopy of astronomical and laboratory sources by Mumma *et al.* (1975). Later review papers were published on spectroscopic applications by Kosichkin and Nadezhdinski (1983), on the applicability of a multimode diode laser in flame absorption spectroscopy by Ng *et al.* (1990), on the transient absorption spectroscopy by Beckwith *et al.* (1987), on the atomic absorption spectroscopy by Hergenroeder and Niemax (1988), on the quantum noise limits in FM spectroscopy by Carlisle *et al.* (1989), etc.

Laser diodes represent a unique spectroscopic instrumental device because they simultaneously supply several important properties:

- Narrow spectral line (10^{-4}–10^{-5} cm^{-1}) sufficient for high-resolution spectroscopy.
- Tunability with simple methods of wavelength control.
- High-speed response sufficient for high temporal resolution measurement (down to a nano-second range).
- Small-size emission spot providing local resolution in the micrometer range (limited by the wavelength).
- Availability of laser diode materials for every wavelength in the range from visible to far IR emission (46 μm).

Owing to these advantages the long-wavelength laser diode has become one of the most important instruments in atomic and molecular spectroscopy. Many of the new scientific results in spectroscopy during the last decades were obtained using IR tunable laser diodes. High-resolution spectroscopy gave a large amount of new information concerning the structure of absorption spectra of technically important species (air pollutants, biologically active substances, ozone molecules, car exhaust components and other pollutants, drugs, isotope-pure substances, etc.).

An example of IR high-resolution spectroscopy of overcooled gaseous hexafluorides was described by Baronov *et al.* (1981). The laser spectrometer with PbSnTe/PbTe and PbSnSe/PbSe laser diodes was employed to perform studies of absorption spectra of SF_6, WF_6, and UF_6. In order to eliminate hot-bands superimposed on fundamental transition bands, the gaseous mixture was overcooled in a supersonic jet. The probe gas constituted 1% of the mixture in He. The rotation temperature of molecules drops to 40 K in the jet. The isotopic structure of the $\nu3$ band is identified for WF_6 (four Q-branches) and for UF_6 (two Q-branches of $\nu3$ band for U^{238} and U^{235} hexafluorides). The isotopic shift was 0.650 ± 0.005 cm^{-1} in the case of UF_6. The fine structure of the branch of $U^{238}F_6$ had been resolved to identify the spectral peak near the 16 μm wavelength with a bandwidth of ~0.05 cm^{-1}. The asymmetric shape of the branch was confirmed, as was found earlier in connection with the technology of uranium isotope enrichment, as discussed by Jensen *et al.* (1976).

There have been many works concerning spectral identification of various gaseous substances. The use of tunable laser diodes for the detection of air pollutants was reported by Hinkley and Kelley (1971) and developed in numerous works. Ku *et al.* (1975) demonstrated long-path monitoring of CO in the atmosphere with a tunable laser diode system. The laboratory sensitivity to some species was reported by Reid *et al.* (1978) for the multiple-path cell with a 200-m-long total optical path. The absorption coefficient of 3×10^{-6} cm^{-1} was estimated as a limit of detectivity of this spectroscopic sensor, which corresponds to impurity detectivity as low as 10^{-9} for SO_2 and N_2O and 10^{-10} for NH_3 pollutants. The sensitivity of the transient absorption spectroscopy was discussed by Beckwith (1987) and the absorbency detection limit was estimated at 10^{-4}–10^{-5}.

Spectroscopic sensing of humidity, impurities, and pollutants has a great practical significance, and the laser-diode–based systems have good perspectives, especially if they use RT-operating lasers. The areas of their application are the automobile industry; power plants; chemical, pharma-chemical, and semiconductor manufactures; and environmental and geophysical control.

Bloch *et al.* (1990) discussed the monitoring problem of global distribution of the ozone using tunable PbSnTeSe laser diodes. They concluded that laser-diode based spectroscopic sensors may be successfully used for detection of O_3 and other species related to ozone circulation in atmosphere in the 7–20 μm wavelength range. The bandwidth of ozone lines at low pressure is about $(1–5)\times10^{-3}$ cm^{-1}, therefore typical laser emission linewidth is suitable for the detection measurements. These measurements may be performed at long-path condition, and high speed scanning of laser emission wavelength allows the obtaining of absorption data in a few microseconds.

An exciting field of the laser diode application is medical diagnostics. In the review paper of Preier (1990), an example of medical diagnostics described was use of laser-diode spectroscopic sensing of the isotopic ratio. The diagnostic tool is based on the possibility of detecting metabolic malfunctions by measuring the temporal development of the isotopic ratio in CO_2 in the breath air after feeding the patient with isotopic C^{13}-labeled food or drugs. The ratio can be determined very accurately by laser spectroscopy with measurement of magnitudes of the isotopic-shifted absorption lines. It is stated that using this stable isotope method it was possible to get, for instance, some information about the digestion of fats of premature babies by injection of C^{13}-enriched triolin. This breath analysis method using stable isotopes has several advantages: It is noninvasive, nonradioactive, and nonpoisonous, free of any side effects.

The commercial applications of long wavelength lasers for communication purposes are likely to materialize in the near future. Atmospheric windows at 3–5 μm and 8–14 μm are available for open optical transmission lines based on the semiconductor lasers. However, development in this field occurs rather slowly in expectation of cheaper and RT operating laser diodes. On the other hand, it was pointed out that some novel fiber materials, like sulfide and fluoride glasses, may have extremely low losses in the range of 2.4 μm, and an absorption level of about 10^{-3} dB/km was predicted. In practical cases, optical losses include also various mechanisms of light scattering. Overall losses of about 1 dB/km were measured in the zirconium fluoride-based fiber at a narrow window near 2.56 μm, as reported by France *et al.* (1986). This result holds promise for improvements of the performance characteristics of future fiber communication lines. IR diode lasers described in this review may appear to be most suitable components for these lines providing broadband-modulated laser emission sources at new optimized wavelengths.

As to photoreceivers for long wavelengths, they are elaborated using the same compounds and alloys used in lasers, but with shifted compositions fitted for detection of the emission by an intrinsic absorption.

2.12   Conclusions

The above-described research of long wavelength semiconductor lasers is important with respect to the practical applications in a wide spectral range. There is an active field of competitive developments of laser versions based on quite different operation mechanisms: either interband transition with a bipolar pumping or intraband transition with a unipolar pumping. The former mechanism is more traditional, with a great experience of commercial production of GaAs- and InP-based devices and large-scale applications. The tunneling-based device of the latter unipolar type (the quantum-cascade laser) is rather new. At present, it is known practically by publications from only one laboratory. It is possible to guess that this new type of laser can be a real alternative to some conventional mid-IR semiconductor lasers. An impressive progress in RT-operating QCLs in the mid-IR gives a hope for a breakthrough to the large-scale laser applications of these lasers. An important advantage of the QCLs is that their technology deals with well-developed materials like InAlGaAs/InP, and thus it does not require looking for very new and sophisticated material technologies.

The progress in conventional long-wavelength laser diodes was linked with the quaternary system of InGaSbAs, which includes ternary alloys of InAsSb and GaInSb, all suitable for heterostructure laser applications. It was argued that the alloy system InGaSbAs lattice matched to GaSb appears to be most suitable for RT operation in the wavelength range of 2.0–2.7 μm. Its advantages are:

1. The active material InGaSbAs may be reproducibly grown by epitaxial technologies with satisfactory emission properties for both injection and electron-beam pumped lasers.

2. The cladding material GaAlSbAs is chosen as being lattice-matched to both active material and to GaSb substrate and as providing sufficient optical and electron confinements for DH laser structures.

The improvement in laser performances in the framework of this hetero-system was achieved with MBE and MOCVD, allowing the carrying out of the crystal growth farther from the chemical equilibrium than in the LPE process. Due to this, some breakthrough regarding the immiscibility gap of compositions was achieved.

The actual aim in a further development is the improvement of laser structure with InSbAsP active materials lattice-matched to GaSb or InAs, which are suitable for emission in the spectral range of 2.5–3.5 μm. The low-temperature laser action has been demonstrated already. In addition, there is the hope of finding new alloy combinations for IR spectral ranges without precise lattice-matching (strained heterostructure). Those may reliably operate if the strained layer is sufficiently thin. The concept of strained structures will give a wide choice of suitable materials.

The lead-salt lasers passed a long way of research and development. At present some of them approach the RT operation. Impressive improvements were achieved using the lattice-matched heterostructures. Most encouraging results may be summarized as follows:

1. Obtaining the longest wavelength laser emission in the range 46–49 μm with the quaternary heterosystem PbSnTeSe at LT.

2. Obtaining the extremely low threshold current density of 3.9 A/cm^2 at LT in the quaternary PbEuTeSe laser heterosystem.

3. Development of mode selective structures for the extended mode tuning range, particularly DFB and DBR devices.

4. Obtaining near-RT pulse operation (280–290 K) in some DH lasers and obtaining the CW operation above 200 K.

In addition to these results, there are interesting achievements in the laser action in far-IR, based on the intraband transitions in valence band of Ge. This allows the obtaining of laser emission from the low-temperature operating devices in the range beyond 100 μm.

Acknowledgments

In conclusion, the author expresses his appreciation to Dr. A. D. Britov, Dr. O. I. Davarashvili, and Dr. A. E. Yunovich for helpful discussions and to E. I. Naidich for the assistance in manuscript preparation.

References

Abrikosov, N. Kh.; Bankina, V. F.; Poretskaya, L. V.; and Shelimova, L. E. (1969). *Semiconducting II–VI, IV–VI and V–VI Compounds.* New York: *Plenum.*

Adachi, S. (1987). *J. Appl. Phys.*, **61**, 4869.

Aggarwal, R. L.; Lax, B.; Chase, C. E.; Pidgeon, C. R.; Limbert, D., and Brown, F. (1971). *Appl. Phys. Lett.*, **18**, 383.

Aidaraliev, M.; Zotova, N. V.; Karandashev, S. A.; Matveev, B. A.; Stus', N. M.; and Talalakin, G. N. (1987a). *Pis'ma Zh. Tekh. Fiz.*, **13**, 329.

Aidaraliev, M.; Zotova, N. V.; Karandashev, S. A.; Matveev, B. A.; Stus', N. M.; and Taialakin, G. N. (1987b). *Sov. Techn. Phys. Lett.*, **13**, 232.

Aidaraliev M.; Zotova, N. V.; Karandashev, S. A.; Matveev, B. A.; Stus' N. M; and Talalakin, G. N. (1988). *Sov. Techn. Phys. Lett.*, **14**, 704.

Aidaraliev, M.; Zotova, N. V.; Karandashev, S. A.; Matveev, B. A.; Stus', N. M.; and Taialakin, G. N. (1989). *Sov. Tech. Phys. Lett.*, **15**, 600.

Aidaraliev, M.; Zotova, N. V.; Karandashev S. A.; Matveev, B. A.; Stus' N. M.; and Talalakin, G. N. (1993). *Sov. Phys. Semicond.*, **27**, 21.

Akiba, S.; Matsushima, Y.; Iketani, T.; and Usami, M. (1988). *Electron. Lett.*, **24**, 1071.

Akimova, I. V.; Bochkarev, A. E.; Dolginov, L. M.; Drakin, A. E.; Druzhinina, L. M.; Eliseev, P. G.; Sverdlov, B. N.; and Skripkin, V. A. (1988). *Zh. Tekh. Fiz.*, **58**, 701.

Aleksandrov, S. N.; Vasil'ev, V. I; Dimov, F. I.; Kuchinski, V. I.; Lasutka, A. S.; Mishurny, V. A.; and Smirnitski, V. B. (1984). *Pis'ma Zh. Tekh. Fiz.*, **10**, 1081.

Andaspaeva, A. A.; Baranov, A. N.; Gel'mont, B. L.; Dzhurtanov, B. E.; Zegrya, G. G.; Imenkov, A. N.; Yakovlev, Yu. P.; and Yastrebov, S. G. (1991). *Sov. Phys. Semicond.*, **25**, 240.

Anderson, W. W. (1977). *IEEE J. Quantum Electron.*, **OE-13**, 532.

Andronov, A. A. (1979). *Pis'ma Zh. Exp. Teor. Fiz.*, **30**, 585.

Andronov, A. A.; Nozdrin, Yu. N.; and Shastin, V. N. (1986). *Izvestia AN SSSR, Ser. Phys.*, **50**, 1103.

Andronov, A. A.; Zverev, I. V.; Kozlov, V. A.; Nozdrin, Yu. N.; Pavlov, S. A.; and Shastin, V. N. (1984). *Sov. Phys.-JETP*, **40**, 804.

Anisimova, I. D.; Ivashneva, N. A.; and Mozzhorin, Yu. D. (1970). *Sov. Phys., Semicond.*, **3**, 1412.

Antcliffe, G. A.; Parker, S. G.; and Bate, R. T. (1972). *Appl. Phys. Lett.*, **21** (10), 505.

Arias, J. M.; Zandian, M.; Zucca, R.; and Singh, J. (1993). *Semicond. Sci. Technol.*, **8**, S 255.

Arushanov, E. K.; Kulyuk, L. L.; Luk'yanova, L. N.; Naterpov, A. N.; Radautsan, S. I.; and Shanov, A. A. (1982). *Kvant. Elektron.*, **9**, 1926.

Benoit a la Guillaume, C., and Debever, J. M. (1964). *Solid State Commun.*, **2**, 145.

Benoit a la Guillaume, C., and Debever, J. M. (1965a). *Solid State Commun.*, **3**, 19.

Benoit a la Guillaume, C.; and Debever, J. M. (1965b). *Radiative Recomb. Semicond. Paris: Dunod*, 11–13.

Baranov, A. N. (1991). *Appl. Phys. Lett.*, **59**, 2360

Baranov, A. N.; Danilova, T. N.; Dzhurtanov, B. E.; Imenkov, A. N.; Konnikov, S. G.; Litvak, A. M.; Usmanski, V. E., and Yakovlev, Yu. P. (1988a). *Pis'ma Zh. Tekh. Fiz.*, **14**, 1671.

Baranov, A. N.; Danilova, T. N.; Dzhurtanov, B. E.; Imenkov, A. N.; and Yakovlev, Yu. P. (1990). *Tezisy All-Union Conf. on Solid-State Electron. Dev.*, Leningrad.

Baranov, A. N.; Danilova, T. N.; Ershov, O. G.; Imenkov, A. N.; and Yakovlev, Yu. P. (1991). *Pis'ma Zh. Techn. Fiz.*, **17**, 54.

Baranov, A. N.; Dzhurtanov, B. E.; Imenkov, A. N.; Rogachev, A. A.; Shernyakov, Yu. M.; and Yakovlev, Yu. P. (1986a). *Fizika i Tekhnika Poluprov.*, **10**, 2217.

Baranov, A. N.; Dzhurtanov, B. E.; Imenkov, A. N.; Shernyakov, Yu. M.; and Yakovlev, Yu. P. (1986b). *Pis'ma Zh. Tekh. Fiz.*, **12**, 557.

Baranov, A. N.; Grebenschikova, E. A.; Dzhurtanov, B. E.; Danilova, T. N.; Imenkov, A N.; and Yakovlev, Yu. P. (1988b). *Pis'ma Zh. Tekh. Fiz.*, **14**, 1839.

Baranov, A. N.; Gusseinov, A. A.; Rogachev, A. A.; Titkov, A. N.; Cheban, B. A.; and Yakovlev, Yu. P. (1988c). *Pis'ma Zh. Tekh. Fiz.*, **48**, 342.

Baranov, A. N., Imenkov, A. N., Mikhailova, M. P., Rogachev, A. A., Yakovlev, Yu. P. (1989). *SPIE Proc.*, **1048**, 188.

Baronov, G. S.; Britov, A. D.; Karavaev, S. M.; Karchevski, A. I.; Kulikov, S. Yu.; Merzlyakov, A. N.; Sivachenko, S. D.; and Shcherbina, Yu. I. (1981). *Kvant. Elektron.*, **8**, 1573.

Baryshev, N. S. (1961). *Sov. Phys.-Solid State*, **3**, 1037.

Baryshev, N. S. (1962). *Fiz. Tverd. Tela*, **3**, 1428.

Basov, N. G.; Dudenkova, A. V.; Krasilnikov, A. I.; Nikitin, V. V.; and Fedoseev, K. P. (1966). *Fiz. Tekh. Poluprov.*, **8**, 1050. *Sov. J. Sol. State*, **8**, 847.

Beckwith, P. H; Brown, C. E.; Danagher, D. J.; Smith, D. R.; and Reid, J. (1987). *Appl. Opt.*, **26**, 2643.

Becla, P. (1988). *J. Vac. Sci. Technol.*, **A 6**, 2725.

Belenov, E. M.; Eliseev, P. G.; Oraevski, A. N.; Romanenko, V. I.; Sobolev, A. G.; and Uskov, A. V. (1988). *Kvant. Elektron.*, **15**, 1595.

Berding, M. A.; Sher, A.; Chen, A.-B.; and Miller, W. E. (1988). *J. Appl. Phys.*, **63**, 107.

Bernard, M.; Chipaux, C.; Durraffourg, G.; Jean-Louis, M.; Loudette, J.; and Noblanc, J. (1963). *Compt. Rend.*, **257**, 2984.

Bishop, S. G.; Moore, W. J.; and Swiggard, E. M. (1969). *Appl. Phys. Lett.*, **15**, 12.

Bloch, M. A.; Bychkova, L. P., Davarashvili, O. I.; Dyad'kin, A. P.; Enukashvili, M. I.; Kekelidze, N. N.; Kuritsyn, Yu. A.; Krivtsun, V. M.; Sisakyan, I. N.; and Shotov, A. P. (1990). *Proc. Tbilisi Uni.*, **295**, 188.

Bochkarev, A. E.; Dolginov, L. M.; Drakin, A. E.; Druzhinina, L. V.; Eliseev, P. G.; and Sverdlov, B. N. (1985). *Kvant. Elektron.*, **12**, 1309.

Bochkarev, A. E.; Dolginov, L. M.; Drakin, A. E.; Druzhinina, L. V.; Eliseev, P. G.; Sverdlov, B. N.; and Skripkin, V. A. (1986). *Kvant. Elektron.*, **13**, 2119.

Bochkarev, A. E.; Dolginov, L. M.; Drakin, A. E.; Eliseev, P. G.; and Sverdiov, B. N. (1988a). *11th IEEE Int. Semicond. Laser Conf., Boston. Post-Deadline Pap. PD-8.*

Bochkarev, A. E.; Dolginov, L. M.; Drakin, A. E.; Eliseev, P. G.; and Sverdlov, B. N. (1988b). *Kvant. Elektron.*, **15**, 2171.

Bochkarev, A. E.; Drakin, A. E.; Eliseev, P. G.; and Sverdlov, B. N. (1992) *Trudy FIAN*, 216.

Bondar, S. A.; Borisov, N. A.; Galchenkov, A. N.; Lavrushin, B. M.; Lebedev, V. V.; and Strel'chenko S. S. (1976). *Kvant. Electron.*, **3**, 94.

Bouchut, P.; Destefanis, G.; Chamonal, J. P.; Million, A.; Pelliciari, B.; and Piaguet, J. (1991). *J. Vac. Sci. Technol.*, **b 9**, 1794.

Britov, A. D. (1983). *Semiconductor tunable lasers of middle and far IR range.* Doct. Dissertation, Ioffe Phys.-Techn. Inst., S.-Petersburg.

Britov, A. D.; Karavaev, S. M.; Kalyuzhnaya, G. A.; Gorina, Yu. I.; Kurbatov, A. L.; Kiseleva K. V.; and Starik, P. M. (1976a). *Kvant. Elektron.*, **3**, 2238.

Britov, A. D.; Penin, N. A.; Maksimovski, S. N.; Karavaev, S. M.; Revokatova, I. P.; Kurbatov, A. L.; Aver'yanov, I. S.; Pyregov, B. P.; and Myzina, V. A. (1976b). *Kvant. Elektron.*, **3**, 253.

Brown, M. A., and Porteous, P. (1967). *Sol.-State Electron.*, **10**, 76.

Bruek, S. R. J., and Mooradian, A. (1971). *Appl. Phys. Lett.*, **18**, 229.

Butler, J. F. (1964). *J. Electrochem. Soc.*, **111**, 1150.

Butler, J. F. (1976). *Proc. SPIE*, **82**, 33.

Butler, J. F.; Calawa, A. R.; and Harman, T. C. (1966). *Appl. Phys. Lett.*, **9**, 427.

Butler, J. F.; Calawa, A. R.; Phelan, R. Jr.; Harman, T. C.; Strauss, A. J.; and Rediker, R. H. (1964a). *Appl. Phys. Lett.*, **5** (4), 75.

Butler, J. F.; Calawa, A. R.; Phelan, R. Jr.; Strauss, A. J.; and Rediker, R. H. (1964b). *Solid State Commun.*, **2**, 303.

Butler, J. F.; Calawa, A. R.; and Rediker, R. H. (1965). *IEEE J. Quant. Electron.*, **QE-1** (1), 4.

Butler, J. F., and Harman, T. C. (1967). *Trans. IEEE Electron. Dev.*, **14** (9), 630.

Bychkova, L. P.; Davarashvili, O. I.; Eliseev, P. G.; Saginuri, M. I.; Chikovani, R. I.; and Shotov, A. P. (1982). *Kvant. Elektron.*, **9**, 2140.

Calawa, A. R., Dimmock, J. O.; Harman, T. C.; and Melngailis, I. (1969). *Phys. Rev. Lett.*, **23**, 7.

Caneau C.; Srivastava, A. K.; Dentai, A. G.; Zyskind, J. L.; Burrus, C. A.; and Pollack, M. A. (1986). *Electon. Lett.*, **22**, 992.

Caneau, C.; Srivastava, A. K.; Dentai, A. G.; Zyskind, J. L.; and Pollack, M. A. (1985). *Electron. Lett.*, 815.

Caneau, C.; Zyskind, J. L.; Sulhoff, J. W.; Glover, T. E.; Centanni, J.; Burrus, C. A.; Dentai, A. G.; and Pollack, M. A. (1987). *Appl. Phys. Lett.*, **51**, 764.

Carlisle, C. B.; Cooper, D. E.; and Preier, H. (1989). *Appl. Opt.*, **28**, 2567.

Chashchin, S. P.; Averianov, I. S.; and Baryshev, N. S. (1970b). *Fiz. Tekh. Poluprov.*, **9**, 1794.

Chashchin, S. P.; Averyanov, I. S.; Baryshev, N. S.; Kudryashev, V. A.; Markina, N. P.; and Shuba, Yu. A. (1969a). *Fiz. Tekh. Poluprov.*, **3**, 1259.

Chashchin, S. P.; Baryshev, N. S.; Averyanov, I. S.; and Markina, N. P. (1969b), *Fiz. Tekh. Poluprov.*, **3** (10), 1572.

Chashchin, S. P., Baryshev, N. S., Averyanov, I. S., and Markina, N. P. (1970a). *Fiz. Tekh. Poluprov.*, **4**(8), 1546.

Cherng, M. J.; Stringfellow, G. B.; Kisker, D. W.; Srivastava, A. K.; and Zyskind, J. L. (1986). *Appl. Phys. Lett.*, **48**, 419.

Choi, H. K., and Eglash, S. J. (1991a). *IEEE J. Quant. Electron.*, **QE-27**, 1555.

Choi, H. K. and Eglash, S. J. (1991b). *Appl. Phys. Lett.*, **59**, 1165.

Choi, H. K., and Eglash, S. J. (1992). *Appl. Phys. Lett.*, **61**, 1154.

Choi, H. K.; Eglash, S. J.; and Connors, M. K. (1992) *50th IEEE Annual Device Res. Conf.*, B-6.

Choi, H. K.; Eglash, S. J.; and Connors, M. K. (1993). *Appl. Phys. Lett.*, **63**, 3271.

Choi, H. K.; Eglash, S. J.; and Turner, G. W. (1994a). *Appl. Phys. Lett*, **64**, 2474.

Choi, H. K., and Turner, G. W. (1995). *Appl. Phys. Lett.*, **67**, 332.

Choi, H. K.; Turner, G. W.; and Eglash, S. J. (1994b). *IEEE Photon. Technol. Lett.*, **6**:7.

Choi, H. K.; Turner, G. W.; and Manfra, M. J. (1996). *Electron. Lett.*, **32**, 1296.

Chow, D. H.; Miles, R. H.; Hasenberg; T. C.; Kost, A. R.; Zhang, Y.-H.; Dunlap, H. L.; and West, L. (1995). *Appl. Phys. Lett.*, **67**, 3700.

Chuiu, T. H.; Zyskind, J. L.; and Tsang, W. T. (1987). *J. Electron. Mater.*, **16**, 7.

Davarashvili, O. I.; Dolginov, L. M.; Eliseev P. G.; Zasavitsky, I. I.; and Shotov, A. P. (1977). *Kvant. Elektron.*, **4**, 904.

Dimmock, J. O.; Melngailis, I.; and Strauss, A. J. (1966). *Phys. Rev. Lett.*, **16**(2), 1193.

Dolginov, L. M.; Drakin, A. E.; Druzhinina, L. V.; Eliseev, P. G.; Milvidski, M. G.; Sverdlov, B. N.; and Skripkin, V. A. (1983). *Trudy FIAN SSSR, Nauka Moscow*, **141**, 46.

Dolginov, L. M.; Druzhinina, L. N.; Eliseev, P. G., Kryukova, I. V., Leskovich, V. I.; Milvidski, M. G.; Sverdlov, B. N.; and Chapnin, B. A. (1976a). *Sov. J. Quantum. Electron.*, **6**, 507.

Dolginov, L. M.; Druzhinina, L. V.; Eliseev, P. G.; Kryukova, I. V.; Milvidski, M. G.; and Sverdlov, B. N. (1976b). *Sov. J. Quantum. Electron.*, **6**, 257.

Dolginov, L. M.; Druzhinina, L. V.; Eliseev, P. G.; Kryukova, I. V.; Milvidski, M. G.; Sverdlov, B. N.; and Shevchenko, E. G. (1978a). *IEEE J. Quantum. Electron.*, **13**, 609.

Dolginov, L. M.; Druzhinina, L. V.; Kryukova, I. V.; Lapshin, A. N.; Matveenko, E. V.; and Mil'vidski, M. G. (1978b). *Kvant. Elektron.*, **5**, 26.

Dolginov, L. M.; Druzhinina, L. V.; Eliseev, P. G.; Lapshin, A. N.; Milvidski, M. G.; and Sverdlov, B. N. (1978c). *Sov. J. Quantum. Electron.*, **8**, 416.

Dolginov, L. M.; Kochergin, Yu. N.; Krukova I. V.; Leskovich, V. I.; Matveenko, E. V.; Milvidski, M. G.; and Stepanov, B. M. (1978d). *Sov. Techn. Phys. Lett.*, **4**, 580.

Donnelly, J. P.; Calawa, A. R.; Harman, T. C.; Fo, A. G.; and Lindley, W. T. (1972). *Solid-State Electron.*, **15** (4) 403.

Drakin, A. E.; Eliseev, P. G.; Sverdlov, B. N.; Bochkarev, A. E.; Dolginov, L. M.; and Druzhinina L. V. (1986). *10th IEEE Int. Semicond. Laser Conf., Kanazawa, Conf. Progr. and Abstr. Pap., Tokyo*, 220.

Drakin, A. E.; Eliseev, P. G.; Sverdlov, B. N.; Bochkarev, A. E.; Dolginov, L. M.; and Druzhinina, L. V. (1987). *IEEE J. Quantum. Electron.*, **23**, 1089.

Dutta, B. V., Temkin, H.; Kolb, E. D.; and Sunder, W. A. (1985). *Appl. Phys. Lett.*, **47**, 111.

Dziuba, E. Z.; Niculescu, D.; and Niculescu, N. (1968). *Phys. Stat. Sol.*, **29**, 813.

Eglash, S. J., and Choi, H. K. (1994). *Appl. Phys. Lett.*, **64**, 833.

Eliseev, P. G. (1991). *Optoelectronics - Dev. and Technol.*, **6**, 1.

Emtage, P. R. (1976). *J. Appl. Phys.*, **47**, 2565.

Eng, R. S.; Mooradian, A.; and Fetterman, H. R. (1974). *Appl. Phys. Lett.*, **25**, 453.

Esaki, L. (1985). *MBE and Heterostructures*, ed. by Chang, L. L., and Ploog, K.-M. Nijhoff, Dordrecht.

Esina N. P.; Zotova, N. V.; Matveev, B. A.; Stus', N. M.; Talalakin, G. N.; and Abishev, T. D. (1983). *Sov. Tech. Phys. Lett.*, **9**, 167.

Esina, N. P.; Zotova, N. V.; and Nasledov, D. N. (1967). *Sov. Phys. Sol. State*, **9**, 1036.

Faist, J.; Capasso F.; Sirtori, C.; Sivco, D. L.; Baillargeon, J. N.; Hutchinson, A. L.; Chu, S. N. G.; and Cho, A. Y. (1996). *Appl. Phys. Lett.*, **68**, 3680.

Faist, J.; Capasso, F.; Sivco, D. L.; Sirtori, C., Hutchinson, A. L.; and Cho, A. Y. (1994a). *Electron Lett.*, **30**, 865.

Faist, J.; Capasso, F.; Sivco, D. L.; Sirtori, C.; Hutchinson, A. L.; and Cho, A. Y. (1994b). *Science*, **264**, 553.

Fang, Z. M.; Ma, K. Y.; Cohen R. M.; and Stringfellow, G. B. (1990a). *J. Appl. Phys.*, **68**, 1187.

Fang, Z. M.; Ma, K. Y.; Jaw D. H.; Cohen, R. M.; and Stringfellow, G. B. (1990b). *J. Appl. Phys.*, **67**, 7034.

Feit, Z.; Kostyk, D.; Woods, R. J.; and Mak, P. (1990a). *J. Vac. Sci. Technol.*, **B8**, 200.

Feit, Z.; Kostyk, D.; Woods, R. J.; and Mak, P. (1990b). *Appl. Phys. Lett.*, **57**, 2891.

Feit, Z.; Kostyk, D.; Woods, R. J.; and Mak, P. (1991a). *SPIE Proc.*, **1512**, 164.

Feit, Z.; Kostyk, D.; Woods, R. J.; and Mak, P. (1991b). *Appl. Phys. Lett.*, **58**, 343.

Feit, Z.; McDonald, M.; Woods, R. J.; Archambault, V.; and Mak, P. (1996). *Appl. Phys. Lett.*, **68**, 738.

Feit, Z.; Woods, R.; Kostyk, D; and Jalenak, W. (1989). *Appl. Phys. Lett.*, **55**, 16.

Fesquet, J. (1988). *Ann. Telecommun.*, **43**, 112.

Flatté, M. E.; Grein, C. H.; Ehrenreich, H.; Miles, R. H.; and Cruz, H. (1995) *J. Appl. Phys.* **78**, 4552.

France, B. W.; Carter, S. F.; Moore, M. W.; and Williams, J. R. (1986). *SPIE Proc.*, **618**, 51.

Galeski, F.; Drozd, I. A.; Lebedeva, L. Ya.; Ten, P.; and Yunovich, A. E. (1977). *Fiz. Tech. Poluprov.*, **11**, 568.

Galkin, L. N., and Korolyov N. V. (1953). *Doklady AN SSSR*, **91**, 529.

Garbuzov, D. Z.; Martinelli, R. U.; Menna, R. J.; York, P. K.; Lee, H.; Narayan, S. Y.; and Connolly, J. C. (1995). *Appl. Phys. Lett.*, **67**, 1346.

Garlick, G. F. J., and Dumbleton, M. J. (1954). *Proc. Phys. Soc.* **67b**, 442.

Giles, N. C.; Han, J. W.; Cook, J. W. Jr.; and Schetzina, J. F. (1989). *Appl. Phys. Lett.*, **55**, 2026.

Grein, C. H.; Young, P. M.; and Ehrenreich, H. (1994). *J. Appl. Phys.*, **76**, 1940.

Groves, S. H.; Nill, K. W.; and Strauss, A. J. (1974). *Appl. Phys. Lett.*, **25**, 331.

Gubarev, A. A.; Lavrushin, B. M.; Nabiev, R. F.; Nasibov. A. S.; Sypchenko, M. N.; and Popov, Yu. M. (1991). *Trudy FIAN*, Moscow, v. 202, 158–186.

Gureev, D. M.; Davarashvili, O. I.; Zasavitski, I. I.; Matsonashvili, B. N.; and Shotov, A. P. (1978). *Kvant. Elektron.*, **5**, 2630.

Haase, M. A.; Qiu, J.; DePuydt, J. M.; and Cheng, H. (1991). *Appl. Phys. Lett.*, **59**, 1272.

Hansen, G. L.; Schmit, J. L.; and Casselman, T. N. (1982). *J. Appl. Phys.*, **53**, 7099.

Harman, T. C. (1979). *J. Electron. Mater.*, **8**, 191.

Harman, T. C.; Calawa, A. R.; Melngailis, I.; and Dimmock, J. 0. (1969). *Appl. Phys. Lett.*, **14**, (11), 333.

Hasenberg, T. C.; Chow, D. H.; Kost A. R.; Miles, R. H.; and West L. (1995). *Electron. Lett.*, **31**, 275.

Haug, A. (1983). *J. Phys., C - Sol. St. Phys.*, **16**, 4159.

Haug, A. (1989). *Semicond. Sci. Technol.*, **4**, 803.

Hergenröder, R., and Niemax, K. (1988). *Spectrochemica Acta*, **43B**, 1443.

Hinkley, E. D. (1970). *Appl. Phys. Lett.*, **16**, 351.

Hinkley, E. D.; Harman, T. C.; and Freed, C. (1968). *Appl. Phys. Lett.*, **13**, 49.

Hinkley, E. D., and Kelley, P. L. (1971). *Science*, **171**, 635.

Hinkley, E. D.; Nill, K. W.; and Blum, F. A. (1976). "Infrared spectroscopy with tunable lasers," in *Laser Spectroscopy*, ed. by Walther, H. Berlin: Springer-Verlag.

Horikoshi, Y. (1985). *In: "Semicond. and Semimetals"*, ed. by Tsang, W. T., Academic Press, Orlando, **22C**, 93.

Horikoshi, Y.; Kawashima, M.; and Saito, H. (1981). *Jpn. Appl. Phys.*, **20**, L897.

Horikoshi, Y.; Kawashima, M.; and Saito, H. (1982a). *Jpn. J. Appl. Phys.*, **21**, 77.

Horikoshi, Y., Kawashima, M., and Saito, H. (1982b). *Jpn. J. Appl. Phys.*, **21**, 198.

Ishida, A.; Fujiyasu, H.; Ebe, H.; and Shinohara, K. (1986). *J. Appl. Phys.*, **59**, 3023.

Ishida, A., Muramatsu, K.; Takashiba, H.;and Fujiyasu, H. (1989). *Appl. Phys. Lett.*, **55**, 430.

Ishida, A.; Muramatsu, K.; Ishino, K.; and Fudjiasu, H. (1990). *Semicond. Sci. Techn.*, **5**, 334.

Ishida, A.; Matsuura, S.; Mizuno, M.; and Fujiyasu, H. (1987). *Appl. Phys. Lett.*, **51**, 478.

Jensen, R.-J.; Marinuzzi, J. G.; Robinson, C. P.; and Rackwood, S.G. (1976). *Laser Focus (May)*, 51.

Jiang, Y.; Teich, M. C.; and Wang, W. I. (1991). *J. Appl. Phys.*, **69**:6869.

Junga, F. A.; Cuff, K. F.; Blakemore, J. S.; and Washwell, E. R. (1964). *Phys. Lett.*, **13**, 103.

Kapon, E., and Katzir, A. (1985). *IEEE J. Quant. Electron.*, **21**, 1947.

Kano, H., and Sugiyama, K. (1980). *Electron. Lett.*, **16**, 146.

Kasemset, D., and Fonstad, C. G. (1979). *Appl. Phys. Lett.*, **34**, 432.

Kasemset, D.; Roffer, S.; and Fonstad, C. G. (1980). *IEEE Electron. Dev. Lett.*, **5**, 75.

Katzir, A.; Rosman, R.; Shani, Y.; Bachem, K. H.; Böttner, H.; and Preier, H. M. (1989). *Handb. Solid-State Lasers*. N.Y., Basel, 227.

Kazarinov R. F., and Suris R. A. (1971). *Sov. Phys. Semicond.*, 5, 707.

Kazczewski, G.; Kowalczyk, L.; and Szczerbakow, A. (1981). *Sol. St. Commun.*, **38**, 499.

Kildal, H., and Mikkelsen, J. C. (1974). *Opt. Commun.*, **10**, 306.

Klein, P. B.; Furneaux, J. E.; and Henry, R. L. (1983). *Appl. Phys. Lett.*, **42**, 638.

Kobayashi, N., and Horikoshi, Y. (1980). *Jpn. J. Appl. Phys.*, **19**, L641.

Kobayashi, N.; Horikoshi, Y.; and Uemura, C. (1980a). *Jpn. J. Appl. Phys.*, **19**, L30.

Kobayashi, N.; Horikoshi Y.; and Uemura, C. (1980b). *Jpn. J. Appl. Phys.*, **19**, (Suppl. 3), 333.

Koguchi, N.; Kiyosawa, T.; and Takahashi, S. (1989). *Jpn. J Appl. Phys.*, **28**, 1170.

Koguchi, N.; and Takahashi, S. (1991). *Phys. Lett.*, **58**, (8), 799.

Kosichkin, Yu. V., and Nadezhdinski, A. I. (1983). *Izvestia AN SSSR, Ser. Phys.*, **47**, 2037.

Kowalczyk, L.; Karczewski, G.; and Szczerbakow, A. (1981). *Phys. Semicond. Compound*, **4**, 183.

Kowalczyk, L., and Szczerbakow, A. (1988). *Acta Phys. Polon.*, **A73**, 447.

Krier, A. (1990). *Appl. Phys. Lett.*, **56**, 2428.

Kroemer, H., and Griffiths, G. (1983). *IEEE Electron. Dev. Lett.*, **4**, 20.

Kryukova, I. V.; Leskovich, V. I.; and Matveenko, E. V. (1979). *Kvant. Elektron.*, **6**, 1401.

Ku, R. T.; Hinkley, E. D.; and Sample, J. O. (1975). *Appl. Opt.*, **14**, 854.

Kucera, Z. (1987). *Phys. Stat. Solidi A*, **100**, 659.

Kurbatov, L. N.; Britov, A D.; Aver'yanov, I. S.; Mashchenko, V. E.; Mochalkin, N. N.; and Dirochka, A. I. (1968). *Fiz. Tekh. Poluprov.*, **2**, 1000.

Kurbatov, L. N.; Britov, A. D.; Karavaev, S. M.; Sivachenko, S. D.; Maksimovski, S. N.; Ovchinnikov, I. I.; Rzaev, M. M.; and Starik, P. M. (1983). *Pis'ma Zh. Exp. Teor. Fiz.*, **9**, 424.

Kurbatov, L. N.; Britov, A. D.; Mashchenko, V. E.; and Mochalkin, N. N. (1967). *Fiz. Tekhn. Poluprov.*, **1**, 1108.

Kurbatov, L. N.; Dirochka, A. I.; Sinitsyn, E. V; Lazarev, V. B.; Shevchenko, V. Ya.; and Kozlov, S. E. (1976). *Sov. J. Quantum. Electrons*, **6**, 166.

Kurbatov, L. N.; Karavaev, S. M.; Britov, A. D.; Sivachenko, S. D.; Maksimovski, S. N.; and Starik, P. M. (1986). *Pis'ma Zh. Tekh. Fiz.*, **12**, 422.

Kurbatov, A. L.; Shubin, M. V.; Starik, P. M.; Luchitski, R. M.; Britov, A. D.; and Polyakova, N. D. (1981). *Fiz. Tekh, Polupr.*, **15**(1), 802.

Kurtz, S. R.; Biefeld, R. M.; Allerman, A. A.; Howard, A. J.; Crawford, M. H.; and Pelczynski, M. W. (1996). *Appl. Phys. Lett.*, **68**, 1332.

Lasher, G., and Stern, F. (1964). *Phys. Rev.*, **133A**, 553.

Le, H. Q.; Turner, G. W.; Eglash, S. J.; Choi, H. K.; and Coppeta, D. A. (1994). *Appl. Phys. Lett.*, **64**, 152.

Lee, H.; York, P. R.; Menna, R. J.; Martinelli, R. U.; Garbuzov, D. Z.; Narayan, S. Y.; and Connolly, J. C. (1995). *Appl. Phys. Lett.*, **66**, 1942.

Linden, K. J.; Nill, K. W.; and Butler, J. F. (1977). *IEEE J. Quantum. Electron., QE-13:* **8**, 720.

Lo, W. (1981). *J. Appl. Phys.*, **52**, 900.

Lo, W., and Swets, D. E. (1980). *Appl. Phys. Lett.*, **36**, 450.

MackIntosh, I. M. (1956). *Proc. Phys, Soc.*, **69B**, 115.

Mani, H.; Joullie, A.; Boissier, G.; Tournie, E.; Pitard, F.; Joullie, A.-M.; and Alibert, C. (1988). *Electron. Lett.*, **24**, 1542.

Martinelli, R. U., and Zamerovski, T. J. (1990). *Appl. Phys. Lett.*, **56**, 125.

Martinelli, R. U.; Zamerovski, T. J.; and Longeway, P. A. (1989). *Appl. Phys. Lett.*, **54**, 277.

Melngailis, I. (1963). *Appl. Phys. Lett.*, **2**, 176.

Melngailis, I., and Rediker, R. H. (1966). *J. Appl. Phys.*, **37**, 899.

Melngailis, I., and Strauss, A. J. (1966). *Appl. Phys. Lett.*, **8**, 179.

Meyer, J. R.; Hoffman, C. A.; Bartoli, F. J.; and Ram-Mohan, L. R. (1995). *Appl. Phys. Lett.*, **67**, 757.

Meyer, J. R.; Vurgaftman, I.; Yang, R. Q.; and Ram-Mohan, L. R. (1996). *Electron. Lett.*, **32**, 45.

Mooradian, A. (1976). *Optoelectron. Conf. Proc.*, Munich, 1975, Gilford, 14.

Moy, J. P., and Reboul, J. P. (1982). *Infrared Phys.*, **22**, 163.

Mumma, M.; Kostiuk, T.; Cohen, S.; Buhl, D.; and von Thuna, P. C. (1975). *Space Sci. Rev.*, **17**, 661.

Murav'ev, A. V.; Pavlov, S. G.; and Shastin, V. N. (1990). *Pis'ma Zh. Exp. Teor. Fiz.*, **52**, 959.

Ng, K. C.; Ali, A. H.; Barber, E.; and Winefordner, J. D. (1990). *Appl. Spectroscopy*, **44**, 849.

Nill, K. W.; Calawa, A. R.; Harman, T. C.; and Walpole, J. N. (1970). *Appl. Phys. Lett.*, **16**(10), 375.

Nill, K. W.; Strauss, A. J.; and Blum, F. A.. (1973). *Appl. Phys. Lett.*, **22**, 677.

Nishijima, Y. (1989). *J. Appl. Phys.*, **6**(5), 935.

Nishizawa, J., and Suto, K. (1980). *J. Appl. Phys.*, **51**, 2429.

Patel, C. K. N., and Shaw, E. D. (1970). *Phys. Rev. Lett.*, **24**, 451.

Patel, N., and Yariv, A. (1970). *IEEE J. Quantum. Electron.*, **6**, 383.

Partin, D. L. (1984a). *J. Electron. Mater.*, **13**, 493.

Partin, D. L. (1984b). *Appl. Phys. Lett.*, **45**, 487.

Partin, D. L. (1985). *Superlat. Microstruct.*, **1**, 131.

Partin, D. L. (1988). *IEEE J. Quantum. Electron.*, *QE-24*, **8**, 1716.

Partin, D. L., and Thrush, C. M. (1984). *Appl. Phys. Lett.*, **45**, 193.

Phelan, R. J.; Calawa, A. R.; Rediker, R. H.; Keyes, R. J.; and Lax, B. (1963). *Sol. St. Res. Lincoln Lab. MIT.*, **3**, 12.

Pidgeon, C. R.; Lax, B.; Aggarwal, R. L.; Chase, C. E.; and Brown, F. (1971). *Appl. Phys. Lett.*, **19**, 333.

Preier, H. (1990). *Semicond. Sci. Technol.*, **5**, 12.

Preier, H.; Bleicher, M.; Riedel, W.; and Maier, H. (1976). *Appl. Phys. Lett.*, **28**, 669.

Preier, H.; Bleicher, M.; Riedel, W.; Pfeiffer, H.; and Maier, H. (1977). *Appl. Phys.*, **12**, 277.

Quinn, H. F., and Manley, G. W. (1965). *Solid-State Electron.*, **9**, 907.

Ralston, R. M.; Melngailis, I.; Calawa, A. R.; and Lindley, W. T. (1973). *IEEE J. Quant. Electron.*, **9**(2), 350.

Ralston, R. W.; Walpole, J. N.; Calawa, A. R.; Harman, T. C.; and McVittie, J. P. (1974). *J. Appl. Phys.*, **45**, 1323.

Ravid, A.; Zussman, Z.; Cinader, G.; and Oron, A. (1989). *Appl. Phys. Lett.*, **55**, 2704.

Ravid, A.; Zussman, A.; Sher, A.; and Shapira, Y. (1991). *Appl. Phys. Lett.*, **58**, 337.

Reid, J.; Schewchun, J.; Garside, B. K.; and Ballik, E. A. (1978). *Appl. Optics.*, **17**, 300.

Rosman, R., and Katzir, A. (1982). *IEEE J. Quantum. Electron.*, **18**, 814.

Sattler, J. P.; Weber, B. A.; and Nemarich J (1974). *J. Appl. Phys. Lett.*, **25**, 491.

Schlereth, K. H.; Spanger, B.; Böttner, H.; Lambrecht, A.; and Tacke, M. (1990). *Infrared Phys.*, **30**, 449.

Selivanov, Yu. G., and Shotov, A. P. (1987). *Kratk. Soobshchen. po Fiz.*, *FIAN, Moscow*, **4**, 21.

Shani, Y.; Katzir, A.; Bachem, K.-H; Norton, P.; Tacke, M.; and Preier, H. M. (1986). *Appl. Phys. Lett.*, **48**, 1178.

Shani, Y.; Rosman, R.; Katzir, A.; Norton, P.; Tacke, M.; and Preier, H. M. (1988). *J. Appl. Phys.*, **63**, 5603.

Shen, W. Z.; Shen, S. C.; Tang, W. G.; Zhao, Y.; and Li, A. Z. (1995). *Appl. Phys. Lett.*, **67**, 3432.

Shi, Z.; Tacke, M.; Lambrecht, A.; and Böttner, H. (1995). *Appl. Phys. Lett.*, **66**, 2537.

Shinohara, K.; Nishijima, Y.; Eba, H.; Ishida, A.; and Fujiyasu, H. (1985). *Appl. Phys. Lett.*, **47**, 1184.

Shotov, A. P. and Selivanov, Yu. G. (1990). *Semicond. Sci. Technol.*, **5**, 527.

Shotov; A. P.; Vyatkin, K. V.; and Sinyatynski, A. L. (1980). *Pis'ma Zh. Tekh. Fiz.*, **6**, 983.

Singh, J., and Zucca, R. (1992). *J. Appl. Phys.*, **72**, 2043.

Spanger, B.; Schiessl, U.; Lambrecht, A.; Böttner, H.; and Tacke, M. (1988a). *Appl. Phys. Lett.*, **53**, 2582.

Spanger, B., Schiessl, U.; Lambrecht, A.; Böttner, H.; and Tacke, M. (1988b). *Appl. Phys. Lett.*, **53**, 2582.

Srivastava, A. K.; Caneau, C.; Zyskind, J. L.; and Pollack, M. A. (1986). *10th IEEE Int. Semicond. Laser Conf., Kanazawa., Conf. Progr. and Abstr. Pap., Tokyo*, 218.

Sugimura, A. (1980). *J. Appl. Phys.*, **51**, 4405.

Sugimura, A. (1982). *IEEE J. Quant. Electron.*, **18**, 352.

Suto, K., and Nishizawa, J. (1983). *IEEE J. Quantum Electron.*, **19**, 1251.

Suto, K., and Nishizawa, J. (1986) *IEEE Proc.*, **133**, Pt. J., 259.

Sweeny, M., and Xu, J. (1989). *IEEE J. Quantum Electron.*, **25**, 885.

Takeshima, M. (1972). *J. Appl. Phys.*, **43**, 4114.

Tarry, H. A. (1986). *Electron. Lett.*, **22**, 416.

Titkov A. N.; Cheban B. N.; Baranov, A. N.; Gusseinov, A. A.; and Yakovlev, Yu. P. (1990a). *Fiz. Tekh, Poluprov.*, **24**, 1056.

Titkov, A. N.; Yakovlev, Yu. P.; and Baranov, A. N. (1990b). *Report at 20th Int. Conf. Semicond Phys.*, Thesaloniki Greece.

Tomasetta, L. R. and Fonstad, C. G. (1975). *IEEE J. Quantum Electron.*, **11**, 384.

Tournie, E.; Grunberg, P.; Fouillant, C.; Baranov, A.; Joullie, A.; and Ploog, K. H. (1994). *Solid-State Electron.*, **37**, 1311.

Tournie, E.; Lazzari, J.-L.; Pitard, F.; Alibert, C.; Joullie, A.; and Lambert, B. (1990). *J. Appl. Phys.*, **68**, 5936.

Turner, G. W.; Choi, H. K.; and Le, H. Q. (1995). *J. Vac. Sci. Technol.*, **B 13**, 699.

Van der Ziel, J. P.; Chiu, T. H.; and Tsang, W. T. (1986a). *J. Appl. Phys.*, **60**, 4087.

Van der Ziel, J. P.; Chiu, T. H.; and Tsang, W. T. (1986b). *Appl. Phys. Lett.*, **48**, 315.

Van der Ziel, J. P.; Logan, R. A.; Mikulyak, R. M.; and Ballman, A. A. (1985). *IEEE J. Quant. Electron.*, **21**, 1827.

Vasil'ev, V. I.; Il'inskaya, N. D.; Kuksenkov, D. V.; Kuchinski, V. I.; Mishurny, V. A.; Sazonov, V. V.; Smirnitsky, V. B.; and Faleev, N. N. (1990). *Pis'ma Zh. Tekh. Fiz.*, **16**, 58.

Verie, C., and Granger, R. (1965). *Comp. Rend.*, **261**, 3349.

Walpole, J. N.; Calawa, A. R.; Chinn, S. R.; Groves, S. H.; and Harman, T. C. (1976). *Appl. Phys. Lett.*, **29**, 307.

Walpole, J. N.; Calawa, A. R.; Chinn, S. R.; Groves, S. H.; and Harman, T. C. (1977). *Appl. Phys. Lett.*, **30**, 524.

Walpole, J. N.; Calawa, A. R.; Ralston, R. W.; and Harman, T. C. (1973b). *J. Appl. Phys.*, **44**, 2905.

Walpole, J. N.; Calawa, A. R.; Ralston, R. W.; Harman, T. C.; and McVittie (1973a) *Appl. Phys. Lett.*, **23**, 620.

Watanabe, Y., and Nishizawa, J. (1957). *Jpn. Patent N 273217*.

Wood, R. A.; McNeish, A.; Pidgeon, C. R.; and Smith, S. D. (1973). *J. Phys.*, **C, 6**, L144.

Yoshikawa, M.; Shinohara, K.; and Ueda, R. (1977). *Appl. Phys. Lett.*, **31**, 699.

Yuh, P.-F., and Wang, K. L. (1987). *Appl. Phys. Lett.*, **51**, 1404.

Yunovich, A. E., (1988). *Mechanism of Radiative Recombination via Impurities in A^3B^5 and A^4B^6 Semiconductors*. Doct. Dissertation, Moscow St. Uni.

Yunovich, A. E.; Averyushkin, A. S.; Drozd, I. A.; and Ogneva, V. G. (1979). *Fiz. Techn. Polupr.,* **9**, 1694.

Zandian, M.; Arias, J. M.; Zucca, R.; Gil, R. V.; and Shin, S. H. (1991). *Appl. Phys. Lett.,* **59**, 1022.

Zasavitski, I. I.; Chizhevski, E. G.; and Shotov, A. P. (1978). *Kvant. Elektron.,* **5** (3), 692.

Zasavitski, I. I.; Matsonashvili, B. N.; and Shotov, A. P. (1971). *Zh. Prikl. Spektroskopii,* **15**, 349.

Zasavitski, I. I.; Matsonashvili, B. N.; and Shotov, A. P. (1975). *Pis'ma Zh. Tekh. Fiz.,* **1**, 341.

Zotova, N. V.; Karandashev, S. A.; Matveev, B. A.; Stus', N. M.; and Talalakin G. N. (1983). *Sov. Techn. Phys. Lett.,* **9**, 167.

Zotova, N. V.; Karandashev, S. A.; Matveev, B. A.; Stus', N. M., and Talalakin, G. N. (1986). *Sov. Techn. Phys. Lett.,* **12**, 599.

Zucca, R.; Bajaj, J., and Blazejewski, E. R. (1988). *J. Vac. Sci. Technol.,* **A 6** (4), 2728.

Zucca, R.; Edwall, D. D.; Chen, J. S.; Johnston, S. L.; and Jounger, C. R. (1991). *J. Vac. Sci. Technol.,* **b 9**, 1823.

Zucca, R.; Zandian, M.; Arias, J. M.; and Gil, R. V. (1992a). *SPIE Proc.,* **1634**, 161.

Zucca, R., Zandian, M., Arias, J. M., and Gil, R. V. (1992b). *J. Vac. Sci. Technol.,* **B 10**, 1587.

Chapter 3

Single-Mode and Tunable Laser Diodes

Markus-Christian Amann

Walter Schottky Institute, Technical University of Munich, D-85748 Garching, Germany

3.1 Introduction

After the successful development of the basic technology for GaAs- and InP-based semiconductor lasers (Agrawal and Dutta, 1986), truly single-mode lasers with side mode suppression ratios above 30 dB have become available during the last decade, and high-performance monolithic tunable devices were first presented around 1990 (Koch and Koren, 1990). This development was mainly driven by evolving optical communication techniques, which demand high-performance optical sources (Wagner and Linke, 1990). For wavelengths used in optical communications, essential progress was achieved with the InGaAsP/InP material system at wavelengths around 1.55 μm.

In this chapter the state of the art of monolithically integrated single-mode and electronically wavelength tunable laser diodes is reviewed. The hybrid approaches, in which the laser diode is put into an external cavity and tuning is performed, *e.g.* via rotation of a diffraction grating (Oshiba *et al.*, 1989; Mehuys *et al.*, 1989; Favre and Le Guen, 1991) or by electrostatically deflected cantilevers as top mirrors in vertical cavity surface emitting lasers (Wu *et al.*, 1995), are omitted. The corresponding devices, therefore, usually represent technologically sophisticated integrated photonic circuits comprising two to three different optoelectronic and optic functions such as light amplification, (coarse and fine) wavelength filtering, and optical phase shifting. Throughout this chapter we first start with the underlying concepts and operation principles of the devices and then derive the corresponding laser structures and their technologies. For each device type, experimental results are given demonstrating the state of the art.

The chapter is divided into three main parts treating the single-mode laser diodes, the continuously tunable laser diodes, and, finally, the widely but discontinuously tunable devices. It should be noted that all laser structures considered are single mode or offer at least the potential for a dominating single longitudinal mode operation. This is because, for most of the anticipated applications of wavelength tunable lasers, the single-mode emission represents an indispensible prerequirement.

3.2   Single-mode laser diodes

3.2.1   The Fabry-Perot laser diode

The most simple longitudinal semiconductor laser structure comprises a Fabry-Perot cavity, which is homogeneously filled with the gain medium. A simplified longitudinal view of this laser type is sketched in Fig. 3.1a. Assuming a transversely single-mode waveguide, the cavity modes consist of the set of the longitudinal Fabry-Perot cavity modes as shown in Fig. 3.1b. All cavity modes exhibit equal total loss

$$\gamma_{tot} = \gamma_m + \gamma_i, \tag{3.1}$$

where γ_i stands for the internal mode loss, and the spatially distributed mirror loss

$$\gamma_m = \gamma_{m1} + \gamma_{m2} \tag{3.2}$$

is the sum of the mirror losses γ_{m1} and γ_{m2} at the left (M_1) and right (M_2) mirror, respectively

$$\gamma_{m1,m2} = \frac{1}{2L} \ln \frac{1}{r_{1,2}}. \tag{3.3}$$

L stands for the laser length, and the power reflectivities of the mirrors M_1 and M_2 are denoted by r_1 and r_2.

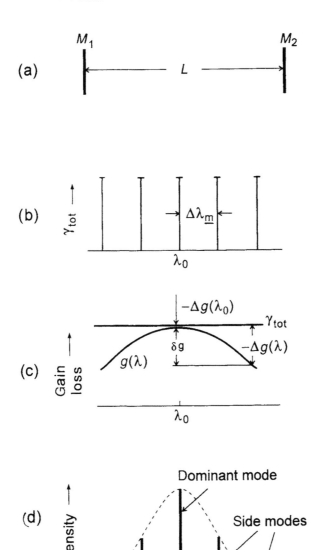

Figure 3.1: Longitudinal view of Fabry-Perot cavity (a), total mode loss spectrum (b), mode gain spectrum (c), and laser spectrum (d).

From the oscillation condition of the laser (Agrawal and Dutta, 1986, Chapter 2.3), the wavelengths of the longitudinal modes are given as

$$\lambda_i = \frac{2n(\lambda_i)L}{i} \; i \, \epsilon N, \tag{3.4}$$

where i is an arbitrary integer denoting the longitudinal mode number (typically 10^3–10^4), n is the real part of the (wavelength dependent) complex cavity index

$$n_c(\lambda) = n(\lambda) + j\frac{g(\lambda)}{2}, \tag{3.5}$$

and $g(\lambda)$ is the power gain of the laser modes. Under stationary conditions, the oscillation condition further requires, that—independently of the optical output power—the total losses are almost exactly compensated by the mode gain

$$g \simeq \gamma_{tot} \; \text{(gain-clamping-mechanism).} \tag{3.6}$$

The longitudinal mode spacings are almost equal (Fig 3.1b) and are given as

$$\Delta\lambda_m = \frac{\lambda_0^2}{2n_g L}, \tag{3.7}$$

where λ_0 is the center wavelength and n_g is the group effective refractive index of the cavity, considering the dispersion of n

$$n_g = \frac{d(k_0 n)}{dk_0}\bigg|_{\lambda_0} = n(\lambda_0) - \lambda_0\frac{dn}{d\lambda}\bigg|_{\lambda_0}. \tag{3.8}$$

Obviously no wavelength selection occurs in the Fabry-Perot cavity; the wavelength dependent mode gain $g(\lambda)$ caused by the active medium merely provides a certain mode selection, as shown in Fig. 3.1c. With a spectral width of several ten nm of the active medium gain in the 1.3–1.5 μm wavelength range and, on the other hand, typical mode spacings $\Delta\lambda_m$ around 0.3–1 nm, only a weak mode selection is obtained, and reproducible and stable single longitudinal mode operation may hardly be achieved this way. A representative multimode spectrum of a Fabry-Perot laser diode is displayed schematically in Fig. 3.1d.

A convenient measure characterizing the spectral purity is the side-mode suppression ratio SSR. It is calculated from the time-averaged power in the laser modes by means of the multimode rate equations (Koch and Koren, 1990). For the dominant mode (mode 0) at wavelength λ_0 and the second strongest mode (mode 1 at $\lambda_1 = \lambda_0 + \Delta\lambda_m$ in the example of Fig. 3.1), these equations read under stationary conditions (Agrawal and Dutta, 1986, Chapter 6.2):

$$\dot{S}_0 = 0 = R_{sp}(\lambda_0) + S_0 v_g \Delta g(\lambda_0) \tag{3.9}$$

$$\dot{S}_1 = 0 = R_{sp}(\lambda_1) + S_1 v_g \Delta g(\lambda_1), \tag{3.10}$$

where $R_{sp}(\lambda_i)$ and S_i $(i = 0, 1)$ are the spontaneous emission rate, and the photon number in the cavity of the ith mode and $v_g = c/n_g$ is the group velocity. The (negative) mode gain difference

$$\Delta g(\lambda) = g(\lambda) - \gamma_{tot} \tag{3.11}$$

is very small as compared with g and γ_{tot} for the dominant mode, as shown in Fig. 3.1c. The side-mode suppression ratio is proportional to the photon density quotient of modes 0 and 1

$$SSR = \frac{S_0}{S_1}. \tag{3.12}$$

Using Einstein's relations, the spontaneous emission rate can be expressed by the mode gain

$$R_{sp}(\lambda) = n_{sp} v_g g(\lambda), \tag{3.13}$$

where n_{sp} is the spontaneous emission factor (≈ 2) considering the incomplete population inversion in the semiconductor (Henry, 1982). Since $g \approx \gamma_{tot}$, R_{sp} can well be approximated by

$$R_{sp}(\lambda_0) \approx R_{sp}(\lambda_1) \approx n_{sp} v_g \gamma_{tot}. \tag{3.14}$$

From Eqs. (3.10) and (3.14) we obtain the photon number S_1 as

$$S_1 = -\frac{n_{sp}\gamma_{tot}}{\Delta g(\lambda_i)}. \tag{3.15}$$

With regard to Fig. 3.1c we may put

$$\Delta g(\lambda_1) = \Delta g(\lambda_0) - \delta g \approx -\delta g, \tag{3.16}$$

where δg is the mode gain difference between the two strongest modes. Since

$$\delta g >> |\Delta g(\lambda_0)|, \tag{3.17}$$

Eq. (3.15) can be rewritten as

$$S_1 \approx \frac{n_{sp}\gamma_{tot}}{\delta g}. \tag{3.18}$$

Considering the relationship between S_0 and the output power P per mirror in mode 0 at equal mirror power reflectivities $r_1 = r_2$ (Agrawal and Dutta, 1986, Chapter 2.6),

$$S_0 = \frac{2P}{h\nu v_g\gamma_m}, \tag{3.19}$$

where $h\nu$ denotes the photon energy, we finally obtain for the side-mode suppression ratio after substitution of S_1 and S_0 by Eqs. (3.18) and (3.19) into Eq. (3.12)

$$SSR = \frac{2P\delta g}{h\nu v_g n_{sp}\gamma_{tot}\gamma_m}. \tag{3.20}$$

It should be noted that the mode gain differs from the material gain in the active region of a laser diode. This is because the optical mode usually extends beyond the active region; *i.e.*, the mode volume is larger than the active region volume. So the mode gain is usually smaller than the active region gain. For illustration, Fig. 3.2 shows a schematic perspective view of a Fabry-Perot laser diode (a) and the principal transverse intensity distribution $P'(x, y)$ (b). As the mode gain can be induced only by that part of the mode energy that is contained within the active region, mode gain g and active region gain g_a are related via the optical confinement factor (or filling factor) $\Gamma(\leq 1)$ (Agrawal and Dutta, 1986, Chapter 2.5.1) as

$$g(\lambda) = \Gamma g_a(\lambda). \tag{3.21}$$

Γ is calculated as

$$\Gamma = \frac{\int_{V_a}P'(x,y)^2 dV}{\int_{V_m}P'(x,y)^2 dV}, \tag{3.22}$$

where V_m and V_a stand for the mode volume and active region volume, respectively.

As a numerical example we now consider a typical InGaAsP/InP Fabry-Perot laser diode at $\lambda_0 = 1.55\ \mu m$ ($h\nu = 0.8\ eV$) with 5 mW optical output power per mirror. The other laser parameters are: $L = 500\ \mu m$,

(a)

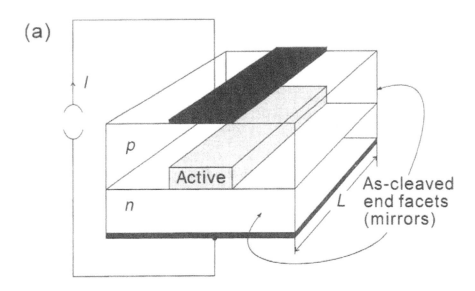

(b)

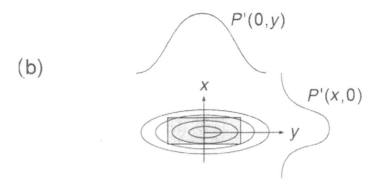

Figure 3.2: Schematic perspective view of laser diode (a) and principal transverse and lateral intensity distributions of laser mode $P'(x, 0)$ and $P'(0, y)$, respectively (b).

$n_g = 4$, optical confinement factor $\Gamma = 0.2$, $n_{sp} = 2$, $r_1 = r_2 = 0.35$ (yielding $\gamma_m = 21$ cm^{-1}), and $\gamma_i = 30$ cm^{-1}. Approximating the active region gain characteristic g_a as parabola around λ_0,

$$g_a(\lambda) = g_a(\lambda_0) - b_2 \, (\lambda - \lambda_0)^2 \, , \qquad (3.23)$$

the gain difference δg is calculated as

$$\delta = \Gamma b_2 \Delta \lambda_m^2 \, , \qquad (3.24)$$

with $b_2 = 0.15$ cm^{-1}/nm^2 in InGaAsP at 1.55 μm wavelength (Westbrook, 1986). Using these parameters we obtain a mode spacing $\Delta \lambda_m$ of 0.6 nm, total losses of $\gamma_{tot} \approx 51$ cm^{-1}, a gain difference δg of 0.011 cm^{-1}, and an *SSR* of 53 or 17 dB ($= 10 \log SSR$), respectively. Since an efficient stationary side-mode suppression ratio (SSR) well above 30 dB is demanded for high-performance optical communications, we can conclude that the side-mode suppression ratio in Fabry-Perot laser diodes is commonly at most of the order 20 dB, making these devices hardly suited for applications where a high spectral purity is needed. Moreover, it should be stressed that in most practical applications, where the lasers are being directly modulated, the effective SSR may be even smaller (Koch and Koren, 1990).

As a consequence, more sophisticated resonator structures are required for the single longitudinal mode operation with a strong suppression of side modes. Only after establishing the single-mode operation is the development of narrow linewidth and wavelength tunable laser structures feasible.

3.2.2   The distributed Bragg reflector (DBR) laser

In fact, large efforts have been made so far to realize high-performance single-mode and narrow linewidth laser diodes, since these devices may yield higher transmission capacity and/or longer transmission distance in optical fiber communications (Kobayashi and Mito, 1988). A detailed introduction to the subject of single-mode laser diodes can be found in the textbooks of Agrawal and Dutta (1986) and Buus (1991).

Principally, a possible method for improving the single longitudinal mode operation is to increase the gain difference between the dominant mode and the side modes by making the total loss γ_{tot} wavelength dependent. This is shown schematically in Fig. 3.3, where an improved mode selection can be achieved if the bending of the loss characteristic $\gamma_{tot}(\lambda)$

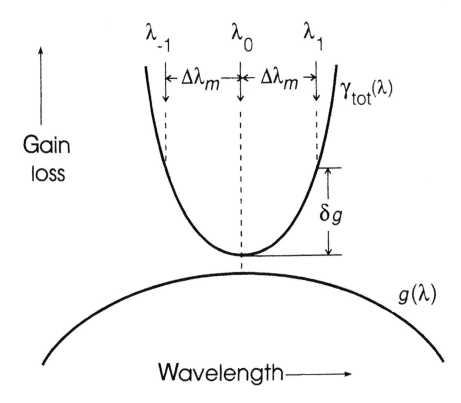

Figure 3.3: Wavelength-dependent gain and loss characteristics for a laser structure with wavelength selective total loss.

is larger than that of the gain curve $g(\lambda)$. In this case, furthermore, the wavelength of minimum loss mainly determines the laser wavelength.

As with the gas lasers, for instance, wavelength selective optical losses can be introduced into the laser diode via the wavelength selectivity of the front-end mirror reflectivities, making the optical feedback wavelength dependent. Particularly by applying interference-type dielectric multilayer mirrors with layers of alternating refractive index, as displayed schematically in Fig. 3.4, a significant wavelength selection may be achieved. This can be understood intuitively by considering the superposition of the reflections occuring each at the refractive index discontinuities. At each discontinuity the contribution to the total reflection at $z = 0$ is labelled, including the effect of the propagation delay. Exactly for those wavelengths, where $k\Lambda$ is an integer multiple of π, constructive

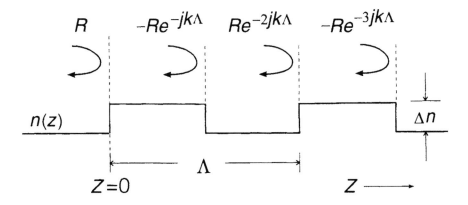

Figure 3.4: Schematic illustration of a Bragg reflector made by periodic refractive index perturbation. Grating period and index discontinuity are labelled by Λ and Δn, respectively.

interference occurs, yielding maximum reflection, while at the other wavelengths constructive and destructive interferences occur simultaneously, leading to smaller total reflection. Note that the sign of the amplitude reflection R (power reflection $r = |R|^2$) changes for subsequent discontinuities due to the change of sign of the refractive index change, which is either Δn for z being an even multiple of the half period $\Lambda/2$ or $-\Delta n$ for z being an odd multiple of $\Lambda/2$. The wavelength λ_B, for which $k\Lambda = \pi$, is denoted as Bragg wavelength, and optimum wavelength selectivity is obtained around λ_B.

As shown in Fig. 3.5, the Bragg grating filters can be arranged either outside the active region at the ends of the laser cavity and act then as wavelength selective reflectors (distributed Bragg reflector: DBR, Fig. 3.5a) (Tohmori and Oishi, 1988; Kano *et al.*, 1989) or can be collocated with the active region along the entire laser cavity, yielding a distributed feedback (DFB) structure (Fig. 3.5b) (Broberg *et al.*, 1984; Dutta *et al.*, 1986; Yamada *et al.*, 1988; Takemoto *et al.*, 1989; Luo *et al.*, 1990; Sasaki *et al.*, 1988). The longitudinal index variation is thereby introduced by periodic thickness variation of the grating layer with refractive index n_1, which covers a semiconductor region with different refractive index n_0. Since these index perturbations occur only within part of the transverse mode profile, their effect on the *effective refractive index* $n_{eff} = k_z/k_0$ of the composite waveguide structure is weakened. As with the $g - g_a$-

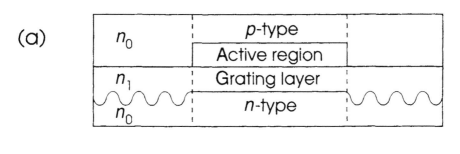

DBR-laser

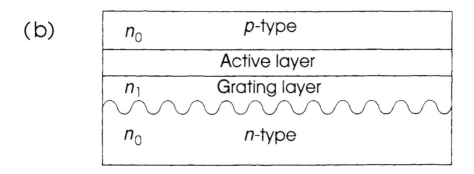

DFB-laser

Figure 3.5: Schematic longitudinal sections of principal DBR (a) and DFB (b) laser structures. The periodic index perturbation is induced by different refractive indexes $n_1 \neq n_0$ on opposite sides of the grating.

relationship (Eq. 3.21), the modulation of n_{eff} is determined by the product of the index difference $n_1 - n_0$ and the confinement factor within the grating area.

First we will consider the DBR laser, which effectively represents a Fabry-Perot laser, whose mirrors are replaced by DBRs so that the mirror reflectivities become wavelength dependent. Accordingly, the characteristics of the DBRs decisively determine the device performance, particularly the spectral purity.

3.2.2.1 Coupled-mode theory for DBR laser

In the DBR and DFB lasers, grating-induced refractive index perturbations lead to a coupling between the forward and backward propagating waves of the particular laser mode. Therefore, the optical feedback is not localized at the cavity end facets but is distributed throughout the entire laser cavity (DFB) or only part of it (DBR). The analysis of these devices, therefore, necessarily requires the investigation of wave propagation in periodic structures. The approach presented here essentially refers to the coupled wave theory of Kogelnik and Shank (1972). A mathematical model for a DBR is shown in Fig. 3.6 with a sinelike z-dependence of n_{eff}. The

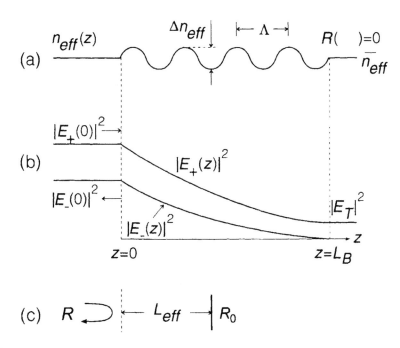

Figure 3.6: Mathematical model for DBR reflector with spatial n_{eff}-modulation (a). The z-dependent intensities of the forward $|E_+|^2$, backward $|E_-|^2$, and transmitted $|E_T|^2$ waves are illustrated (b), and a simple equivalent reflector structure is shown in (c).

DBR extends from $z = 0$ to $z = L_B$, and the z-dependent effective index reads (cf. Fig. 3.6a)

$$n_{eff}(z) = \bar{n}_{eff} + \frac{\Delta n_{eff}}{2} \sin k_g z \,, \qquad (3.25)$$

where $\bar{n}_{eff}$ and Δn_{eff} denote the average effective refractive index and the effective refractive index difference (peak-to-peak). The grating wavevector k_g is related to the grating period Λ as

$$k_g = \frac{2\pi}{\Lambda} \,. \qquad (3.26)$$

Assuming a small index perturbation ($\Delta n_{eff} << \bar{n}_{eff}$), the contribution of Δn_{eff}^2 can be neglected in the square of the z-dependent wave vector $k(z) = k_0 n(z)$, where $k_0 = 2\pi/\lambda$ stands for the vacuum wave vector. This yields

$$k(z)^2 \approx k_0^2 \left(\bar{n}_{eff}^2 + \bar{n}_{eff} \Delta n_{eff} \sin k_g z \right) \,, \qquad (3.27)$$

so that the stationary one-dimensional wave equation reads within the same approximation

$$\frac{d^2 E}{dz^2} + k_0^2 (\bar{n}_{eff}^2 + \bar{n}_{eff} \Delta n_{eff} \sin k_g z) \, E = 0 \,. \qquad (3.28)$$

In the frame of the coupled mode analysis, we decompose the z-dependent electric field strength E into two waves,

$$E(z) = E_+(z) + E_-(z) \,, \qquad (3.29)$$

propagating each into the $+z$-direction (E_+) and the $-z$-direction (E_-). Considering the wavelength range around the Bragg wavelength

$$\lambda_B = 2 \mathcal{R}(\bar{n}_{eff}) \Lambda \qquad (3.30)$$

with Bragg wavevector

$$k_B = \frac{k_g}{2} = \frac{2\pi \mathcal{R}(\bar{n}_{eff})}{\lambda_B} \,, \qquad (3.31)$$

both waves are written as

$$E_+(z) = A(z) \exp(-jk_B z) \qquad (3.32)$$

$$E_-(z) = B(z) \exp(+jk_B z) \,, \qquad (3.33)$$

where $\mathscr{R}(\bar{n}_{eff})$ denotes the real part of $\bar{n}_{eff}$. $A(z)$ and $B(z)$ represent functions that vary slowly along z, while the fast-varying components of E_+ (z) and $E_-(z)$ are taken into account by the exponential functions. By evaluating the second derivative of $E(z)$ with respect to z in the wave Eq. (3.28), therefore, the second derivatives of $A(z)$ and $B(z)$ can be neglected, i.e., $\left|\frac{d^2A}{dz^2}\right|, \left|\frac{d^2B}{dz^2}\right| \ll k_B\left|\frac{dA}{dz}\right|, k_B\left|\frac{dB}{dz}\right|$. We thus obtain

$$\frac{d^2E}{dz^2} = \frac{d^2E_+}{dz^2} + \frac{d^2E_-}{dz^2}, \tag{3.34}$$

where with Eqs. (3.31), (3.32), and (3.33)

$$\frac{d^2E_+}{dz^2} \approx \left\{-2jk_B\frac{dA}{dz} - k_B^2A\right\}\exp(-jk_Bz) \tag{3.35}$$

$$\frac{d^2E_-}{dz^2} \approx \left\{2jk_B\frac{dB}{dz} - k_B^2B\right\}\exp(jk_Bz). \tag{3.36}$$

Using $\sin x = \frac{1}{2j}(\exp(jx) - \exp(-jx))$, the second expression in the wave Eq. (3.28) can be developed as

$$k_0^2\left\{\bar{n}_{eff}^2 + \frac{\bar{n}_{eff}\Delta n_{eff}}{2j}(\exp(j2k_Bz) - \exp(-j2k_Bz))\right\}$$
$$\times \{A\exp(-jk_Bz) + B\exp(jk_Bz)\}.$$

Carrying out the multiplications and neglecting the rapidly varying terms with $\exp(\pm j3k_Bz)$, we obtain

$$\underbrace{\left\{k_0^2\bar{n}_{eff}^2A + jk_0^2\frac{\Delta n_{eff}\bar{n}_{eff}}{2}B\right\}\exp(-jk_Bz)}_{\text{foreward propagating}} +$$

$$+ \underbrace{\left\{k_0^2\bar{n}_{eff}^2B - jk_0^2\frac{\Delta n_{eff}\bar{n}_{eff}}{2}A\right\}\exp(+jk_Bz)}_{\text{backward propagating}}. \tag{3.37}$$

Substituting (3.35), (3.36) and (3.37) into (3.28) yields

$$\underbrace{\left\{(k_0^2\bar{n}_{eff}^2 - k_B^2)A - j2k_B\frac{dA}{dz} + jk_0^2\frac{\Delta n_{eff}\bar{n}_{eff}}{2}B\right\}\exp(-jk_Bz)}_{\text{forward propagating wave}} +$$

$$+ \underbrace{\left\{(k_0^2\bar{n}_{eff}^2 - k_B^2)B + j2k_B\frac{dB}{dz} - jk_0^2\frac{\Delta n_{eff}\bar{n}_{eff}}{2}A\right\}\exp(jk_Bz) = 0.}_{\text{backward propagating wave}} \tag{3.38}$$

This equation is satisfied only if the expressions for the forward and backward propagating waves are each equal to zero. We put

$$k_0 \bar{n}_{eff} = k_B + \Delta\beta, \tag{3.39}$$

where $\Delta\beta$ stands for the deviation of the wave vector $k_0 n_{eff}$ from the wave vector at the Bragg wavelength k_B. From Eqs. (3.39) and (3.31), $\Delta\beta$ can be expressed by the wavelength deviation from the Bragg wavelength $\delta\lambda = \lambda - \lambda_B$ as

$$\Delta\beta = 2\pi \left\{ \frac{\bar{n}_{eff}(\lambda_B + \delta\lambda)}{\lambda_B + \delta\lambda} - \frac{\mathscr{R}(\bar{n}_{eff}(\lambda_B))}{\lambda_B} \right\} \approx jk_0(\lambda_B)\mathscr{I}(\bar{n}_{eff}) - k_0(\lambda_B)\bar{n}_{eff,g} \frac{\delta\lambda}{\lambda_B}, \tag{3.40}$$

where $\bar{n}_{eff,g}$ is the group effective refractive index at λ_B

$$\bar{n}_{eff,g} = \mathscr{R}(\bar{n}_{eff}(\lambda_B)) - \lambda_B \left. \frac{d\mathscr{R}(\bar{n}_{eff})}{d\lambda} \right|_{\lambda = \lambda_B}, \tag{3.41}$$

which takes into account the dispersion of $\bar{n}_{eff}(\lambda)$.

In the wavelength regime around λ_B, therefore, $\Delta\beta$ is relatively small

$$|\Delta\beta| << k_B, \tag{3.42}$$

so we may write

$$k_0^2 \bar{n}_{eff}^2 - k_B^2 \approx 2k_B \Delta\beta, \tag{3.43}$$

and Eq. (3.38) decomposes into

$$\Delta\beta A - j\frac{dA}{dz} + j\kappa B = 0 \tag{3.44}$$

$$\Delta\beta B + j\frac{dB}{dz} - j\kappa A = 0, \tag{3.45}$$

where the coupling coefficient κ is defined as

$$\kappa = \frac{k_0 \Delta n_{eff}}{4}, \tag{3.46}$$

and $k_0 \bar{n}_{eff}$ in the third terms within the curly braces of Eq. (3.38) has been approximated by k_B.

Combining A and B as

$$\vec{E} = \begin{pmatrix} A \\ B \end{pmatrix}, \tag{3.47}$$

the two scalar differential equations (3.44) and (3.45) can be written as one vector differential equation

$$\frac{d}{dz}\vec{E} = \underbrace{\begin{pmatrix} -j\Delta\beta & \kappa \\ \kappa & j\Delta\beta \end{pmatrix}}_{\mathbf{M}} \vec{E}. \tag{3.48}$$

The eigenvalues $s_{1,2}$ of matrix $\mathbf{M}$ are

$$s_{1,2} = \pm s, \tag{3.49}$$

where

$$s = \sqrt{\kappa^2 - \Delta\beta^2} \tag{3.50}$$

and the corresponding eigenvectors read

$$\vec{E_1} = E_0 \begin{pmatrix} 1 \\ (j\Delta B + s)/\kappa \end{pmatrix} \tag{3.51}$$

$$\vec{E_2} = E_0 \begin{pmatrix} 1 \\ (j\Delta B - s)/\kappa \end{pmatrix}, \tag{3.52}$$

with E_0 being a normalization constant with the dimension of the electrical field strength (voltage per length). The general solution of Eq. (3.48) is then obtained as

$$\vec{E} = C_2\vec{E_1}\exp(sz) + C_2\vec{E_2}\exp(-sz), \tag{3.53}$$

where the constants C_1 and C_2 are chosen such that the boundary conditions are met appropriately. It is convenient, however, to eliminate C_1 and C_2 and to express the fields at $z = 0$ by the fields at $z = L_B$. To this end we put

$$\vec{E}(0) = \mathbf{T}\cdot\vec{E}(L_B), \tag{3.54}$$

where $\mathbf{T}$ denotes the transfer matrix of the DBR. $\mathbf{T}$ is easily calculated from Eqs. (3.51)–(3.54) as

$$\mathbf{T} = \begin{pmatrix} \cosh(sL_B) + \frac{j\Delta\beta}{s}\sinh(sL_B) & -\frac{\kappa}{s}\sinh(sL_B) \\ -\frac{\kappa}{s}\sinh(sL_B) & \cosh(sL_B) - \frac{j\Delta\beta}{s}\sinh(sL_B) \end{pmatrix}. \qquad (3.55)$$

3.2.2.2 Purely passive DBR with nonreflecting end mirror

We now consider the case $R(L_B) = 0$ (*cf.* Fig. 3.6a), which corresponds to $B(L_B) = 0$, *i.e.* no reflections occur at the right end of the DBR. We also assume the DBR section to be free of optical loss or gain, so that $n_{eff}(z)$ and $\bar{n}_{eff}$ are purely real. Eq. (3.54) then reads explicitly

$$\begin{pmatrix} A(0) \\ B(0) \end{pmatrix} = \mathbf{T} \cdot \begin{pmatrix} A(L_B) \\ 0 \end{pmatrix}. \qquad (3.56)$$

The amplitude reflectivity at $z = 0$,

$$R = \frac{B(0)}{A(0)}, \qquad (3.57)$$

equals T_{21}/T_{11} in this case and is calculated from Eq. (3.55) as

$$R = -\frac{\kappa \sinh(sL_B)}{j\Delta\beta \sinh(sL_B) + s\cosh(sL_B)}. \qquad (3.58)$$

Exactly at the Bragg wavelength, where $\Delta\beta = 0$ and (with Eq. (3.50)) $s = \kappa$, the amplitude reflection is purely negative real and amounts to

$$R(\lambda_B) = -\tanh(\kappa L). \qquad (3.59)$$

Fig. 3.6b shows the longitudinal intensity distributions of the forward (E_+) and backward (E_-) propagating waves within the DBR. E_T stands for the wave transmitted through the DBR.

Besides the spectral filtering, however, the DBR also increases the effective length of the laser cavity. This can be understood by considering the wavelength dependence of the phase of R in the immediate vicinity of the Bragg wavelength, where the DBR can approximately be described by a wavelength independent reflector R_0 displaced by an effective length L_{eff} from $z = 0$ as displayed in Fig. 3.6c. The DBR reflectivity R in Eq. (3.58) is thus approximated by (Suematsu *et al.*, 1983):

$$R \approx R_0 \exp(-2jk_0\bar{n}_{\text{eff}}L_{eff}), \qquad (3.60)$$

where

$$L_{eff} = \frac{1}{2} \frac{\mathrm{d}\arg(R)}{\mathrm{d}\Delta\beta}\bigg|_{\Delta\beta=0} = \frac{\tanh(\kappa L_B)}{2\kappa} \qquad (3.61)$$

and

$$R_0 = R(\lambda_B)\exp(2jk_B L_{eff}), \qquad (3.62)$$

so that

$$R \approx -\tanh(\kappa L_B)\exp(-2j\Delta\beta L_{eff}). \qquad (3.63)$$

For κL_B-products larger than unity, the effective length is about $1/2\kappa$, so that with typical κ-values between 20 and 50 cm^{-1} L_{eff} amounts to several hundred μm. This is important for the longitudinal mode spacing $\Delta\lambda_m$, because L_{eff} adds to the effective length of the laser cavity, and thereby larger L_{eff} lead to reduced mode spacings. Since the side mode suppression ratio SSR is proportional to the square of the mode spacing $\Delta\lambda_m$ ($\delta g \propto \Delta\lambda_m^2$, cf. Fig. 3.1c), the increased effective cavity length counteracts the selectivity gained by the DBR. Nevertheless, in a properly designed DBR laser, the SSR is significantly enhanced replacing a reflecting end facet by a DBR.

3.2.2.3   Wavelength selectivity of DBR laser

For illustration, the calculated power reflection $r = |R|^2$ of a 400-μm-long passive DBR at 1.55 μm wavelength with $\bar{n}_{eff,g} = 4$ is plotted in Fig. 3.7 versus the wavelength deviation from the Bragg wavelength with the κL_B-product as parameter. As can be seen, a strong wavelength selectivity is achieved for all κL_B-values investigated, and the bandwidth (FWHM) is of the order 1 nm.

Besides the selection of one single longitudinal mode, the Bragg gratings are also very effective in the determination of the absolute laser wavelength. Referring to Eq. (3.30), however, this requires the knowledge of the effective refractive index with the same relative accuracy as the desired wavelength. Since the effective refractive index strongly depends on the composition and thickness of the various layers in the laser structure, a high performance is required during device fabrication. This goal can be achieved by advanced epitaxial techniques such as metal-organic vapor-phase epitaxy (MO VPE) (Meyer *et al.*, 1988) or chemical beam

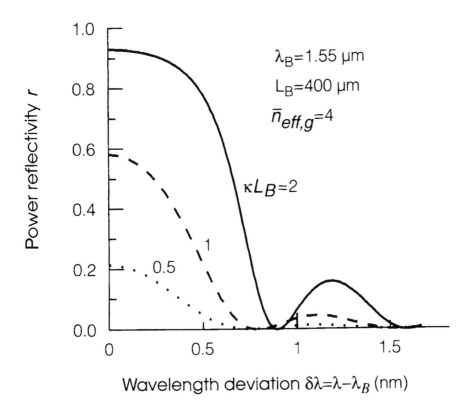

Figure 3.7: Power reflectivity versus wavelength deviation from Bragg wavelength for a 400-μm-long DBR at 1.55 μm wavelength with the κL_B-product as parameter.

epitaxy (CBE) (Tsang, 1990), which provide excellent thickness uniformity and reproducibility.

As the effective index of a laser waveguide is determined by both the transverse and lateral dimensions, the stripe width also needs to be controlled accurately. As a numerical example, the Bragg wavelength of a 1-μm-wide buried heterostructure (BH) type (Tohmori and Oishi, 1988) DBR at $\lambda = 1.55$ μm changes by about 2 nm, or 250 GHz, respectively, at a stripe-width change of only 0.1 μm.

The fabrication of the Bragg grating is commonly done either by holographic techniques (Suematsu *et al.*, 1983) or by means of electron beam lithography (Öberg *et al.*, 1995), and wet or dry etching techniques into an InP or InGaAsP layer, respectively. In a following epitaxial run, the etched corrugation is covered by a compound of different refractive index, yielding the longitudinal modulation of the effective refractive index.

In the DBR lasers, the Bragg grating replaces one or both of the cavity mirrors as displayed in Fig. 3.8a and b. Depending on the grating

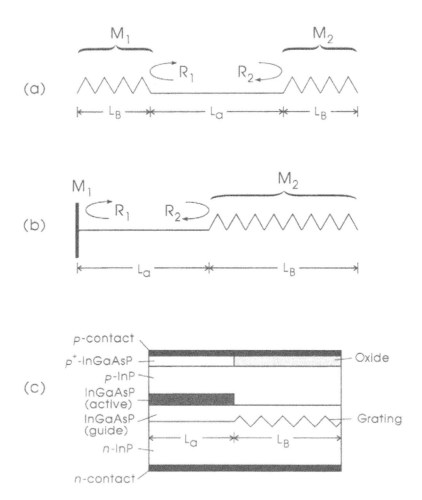

Figure 3.8: Distributed Bragg reflector (DBR) laser: (a) Both mirrors replaced by Bragg gratings. (b) One mirror replaced by a Bragg grating. (c) Schematic longitudinal view of InGaAsP/InP DBR laser.

reflection and active region length L_a, wavelength selective mirror losses are thus achieved preferring the cavity modes near the Bragg wavelength.

The schematic longitudinal cross section of an InGaAsP/InP DBR structure with one mirror (M_2) replaced by a DBR is shown in Fig. 3.8(c). In this device, the Bragg grating is made between InP and an InGaAsP waveguiding layer with a band gap wavelength λ_g *(WG)* shorter than the laser wavelength λ_0 *(e.g.*, λ_g *(WG)* $= 1.3$ μm versus $\lambda_0 = 1.55$ μm) and with a refractive index n_1 larger than that of InP (≈ 3.4 versus 3.17). Note that λ_g is related to the band gap energy E_g by $\lambda_g = hc/E_g$, with h and c denoting Planck's constant and the vacuum speed of light. Accordingly, this layer acts as a passive optical waveguide in the surrounding InP, supporting the laser mode. The active region, consisting of λ_g(active) $= 1.55$ μm InGaAsP, is placed at one end of the cavity. The laser current is confined longitudinally to the active region by an oxide isolation, and the lateral device structure is usually of the BH type (Tohmori and Oishi, 1988). Typically the length of the DBR is adjusted between 300 and 1000 μm with κL-products between 1 and 3, and the gain region is between 200 and 500 μm long.

Considering the structure of Fig. 3.8c (only one DBR), the mirror loss γ_m becomes wavelength dependent, because with Eq. (3.2) γ_m is the sum of the left and right mirror losses γ_{m1} and γ_{m2}. It should be noted that the averaging of the mirror losses *(cf.* Eq. (3.3)) must be done over the active region length L_a, which in the case of the DBR laser differs from the total laser length L. Assuming $L_a = L_B = 400$ μm, $\kappa L_B = 1$ and $n_g = 4$, the wavelength-dependent loss of the right mirror (M_2) $\gamma_{m2}(\lambda)$ is plotted in Fig. 3.9. For comparison, the mode gain characteristic $g(\lambda)$ for InGaAsP at 1.5 μm ($\Gamma = 0.2$) is also shown, revealing the paramount wavelength selectivity of the Bragg reflector.

In the past, single-mode DBR lasers were fabricated successfully (Tohmori *et al.*, 1985); however, a reproducible single-mode operation is not easily achievable. This is because Bragg wavelength λ_B and comb mode spectrum λ_i (cf. Fig. 3.1b) are not synchronized, so that the relative position of the comb modes with respect to λ_B varies randomly from device to device. Optimum single-mode operation, however, occurs only if one of the comb mode wavelengths equals λ_B. Otherwise, reduced side-mode suppression is obtained and, in the worst case, if λ_B is detuned from the comb modes by $\Delta\lambda_m/2$ two longitudinal modes oscillate with equal amplitude. Furthermore, an inherent problem in the DBR laser is that active region and DBR form two different dielectric waveguides, which

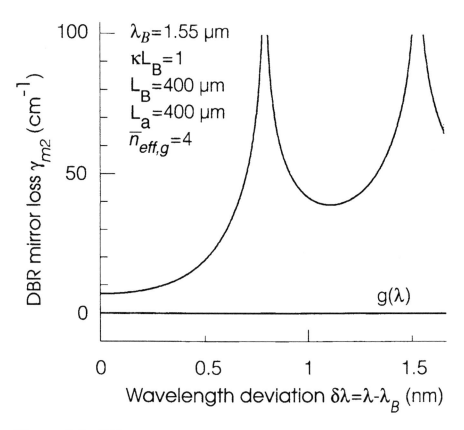

Figure 3.9: DBR mirror loss versus the wavelength deviation from the Bragg wavelength. For comparison, the spectral dependence of the mode gain $g(\lambda)$ for $\Gamma = 0.2$ is also shown putting $g(\lambda_B) = 0$.

are coupled together as indicated by the broken lines in Fig. 3.5a. This not only leads to possible coupling losses (Suematsu *et al.*, 1983) and parasitic reflections but also makes the fabrication technology usually more complex than in case of the DFB lasers with their quasi-homogeneous longitudinal structure. Finally, the total length of DFB lasers is usually smaller than for the DBR devices, because the laser gain as well as the distributed feedback act along the entire cavity and thus make a more efficient use of the laser area; consequently, more lasers can be made from a wafer so that a larger device yield can be obtained during fabrication.

As a consequence, the DBR lasers have not gained importance as commercial single-mode devices, even though they were developed in parallel with the DFB lasers; however, the fundamental DBR laser structure represents the basis for various wavelength tunable laser diodes as well as for optoelectronic integrated circuits (OEICs).

3.2.3 Distributed feedback (DFB) laser diodes

3.2.3.1 Coupled mode theory for DFB laser

As displayed in Fig. 3.10, the simplest DFB laser structure consists of a Bragg grating, nonreflecting end mirrors ($R_1 = R_2 = 0$), and a homogeneous longitudinal gain distribution. The threshold gain for this so called "pure" DFB laser can be calculated by means of Eq. (3.58) taking $L_B = L$ and requiring that the DFB laser emits light even without feeding it with external light. Referring to Fig. 3.6b, this means that $E_-(0) \neq 0$ and $B(0) \neq 0$ even for $E_+(0) = 0$ and $A(0) = 0$. Obviously, the DFB laser represents an active device providing optical gain within the periodic structure. Mathematically, the presence of optical (power) gain g manifests itself as a positive imaginary part of $\bar{n}_{eff}$ as $g = 2k_0 \mathcal{I}(\bar{n}_{eff})$, where $\mathcal{I}(\bar{n}_{eff})$

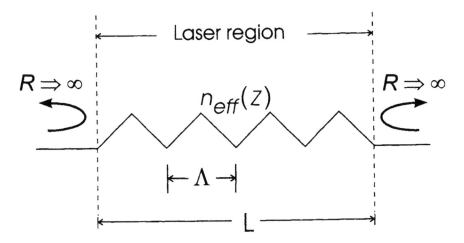

Figure 3.10: Idealized "pure" DFB laser with nonreflecting end facets.

denotes the imaginary part of $\bar{n}_{eff}$. Around the Bragg wavelength we may put $k_0(\lambda) \simeq k_0(\lambda_B)$, so that

$$g \simeq 2k_0(\lambda_B)\mathcal{I}(\bar{n}_{eff}). \tag{3.64}$$

According to the definition of the Bragg grating reflectivity in Eqs. (3.57) and (3.32), the reflectivity R (*cf.* Fig. 3.10) of a DFB laser can be considered to be infinite ($R \rightarrow \infty$). Eq. (3.58) then yields

$$j\Delta\beta \sinh(sL) + s\cosh(sL) = 0. \tag{3.65}$$

To make the results independent of the laser length, it is advantageous to consider the normalized quantities $\Delta\beta L$ and sL as functions of κL. We therefore multiply Eq. (3.65) with L on both sides and eliminate $\Delta\beta$ with Eq. (3.50) in order to get an eigenvalue equation for the product sL as function of the κL-product

$$sL = \pm j\kappa L \sinh(sL), \tag{3.66}$$

yielding an infinite set of complex solutions for sL. Each solution for sL corresponds to a longitudinal mode of the DFB laser. With Eq. (3.50) one obtains the corresponding $\Delta\beta L$-values of the DFB laser modes, which are complex as well.

Inserting $\Delta\beta L$ into Eq. (3.40) and equating real and imaginary parts yields with Eq. (3.64) for each mode the threshold gain g_{th} and wavelength (as $\lambda = \lambda_B + \delta\lambda$)

$$g_{th} = \frac{2\mathcal{I}(\Delta\beta L)}{L} \tag{3.67}$$

$$\delta\lambda = -\frac{\lambda_B}{k_0(\lambda_B)\bar{n}_{eff,g}L}\mathcal{R}(\Delta\beta L), \tag{3.68}$$

In terms of the longitudinal mode spacing of an equally long Fabry-Perot laser, $\delta\lambda$ can be expressed as

$$\delta\lambda = -\Delta\lambda_m \frac{\mathcal{R}(\Delta\beta L)}{\pi} \tag{3.69}$$

with

$$\Delta\lambda_m = \frac{\lambda_B^2}{2\bar{n}_{eff,g}L}, \tag{3.70}$$

according to Eq. (3.7).

Fig. 3.11 shows numerical calculations of g_{th} and $\delta\lambda/\Delta\lambda_m$ for the 20 lowest-order modes of the pure DFB laser for κL products of 0.5 (a), 2 (b), and 5 (c). As can be seen immediately, the modal threshold gain is smallest for the two modes nearest to the Bragg wavelength, where $\delta\lambda = 0$; at the Bragg wavelength itself, however, no lasing mode exists. The two lowest threshold modes are separated by the so-called stop band of width $\Delta\lambda_s$,

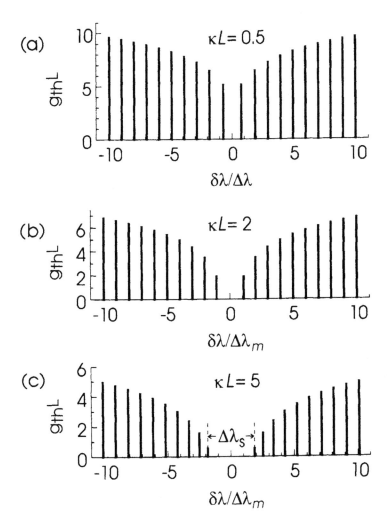

Figure 3.11: Threshold gain-length product $g_{th}L$ and normalized wavelength deviation $\delta\lambda/\Delta\lambda_m$ of DFB laser modes with κL-product of 0.5 (a), 2 (b), and 3 (c).

which markedly increases with increasing κL-product. Threshold gain and mode discrimination strongly depend on κL. The figure also clearly shows that far away from the Bragg wavelength the mode spacings approach the Fabry-Perot mode spacing $\Delta\lambda_m$.

Unfortunately, the longitudinal modes of this pure DFB laser are degenerate in pairs with respect to modal threshold gain, so that the modes spaced equally from λ_B attain the same threshold gain. On the other hand, however, the threshold gain discrimination is very effective against the modes with larger separation to λ_B. Taking, for instance, $\kappa L = 2$, the $g_{th}L$-values for the two mode pairs nearest to λ_B are ≈ 2 and ≈ 3.5, respectively. For a laser length of 400 μm, we obtain g_{th} of 50 cm^{-1} and 87.5 cm^{-1}, respectively, yielding a threshold gain difference of as much as 37.5 cm^{-1}. Compared with the Fabry-Perot laser numerical example in Sect. 3.2.1 ($\delta g \approx 0.01$ cm^{-1}), this corresponds to an improvement by more than 3 orders of magnitude.

The longitudinal intensity distributions of two modes with equal threshold gain spaced equally from λ_B are equal as well. For the two modes nearest to λ_B, which will be the dominant laser modes due to their minimum threshold gain, the longitudinal intensity distributions are sketched schematically in Fig. 3.12 for the two limiting cases of small (a) and large (b) κL-product. At small κL-product (<1), the longitudinal intensity distribution exhibits maximum power at the end facets (Fig. 3.12a), so that the effective reflectivity is small, and, consequently, the total loss is large. On the other hand, with large κL-product (>2) the longitudinal intensity distribution peaks in the cavity center (Fig. 3.12b), yielding high effective reflection and low total loss. Accordingly, these field distributions explain well the κL-dependence of the mode threshold gains as calculated in Fig. 3.11. In the operation regime far above threshold, *i.e.*, at high optical power, the longitudinal intensity distribution is important because an inhomogeneous intensity profile means a spatially inhomogeneous stimulated emission rate. This in turn leads to a power-dependent longitudinal redistribution of the carrier density and optical gain, the so-called spatial "hole-burning," which makes the laser characteristics power dependent.

Independently of the κL-product, however, the degeneracy in g_{th} of the modes placed symmetrically around λ_B makes this idealized "pure" DFB laser unsuited as a single-mode laser.

One approach to break the threshold gain degeneracy in DFB lasers is to use a Bragg grating with a quarter-wave shift (*i.e.*, $\Lambda/2$) near the

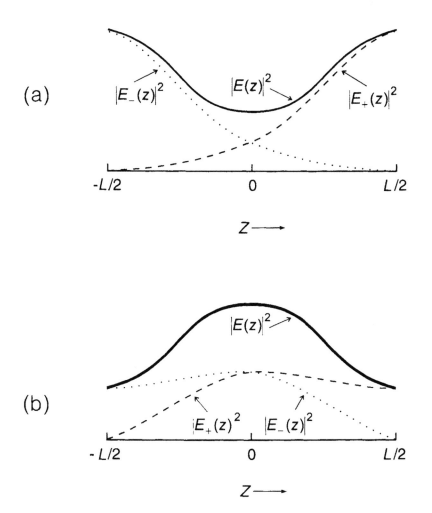

Figure 3.12: Longitudinal intensity distribution in the idealized "pure" DFB laser for small (a) and large (b) κL-product.

cavity center (Utaka *et al.*, 1986). The corresponding grating structure is shown in Fig. 3.13a. Again, the end facets should be nonreflecting. The solutions of the coupled-mode equations for this cavity structure yield the modal threshold gain spectrum shown schematically in Fig. 3.13b. As can be seen, a low-threshold mode now appears exactly at the Bragg wavelength, while the other modes are again each degenerated in pairs with respect to g_{th}. A large threshold gain difference between the mode

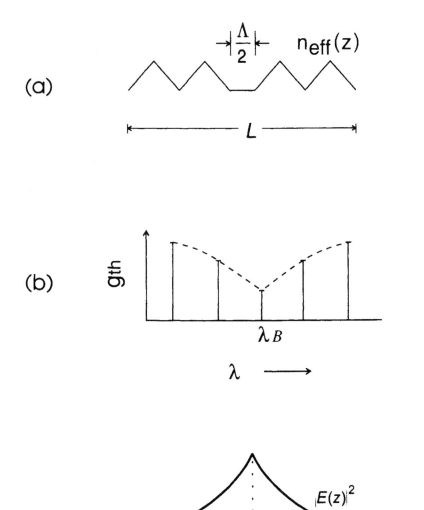

Figure 3.13: DFB laser with $\lambda/4$ phase-shifted Bragg grating and nonreflecting end facets: (a) effective refractive index profile; (b) modal threshold gain spectrum; and (c) longitudinal intensity distribution.

at λ_B and the neighboring modes can be achieved simultaneously with a low threshold gain in this way. The low threshold gain manifests itself in the longitudinal intensity profile (Fig. 3.13c), revealing a strong concentration of the mode power in the cavity center.

As outlined already in the early DFB laser theory (Kogelnik and Shank, 1972), a purely imaginary refractive index grating would also break the gain degeneracy and provide a low-threshold mode exactly at the Bragg wavelength. The corresponding so-called "gain-coupled" DFB lasers usually comprise a longitudinally modulated loss grating that yields the imaginary refractive index perturbation. Recently, these devices have been realized showing the expected performance with respect to single-mode operation, spectral linewidth, and reduced reflection sensitivity (Luo *et al.*, 1990; Borchert *et al.*, 1991).

From the practical point of view, however, the threshold gain degeneracy of the "pure" DFB laser structure plays a less important role. This is because the experimental DFB lasers, as shown schematically in Fig. 3.14, always exhibit reflections at the end facets ($R_1 \neq 0$, $R_2 \neq 0$) if no antireflection coatings are applied. Consequently, the reflectivity of the end facets has to be included in a realistic DFB laser modelling. Depending

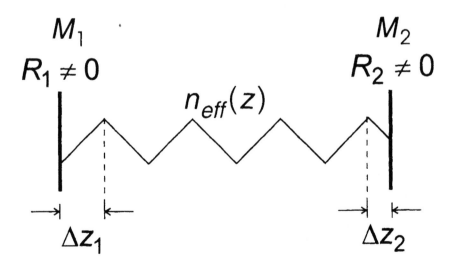

Figure 3.14: Realistic DFB laser structure with as-cleaved reflecting end facets ($r \approx 0.35$), random facet-to-grating position, and non–phase-shifted Bragg grating.

on the relative position of the facets with respect to the Bragg grating, moreover, strong changes of the g_{th} spectra occur, including deterioration of the symmetry, changes of the stop-band width, and breaking of the g_{th}-degeneracy. Since the relative position or phase difference between the end facet and grating changes essentially even if the facet position varies by only 50 nm (*i.e.*, $\Lambda/4$), it is technologically impossible at present to establish a well-defined facet-to-grating phase. In DFB lasers with reflecting end facets, therefore, the effect of the facet-to-grating phase is a purely random function even for devices from the same wafer, so that the analysis of mode discrimination merely yields statistical data (Buus, 1986). The corresponding calculations show that with κL-products around 2, usually more than 60% of uncoated DFB lasers provide a dominating single-mode operation with threshold gain differences large enough (> 5 cm^{-1}) to assure an SSR far beyond 30 dB. This percentage can be increased markedly by applying a high-reflective coating onto one end facet and an antireflection coating onto the other one. Under stationary conditions, SSR-values above 35 dB are commonly achieved (see, *e.g.*, Sasaki *et al.*, 1988, and Tanbun-Ek *et al.*, 1990), so that the target of a single-mode laser diode can well be achieved today with the DFB lasers. Employing the gain-coupled DFB lasers, moreover, the SSR still can be improved up to about 45 dB (Borchert *et al.*, 1991).

3.2.3.2   Spectral linewidth

In many applications the spectral linewidth of the lasing mode is of great importance. As mode competition in multimode lasers leads to significant linewidth broadening, a large SSR is revealed as a prerequisite for the laser performance. With SSR values well above 30 dB, the linewidth broadening by mode competition can be neglected. Then the diode laser linewidth $\Delta\nu$ is dominated by the phase noise due to the spontaneous emission and can thus be well approximated by the Schawlow-Townes-Henry linewidth $\Delta\nu_{STH}$ (Henry, 1982)

$$\Delta\nu \simeq \Delta\nu_{STH} = \frac{R_{sp}}{4\pi S_0}(1 + \alpha^2). \tag{3.71}$$

Here R_{sp} is the spontaneous emission rate into the lasing mode, S_0 is the photon number of the lasing mode in the cavity, and the linewidth enhancement factor (gain-phase coupling coefficient) is defined as

$$\alpha = -\frac{\partial\mathcal{R}(n_a)/\partial N_a}{\partial\mathcal{I}(n_a)/\partial N_a}, \tag{3.72}$$

with n_a and N_a being the refractive index and carrier density in the gain medium (active region). Using Eqs. (3.14) and (3.19), $\Delta\nu_{STH}$ of a single mode Fabry-Perot laser diode can be written as a function of the output power per facet P as (Henry, 1982)

$$\Delta\nu_{STH} = \frac{v_g^2 h \, \nu n_{sp} \gamma_m \gamma_{tot}}{8\pi P} (1 + \alpha^2).$$ (3.73)

The occurence of the α-factor is a consequence of the asymmetry of the semiconductor gain curve with respect to wavelength. Therefore in contrast to, e.g., gas lasers, in a diode laser operating at a wavelength near the gain maximum, i.e., the usual operation regime, changes of the carrier densities (inversion) change the material gain as well as the index. With typical α-factors of 3–7 in bulk InGaAsP operating near 1.55 μm, the linewidth enhancement due to the gain-phase coupling is up to about a factor of 50.

Eq. (3.73) cannot easily be adopted to estimate the spectral linewidth of a DFB laser. This is because there is no simple way to calculate the mirror losses of a DFB laser. Owing to the inhomogeneous intensity distribution within the DFB laser cavity (cf. Fig. 3.12), the P versus S_0 relationship is also different from that of the Fabry-Perot laser (Eq. 3.19). However, DFB lasers with as-cleaved end facets ($r_1 = r_2 \simeq 0.35$ for InGaAsP lasers) and κL-values up to about one can approximately be viewed as single-mode Fabry-Perot type lasers, and the effect of the DFB can be neglected in the linewidth calculation (even though the DFB provides the single-mode operation, which is an indispensible prerequisite for a narrow spectral linewidth). In this common case, Eq. (3.73) may still be applied with reasonable accuracy by inserting the loss parameters of the Fabry-Perot cavity.

As with the laterally gain-guided laser diodes (Petermann, 1979), the complex longitudinal field distribution in DFB lasers implies that a longitudinal K-factor of up to order two has to be added to the linewidth formula (Wang et al., 1987). As a further consequence of the complex longitudinal field distribution, an effective α-factor (Furuya, 1985) has to be used for the linewidth calculation, which may differ somewhat from the material α-factor (Amann, 1990; Pan et al., 1990).

Linewidth calculations for DFB laser without reflecting end facets (Kojima et al., 1985) show a much stronger dependence on L than for the Fabry-Perot laser, because the focussing of the intensity distribution in the cavity center for large κL (Fig. 3.12b) strongly changes the P versus

S_0 relationship, *i.e.*, smaller P than in the Fabry-Perot laser for equal S_0. In the limiting case of large κL, correspondingly, the linewidth of DFB lasers without reflecting facets is proportional to $\kappa^{-2} L^{-3}$. Depending on κL, the length dependence of the linewidth in DFB lasers is therefore stronger than $1/L$ as in the Fabry-Perot laser.

Despite these differences with respect to the Fabry-Perot laser, in the DFB lasers the spectral linewidth is also reciprocal to the output power and essentially proportional to $(1 + \alpha^2)$, with α denoting the material α-factor. Thus the linewidth is usually plotted versus the reciprocal optical power as shown schematically in Fig. 3.15. In the validity range of Eq.

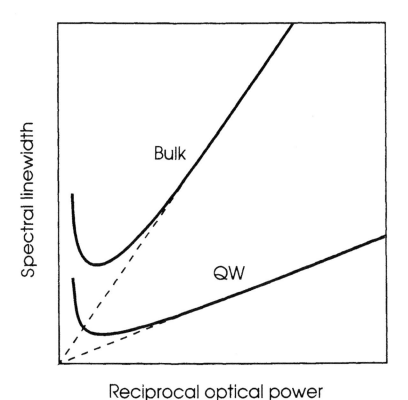

Figure 3.15: Spectral linewidth versus reciprocal optical power of a single-mode laser diode. The effect of the active region structure (bulk or quantum well) is indicated. At large optical power the measured linewidth rebroadens and Eq. (3.71) no longer applies.

(3.71), straight lines approaching the origin for infinite optical power should be obtained. The slope of these lines is proportional to $(1 + \alpha^2)$, so that properly designed quantum well (QW) laser diodes with their smaller α-factor yield smaller slope as well as smaller spectral linewidth. In practice, one usually finds a deviation from this ideal relationship at large optical power yielding a strong rebroadening of the laser linewidth.

The realization of narrow-linewidth DFB lasers therefore indispensably requires the use of long devices with a large optical output power and small α-factor. While the latter can be achieved by the material design of the active region (Ohtoshi and Chinone, 1989; Dutta *et al.*, 1990; Tiemeijer *et al.*, 1991; Kano *et al.*, 1993), the realization of long devices with high output power in a single mode requires a careful cavity design (Okai, 1994) because of spatial hole burning, which rebroadens the line (Pan *et al.*, 1990).

3.2.3.3 Experimental results

Schematical longitudinal and cross-sectional end views of a representative InGaAsP/InP DFB laser (Wolf *et al.*, 1991) for 1.55 μm wavelength with the metal-clad ridge-waveguide (MCRW) (Baumann *et al.*, 1988) structure are shown in Fig. 3.16. This laser is made by single-step epitaxy on a corrugated InP substrate. Differing from the most simple DFB laser structure, an n-InP spacer layer was introduced between active and grating layers. Therewith both a small optical confinement in the active layer and a small coupling coefficient κ can be achieved. The relevant technological parameters and refractive indexes are listed in Table 3.1. The first-order Bragg grating with a pitch of 241 nm is made into the n-InP substrate using holography and wet chemical etching. The typical grating height after epitaxy is 30 nm, yielding κ of about 22 cm^{-1}. The laser length is varied from 400 to 800 μm, so the κL-product is between 1 and 2. Applying a stripe width of 3 μm and an effective index step of 0.04–0.07 provides low threshold current and stable TE-polarized emission. The narrow linewidth operation is obtained by a -20 nm detuning of the Bragg wavelength (1.55 μm) relative to the active region gain peak (1.57 μm) (Ogita *et al.*, 1988) in order to reduce the α-factor and by using as-cleaved devices with κL-products around 1–2 to obtain small mirror losses and reduced spatial hole burning. With respect to small intervalence band absorption (IVBA) (Childs *et al.*, 1986), which essentially determines the internal losses in long-wavelength laser diodes, the confinement factor

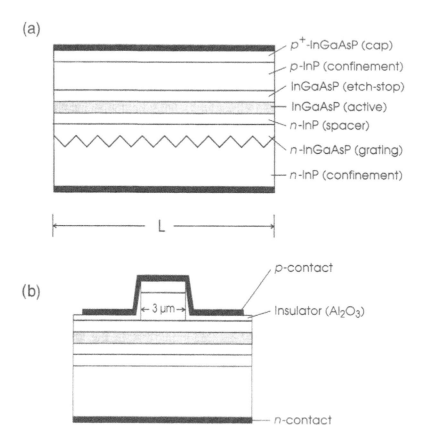

Figure 3.16: Schematic longitudinal (a) and cross-sectional end views (b) of InGaAsP/InP metal-clad ridge-waveguide (MCRW) DFB laser for 1.55 μm wavelength.

in the active region is adjusted to below 15% by choosing an active layer thickness of 0.08–0.1 μm and an 0.15-μm-thick spacer layer.

A typical mode spectrum of a 400-μm-long laser is shown in a logarithmic scale in Fig. 3.17. This device has a threshold current around 50 mA at room temperature, and the spectrum was measured at a current of 150 mA, corresponding to an optical output power around 8 mW. As can be seen very clearly, the emission occurs in one single mode, and the neighboring longitudinal modes are suppressed by more than 30 dB. Note that the width of the modes is determined by the resolution of the spectrometer.

Layer	Function	λ_g (μm)	Thickness (μm)	Doping (cm^{-3})	Refractive Index
n-InP	substrate	0.92	80	$2 \cdot 10^{18}$	3.15
n-InGaAsP	grating	1.12	0.09 ± 0.015	$5 \cdot 10^{17}$	3.29
n-InP	spacer	0.92	0.15	$5 \cdot 10^{17}$	3.17
InGaAsP	active	1.57	0.09	undoped	3.55
p-InGaAsP	etch-stop	1.30	0.13	$3 \cdot 10^{17}$	3.39
p-InP	confinement	0.92	1.5	$5 \cdot 10^{17}$	3.17
p^+-InGaAsP	cap	1.30	0.2	$\approx 5 \cdot 10^{19}$	—

Table 3.1: Parameters of InGaAsP-InP MCRW DFB laser for 1.55 μm. The band gap wavelength λ_g is related to the band gap energy E_g as: $\lambda_g = hc/E_g$, where h is Planck's constant.

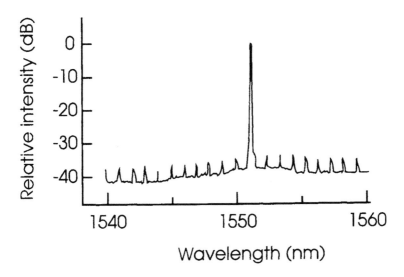

Figure 3.17: Longitudinal mode spectrum of a 400 μm long InGaAsP/InP MCRW DFB laser at 150 mA laser current.

The spectral linewidth of the MCRW DFB laser is plotted in Fig. 3.18 against the reciprocal optical power. The rectangles and circles correspond to 400 and 800 μm laser length, respectively. As can be seen, according to Eq. (3.73) the linewidth monotonically decreases with optical power up

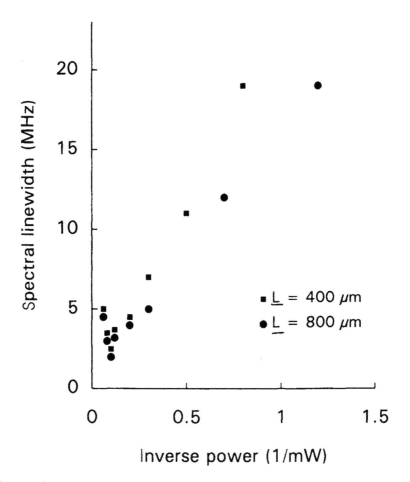

Figure 3.18: Spectral linewidth versus reciprocal optical power of 400- and 800-μm-long InGaAsP/InP MCRW DFB lasers.

to about 10 mW, approaching minimum values of typically 2 MHz (L = 800 μm) and 2.5 MHz (L = 400 μm). Minimum linewidths of selected samples of these rather simple DFB laser devices are as low as 1.5 MHz. The linewidth rebroadening at optical power above 15 mW, which is typical for most InGaAsP/InP DFB lasers, can be attributed to spatial hole burning (Pan *et al.*, 1990).

Extensive work has been performed to reduce the α-factor of the InGaAsP/InP DFB lasers. As mentioned above, one approach is the detuning of the laser wavelength relative to the gain maximum. This technique became possible after the development of the DFB and DBR devices, which allow forcing the laser operation near the Bragg wavelength, which must not necessarily coincide with the wavelength of maximum material gain. As an example, if the gain maximum of a DFB laser lasing at $\lambda =$ 1.52 μm is adjusted to about 1.56 μm, the α-factor reduction yields a narrowing of the linewidth by about a factor of 2 (Ogita *et al.*, 1988).

A further significant linewidth reduction by reducing α can be obtained with properly designed QW structures (Kojima and Kyuma, 1990). Particularly in the case of strained QW structures (Ohtoshi and Chinone, 1989; Dutta *et al.*, 1990), α-factors as low as 1–2 can be obtained and sub-MHz linewidth has been obtained.

Quite recently, an ultra-narrow spectral linewidth has been demonstrated with the corrugation-pitch-modulated DFB lasers, in which the spatial hole burning is effectively reduced by a sophisticated chirped grating structure (Okai, 1994; Okai *et al.*, 1994). As a consequence, the spectral rebroadening at large optical power is prevented, and the spectral linewidth can be as small as 3.6 kHz for a 1.5-mm-long laser with a compressively strained MQW active region operated at 55 mW optical output power.

3.3 Wavelength tunable laser diodes

In many applications, the electronic wavelength tuning of a single-mode laser diode is highly desired. This is particularly true for the coherent optical communication technique (Koch and Koren, 1990; Kotaki and Ishikawa, 1991), where the wavelength tunability is crucial for the local oscillator function (Chawki *et al.*, 1990; Noe *et al.*, 1992). But, also on the transmitter side, modulation schemes based on frequency modulation would gain from the availability of these devices. Besides the coherent optical transmission technique, wavelength division multiplexing applications (Amann, 1994; Reichmann *et al.*, 1993; Willner *et al.*, 1992) and wavelelength conversion (Yasaka *et al.*, 1995; Durhuus *et al.*, 1993) also require light sources with different wavelengths spaced by some nanometers. Furthermore, optical sensing techniques, such as range sensing by frequency modulated continuous wave (FMCW) radar methods (Economou *et al.*, 1986; Slotwinski *et al.*, 1989; Uttam and Culshaw, 1985; Strzelecki *et al.*, 1988), fiber measurement (Ebberg and Noe, 1990), spectroscopy

(Schneider *et al.*, 1995), reflectometry (Lee *et al.*, 1995; Huang and Carter, 1994), beam-steering (Rosenberger *et al.*, 1993), anemometrie (Koelink *et al.*, 1992; Dopheide *et al.*, 1988), and short-pulse generation (Schell *et al.*, 1993) can be done with such devices. Accordingly, after the development of high-performance DFB and DBR single-mode laser diodes, extensive research has been carried out worldwide in recent years to make these devices electronically tunable.

For most of these applications, a high spectral purity and a continuous tuning behavior are required. From a practical point of view, a clear separation of the control functions for wavelength setting and output power control is thereby needed. The demand for a continuous tuning specifically implies that the wavelength versus current (or voltage) characteristic is monotonic and smooth, allowing the unambiguous access to all wavelengths within the tuning range. A narrow spectral linewidth of at most several ten MHz is also essential for most applications. As an example, system experiments have shown that a bit error rate below 10^{-9} is achieved in a coherent digital optical transmission system with frequency shift keying (FSK) at bit rate of 100 Mb/s and a frequency deviation of 1.5 GHz only if the laser linewidth is below 25 MHz (Wagner and Linke, 1990). Equally important is a narrow linewidth for range-sensing applications in order to obtain a high resolution and a sufficiently large coherence length (Uttam and Culshaw, 1985; Dieckmann and Amann, 1995), which determines the maximum measurement range of the FMCW radar. Besides a narrow spectral linewidth, a high spectral purity also means a strong suppression of side modes in the spectrum as characterized by an SSR in excess of typically 30 dB.

3.3.1 Physical mechanisms for electronic wavelength tuning in laser diodes

The basic physical mechanism underlying the electronic wavelength tuning is the refractive index control either by carrier injection (plasma effect) (Westbrook, 1986; Okuda and Onaka, 1977; Bennet *et al.*, 1990) or by exploiting electro-optical effects such as the Quantum Confined Stark Effect (QCSE) (Miller *et al.*, 1984; Zucker *et al.*, 1988; Susa and Nakahara, 1992), rather than by spectrally shifting the gain curve of the active medium (Lau, 1990). This is because, in most of the applications cited above, the wavelength tuning must be smooth and continuous, a requirement that is not met if the gain curve is spectrally shifted along the discrete mode spectrum of the laser cavity. Indeed, referring to Fig. 3.1d, the spectral shift of the gain curve results in successive mode jumps and would not allow access to all wavelengths within the tuning range. In the case of the refractive index control, however, the cavity mode spectrum can be tuned itself in a smooth manner (Coldren and Corzine, 1987).

Up to now the largest refractive index changes and tuning ranges have been achieved by using the plasma effect of the free carriers injected into double heterostructures. The refractive index change Δn caused by the injected electron-hole plasma is mainly due to the polarization of the free carriers, but the spectral shift of the absorption edge also yields a contribution to Δn (Soref and Lorenzo, 1986; Weber, 1994). The dominating effect of the carrier polarization on Δn is given by (Soref and Lorenzo, 1986)

$$\Delta n = -\frac{e_0^2 \lambda^2}{8\pi^2 c^2 n \epsilon_0} \left(\frac{\Delta N}{m_e} + \frac{\Delta P}{m_h} \right), \qquad (3.74)$$

where $\Delta N (\Delta P)$ and $m_e (m_h)$ denote the density and effective mass of the injected electrons (holes) and n is the refractive index of the semiconductor. As a numerical example with $\lambda = 1.5$ μm, $n = 3.3$, $m_e = 0.05 m_0$, $m_h = 0.5 m_0$, one obtains for an ambipolar injection of $\Delta N = \Delta P = 3 \cdot 10^{18}$ cm^{-3} an index change of -0.021. In addition to the plasma effect, the injection of an electron-hole plasma changes the spectral shape of the band-to-band optical absorption via band gap shrinkage and band filling (Weber, 1994). Therefore, maximum refractive index changes up to about -0.04 can be achieved at 1.5 μm wavelength with $\Delta N = \Delta P = 3 \cdot 10^{18}$ cm^{-3} injected into InGaAsP with a band gap wavelength $\lambda_g (= hc/E_g)$ of 1.3 μm. Besides the refractive index change, the carrier injection also causes additional optical losses. The main role is played by the IVBA (Childs *et al.*, 1986) and not by the absorption due to the free carrier movement (Soref and Lorenzo, 1986). At 1.5 μm wavelength, the IVBA in InP and InGaAsP is 20–40 cm^{-1} for $\Delta N = \Delta P = 10^{18}$cm^{-3} (Asada *et al.*, 1984; Casey and Carter, 1984).

Since the injected electron-hole pairs recombine, a sustained current must be applied to the tuning region. This means not only parasitic heat generation that, among other things, reduces the laser power and efficiency, but also a relatively large time constant (several ns) due to the electron-hole recombination process. For a fixed current, the injected carrier density and, consequently, the refractive index and wavelength changes are proportional to the carrier lifetime, while the tuning speed is proportional to the reciprocal carrier lifetime. For a given device structure, therefore, the product of tuning range and tuning speed is fixed, and both laser parameters cannot be optimized independently.

In case of the QCSE, the refractive index changes are typically only of the order 10^{-3}–10^{-2}, depending on how closely the band gap wavelength of the quantum well structure used for tuning matches to the laser wavelength (Zucker *et al.*, 1988; Yamamoto *et al.*, 1985). In addition, the optical confinement achievable in these quantum wells is distinctly smaller than in case of the bulk semiconductor structures used for the plasma effect. As a consequence, the largest tuning ranges have been achieved so far with the plasma effect. On the other hand, for the QCSE, the semiconductor *pn*-junctions are reversely biased so that essentially no current flows, and heat generation does not occur. Furthermore, no carrier lifetime limitation exists for the tuning speed.

In almost every application, the continuous wavelength tuning is preferred, even though it is not always the required tuning mode. This is because of the simplicity and unambiguity of the wavelength setting (Coldren and Corzine, 1987) and because any wavelength within the tuning range can be accessed. In this wavelength tuning scheme, the laser emits in the same longitudinal mode within the entire tuning range, and mode changes or jumps do not contribute to the tuning effect. As a consequence, the tuning range $\Delta\lambda$ cannot be larger than that of a longitudinal laser mode. The latter is determined by the real part of the maximum change $\delta\bar{n}_{eff}$ of the effective refractive index $\bar{n}_{eff}$ as achievable by the electronic control current or voltage. Considering, for example, a DFB laser operating exactly at the Bragg wavelength ($\lambda_0 = \lambda_B$), the tuning range for continuous tuning is given with Eq. (3.30) as

$$\frac{\Delta\lambda}{\lambda_0} = \frac{|\mathscr{R}(\delta\bar{n}_{eff})|}{\bar{n}_{eff,g}}. \tag{3.75}$$

In the less favorable discontinuous tuning mode, on the contrary, tuning is done essentially by mode jumps from one longitudinal mode to the other one, so that this limitation does not exist. Accordingly, the ultimate limitation of the tuning range in the discontinuous tuning scheme is set by the spectral width of the active region gain function.

3.3.2   Continuously wavelength tunable laser diodes

Ideally, continuously tunable laser diodes should exhibit two electronic controls (current[s] and/or voltage[s]) that allow for the independent adjustment of the output power and the wavelength (Coldren and Corzine, 1987), respectively, as illustrated schematically in Fig. 3.19a. Thereby the wavelength control current I_t should affect only the wavelength but not the output power (Fig. 3.19b), while the output power control current I_a should influence exclusively the output power and not the wavelength

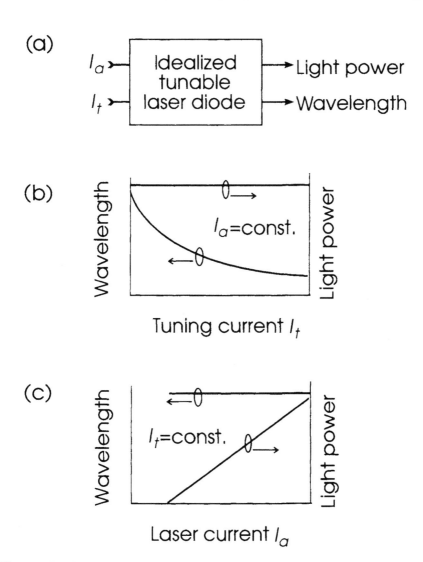

Figure 3.19: Idealized tunable laser diode (a). Output power and wavelength can be controlled independently and unambiguously by currents I_t (b) and I_a (c), respectively.

(Fig. 3.19c). The wavelength versus tuning current characteristic should be unambiguous and smooth, as plotted schematically in Fig. 3.19b. In practice, however, a complete separation between power and wavelength control can hardly be achieved because the tuning function, as done, *e.g.*, by carrier injection, also introduces optical losses that reduce the output power and also because changes of the laser current induce temperature changes that correspondingly affect the emission wavelength (Amann, 1995). Nevertheless, a separation of the two functions as far as possible should be aimed at in the laser development, because it significantly improves the device handling convenience and suitability.

As mentioned above, the largest tuning ranges may be achieved with the free-carrier plasma effect. Considering that the optical confinement in the tuning region is less than unity and noting that $\bar{n}_{eff} \approx n$, the maximum electronic tuning range (excluding thermal heating) is therefore restricted by Eq. 3.75 to values less than about 15 nm at 1.5 μm wavelength (Kotaki and Ishikawa, 1991). In spite of this restriction on the tuning range, the continuously tunable lasers are well suited for coherent optical communications and also for densely spaced wavelength division multiplexing (DWDM), where with channel spacings of 1–2 nm a moderate number (4–8) of channels can be covered by a single device.

Different structures have been developed so far to achieve a wide continuous tuning and to maintain a narrow spectral linewidth. Among these are the multisection DBR devices (Murata *et al.*, 1987; Oeberg *et al.*, 1991; Koch *et al.*, 1988b) and the TTG (Tunable Twin-Guide) laser (Amann *et al.*, 1989; Illek *et al.*, 1990b), in both of which tuning is performed by index changes in passive regions, and multisection DFB lasers, in which a longitudinally varying bias to the active region induces the wavelength tuning (Okai *et al.*, 1989; Kuindersma *et al.*, 1990; Wu *et al.*, 1990). Thereby all these lasers represent integrated optoelectronic devices, each comprising an active region, a Bragg grating filter, and a tuning region. These three basic components provide the optical gain, single mode selection, and electronic tuning function, respectively.

So far, the combination of these three functions has been done either by the longitudinal or the transverse integration technique as shown schematically in Fig. 3.20. Evidently, the longitudinally integrated structures evolve from the DBR laser while the transversely integrated devices resemble the DFB laser structure. Therefore, essentially different tuning behavior is obtained, even though the operation principle is rather similar. Regarding the longitudinal integration, the extremely large longitudinal

(a) Longitudinal integration

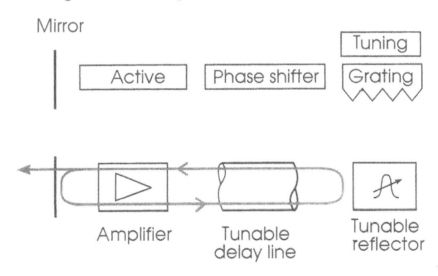

(b) Transverse integration

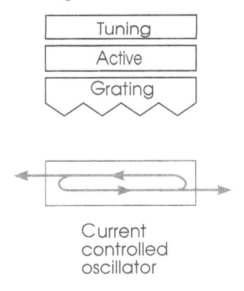

Figure 3.20: Longitudinal (a) and transverse (b) integration of electronically tunable laser diodes.

mode number (10^3–10^4, *cf.* Sect. 3.2.1) of laser diodes means that the longitudinal mode discrimination is relatively weak. So mode changes (jumps) and the resulting wavelength steps may easily occur during tuning. This can be considered as an immediate consequence of the absence of synchronization between Bragg wavelength and comb mode spectrum in the DBR lasers, as already discussed in Sect. 3.2.2. According to Fig. 3.20a, however, they can be synchronized by integrating a phase shift section within the composite laser cavity. In this way the optical cavity length, *i.e.*, the product of effective refractive index and effective cavity length (*cf.* Eq. 3.4), can be matched to the Bragg wavelength such that within a certain wavelength range (continuous tuning range) the same longitudinal mode is placed exactly at the Bragg wavelength.

On the other hand, diode lasers usually operate in the fundamental transverse mode with the higher-order transverse modes being cut off. So transverse mode jumps are principally excluded, and the tuning behavior of the transversely integrated devices is inherently continuous. The latter is a consequence of the clearly defined single-mode selection of a properly designed DFB laser, which is not deteriorated by tuning of the Bragg wavelength.

With no clear separation of the fundamental functions, the multisection DFB lasers represent a special case of a longitudinally integrated tunable laser. The tuning behavior, therefore, is closely related to that of the tunable DBR lasers.

A detailed comparison of the longitudinal and transverse integration schemes (Amann and Thulke, 1990) predicts that the largest continuous tuning ranges are achieved by the transversely integrated lasers. Up to now this has been well confirmed in practice, revealing up to a factor of 2–3 larger continuous tuning ranges for the transversely integrated lasers. On the other hand, however, the longitudinally integrated devices have proved superior so far with respect to other important laser characteristics, such as optical power and spectral linewidth.

3.3.2.1 Longitudinally integrated structures

The longitudinal sections of tunable DFB and DBR laser diode structures (Koch and Koren, 1990; Kotaki and Ishikawa, 1991) are displayed schematically in Fig. 3.21. The technologically simplest approach is a multi-electrode DFB laser, the top contact of which is longitudinally separated into two or three individually biased sections (Fig. 3.21a). By differently

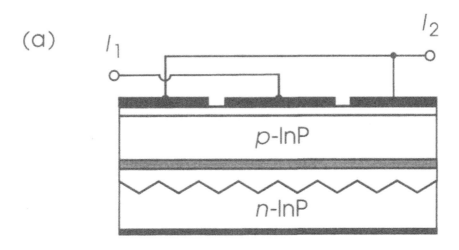

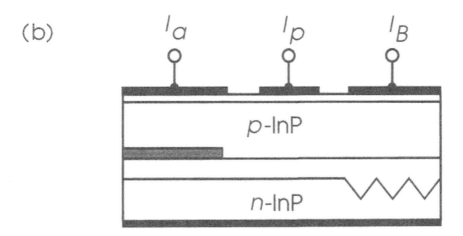

Figure 3.21: Schematic longitudinal sections of multisection tunable DFB laser (a) and three-section tunable DBR laser (b) for 1.55 μm wavelength. The active regions are marked black.

biasing the laser sections, a wavelength change can be obtained. Besides the thermal tuning, it seems surprising at first glance that an inhomogeneous bias causes significant laser wavelength changes, because the gain-clamping mechanism (Eq. 3.6) fixes the average carrier density. In case of equal phase-amplitude coupling coefficient α in all sections, consequently, the constant mode gain would keep the average effective index and the laser wavelength constant as well. While this argument is true for the Fabry-Perot laser diode, it does not apply for the DFB laser, in which the complex fields lead to a position-dependent effective α-factor even if the material α-factor is constant (Amann and Borchert, 1992). As a consequence, the clamping of the mode gain does not mean a fixing of the laser wavelength at the same time.

Theoretical investigations (Kusnetzow, 1988; Tohyama *et al.*, 1993) of the tuning mechanism of multielectrode DFB lasers have revealed that the wavelength tuning by nonuniform current injection is a rather complex mechanism, comprising also spatial hole-burning effects (Pan *et al.*, 1990), thermal heating (Okai *et al.*, 1992), and, for the case of a quantum-well active layer, the gain-levering effect (Okai *et al.*, 1989; Lau, 1990). Theoretically, the occurence of an inhomogeneous effective α-factor inevitably leads to carrier shot noise, which particularly manifests itself in a broadened spectral linewidth (Tromborg *et al.*, 1991). It should also be noted that multisection DFB lasers may become unstable, yielding self-pulsations (Bandelow *et al.*, 1993).

In spite of the tendency for mode jumps, continuous wavelength tuning may be performed by a careful mutual adjustment of the laser currents. Owing to the random mirror phases and the inevitable waveguide perturbations, the longitudinal field and carrier density distributions are different for each DFB laser (even for devices from the same wafer), so that the tuning behavior, particularly the I_1/I_2-ratio for continuous tuning, differs from device to device. Accordingly, a large number of measurements are required to select and characterize the suited lasers.

An essential disadvantage in the practical application of tunable DFB lasers is that the controls of output power and wavelength are not separated, so that both parameters are affected similarily by all control currents. This is in strong contrast to the idealized tunable laser shown in Fig. 3.19. On the other hand, however, these lasers yield the smallest spectral linewidths of the (monolithically) tunable laser diodes. With tuning ranges around 1.3 nm and 1.9 nm, spectral linewidths of 98 kHz and 900 kHz, respectively, were achieved with three-section DFB lasers (Okai and Tsuchiya, 1993; Kotaki *et al.*, 1989). Moreover, placing an electrically tunable phase-control section in the center of a two-section DFB laser proved efficient in keeping the spectral linewidth constant throughout the tuning range (Numai *et al.*, 1988).

The three-section distributed Bragg reflector (3S DBR) laser, as shown in Fig. 3.21b, provides a more convenient handling, because an effective separation between the power control and tuning function is achieved: Current I_a mainly determines the power, while both currents I_p and I_B essentially control the emission wavelength. The operation principle of this device is illustrated in Fig. 3.22. The three device functions of amplification, phase shifting via the optical cavity length, and tunable reflection by the Bragg grating (*cf.* Fig. 3.20a) are controlled by currents $I_a, I_p,$ and I_B, respectively. Changing exclusively I_B or I_p yields a discontin-

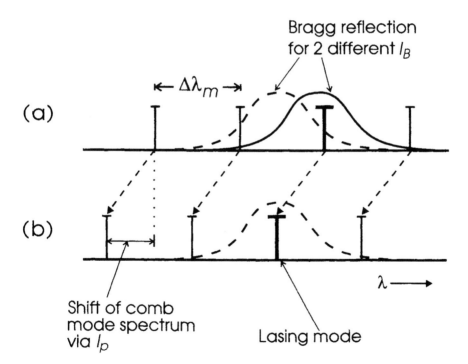

Figure 3.22: Operation principle of tunable three-section DBR laser: (a) longitudinal mode spectrum and total cavity gain curve for different Bragg wavelength but constant comb mode spectrum (discontinuous tuning); (b) synchroneous relative changes of Bragg wavelength and comb-mode spectrum (continuous tuning).

uous tuning by mode jumping, because either solely the Bragg wavelength or the optical cavity length is changed. This is shown for the case of a solitary Bragg wavelength change via I_B in Fig. 3.22a. Considering the solid curve, lasing occurs in the longitudinal cavity mode with a wavelength equal to that of the Bragg reflection peak (*i.e.*, λ_B). Increasing I_B (broken reflection curve), however, λ_B changes to smaller wavelengths and no longer matches the initial lasing mode. For a detuning larger than $\Delta\lambda_m/2$, a mode jump occurs to the neighboring longitudinal mode with a smaller wavelength. Applying only I_p, on the other hand, would reduce the effective index in the phase section and shrink the optical cavity length. This yields a shift of the comb-mode spectrum toward smaller wavelengths by keeping λ_B constant. Again, therefore, comb-mode spectrum and reflection peak become mismatched, and mode jumps will occur.

By a proper mutual adjustment of I_B and I_p, however, a continuous wavelength tuning can be obtained, provided that the relative changes of the Bragg wavelength exactly equal the relative changes of the optical cavity length. This device operation scheme is sketched in Fig. 3.22b, showing laser operation in the same longitudinal mode for both biasing conditions.

It should be stressed that, due to thermal heating, a wavelength change also occurs during I_a changes. Thus the emission wavelength of the tunable three-section DBR laser is a function of all three control currents.

This is demonstrated experimentally (Murata *et al.*, 1987) in Fig. 3.23, showing the laser wavelength as functions of each of the control currents while keeping the other two control currents fixed. As can be seen, the wavelength critically depends on all currents, and wavelength jumps limit the continuous tuning range for one-current tuning to below about 1 nm. By adjusting I_p and I_B, the continuous tuning of this laser can be extended to about 3 nm. In the discontinuous tuning scheme, on the contrary, up to a 5.8 nm tuning range is achieved as indicated in the figure.

The combined effect of I_p and I_B on the longitudinal modes is illustrated schematically in Fig. 3.24a. This diagram shows the lasing areas for the relevant longitudinal modes and indicates the boundaries where mode jumps occur. The mode order is labelled assuming that mode N is lasing initially at $I_B = I_p = 0$. Due to the nonlinear recombination of the carrier injection into the Bragg and phase-shift section, the mode boundaries differ from straight lines. Continuous tuning requires a con-

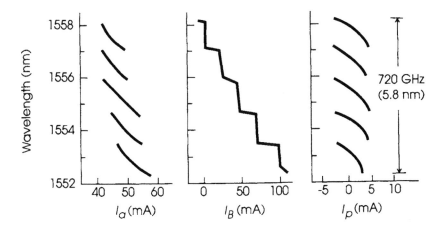

Figure 3.23: Effect of the three control currents on the wavelength of an experimental three-section DBR laser. (After Murata *et al.*, 1987).

stant mode order and can, for instance, be accomplished along the broken line. Effectively, a constant ratio I_p/I_B exists along this line, which can easily be realized by using a splitting network for the total tuning current $I_t = I_B + I_p$, as shown in Fig. 3.24b. In this way only one tuning current needs to be controlled, and continuous tuning up to 3.8 nm has been demonstrated (Ishida *et al.*, 1994).

A lot of work has been performed (Murata *et al.*, 1988; Kotaki *et al.*, 1988; Koch *et al.*, 1988a) to optimize the three-section tunable DBR laser yielding discontinuous tuning up to about 9 nm and maximum continuous tuning ranges around 4.4 nm, which fairly well corresponds to the theoretical limit (Pan *et al.*, 1988).

The separation of the laser-active region from the (passive) Bragg grating region in this device allows the extension of the tuning range by a strong heating of the Bragg section, while keeping the temperature in the gain section constant. In this way, large tuning ranges up to 22 nm have recently been realized in the quasi-continuous (*i.e.*, stepwise continuous) tuning mode (Oeberg *et al.*, 1991).

In many applications the dynamics of the tuning is important. Analytical treatments (Braagaard *et al.*, 1994; Zatni and LeBihan, 1995) showed that sub-ns wavelength switching times can be achieved and that either a pure AM or a pure FM modulation can be performed. Experimentally, a flat FM response up to 1 GHz was demonstrated (Ishida *et al.*, 1989).

(a)

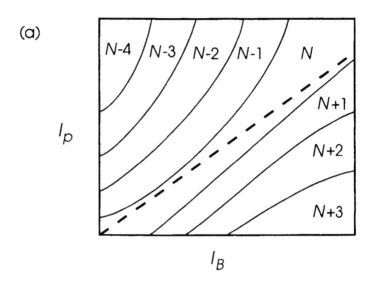

(b)

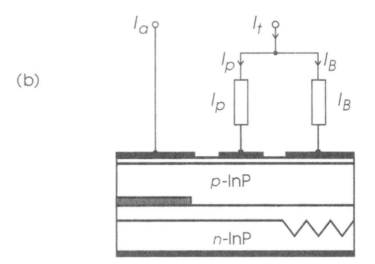

Figure 3.24: (a) Operation regimes for longitudinal modes in the I_B–I_p-plane. Mode and wavelength jumps occur at the mode boundaries (solid curves). Continuous tuning with single mode operation in mode N is obtained along the broken line. (b) Splitting network for the currents into the phase shift and Bragg sections, enabling continuous tuning with only one wavelength control current (I_t) along the broken line.

Using a four-step MOVPE process and a semi-insulating InP:Fe current-blocking structure, an AM modulation bandwidth of 9 GHz and a quasi-continuous tuning range of 9.1 nm were demonstrated with recent 3S DBR lasers (Stoltz *et al.*, 1993). The switching time for the transient between two successive modes can be as small as 500 ps (Delorme *et al.*, 1993; Zhang and Cartledge, 1995; Zhang and Cartledge, 1993).

3.3.2.2 Transversely integrated structures

The operation principle of the transversely integrated TTG DFB laser is schematically sketched in Fig. 3.25. The TTG laser essentially represents

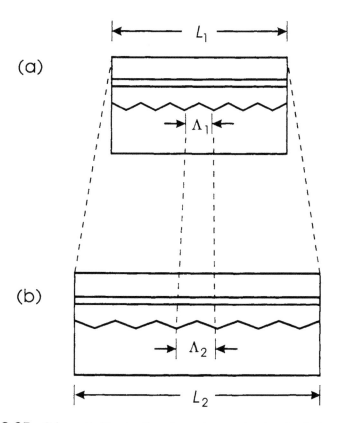

Figure 3.25: Schematic illustration of a tuning mechanism in the transversely integrated tunable twin-guide (TTG) DFB laser. The homogeneous longitudinal stretching of the optical cavity length yields the inherent continuous tuning of the Bragg and laser wavelength.

a DFB laser with an electronically tunable Bragg wavelength. Stable single-mode operation may be provided, for instance, by a quarter-wavelength shifted grating, an antireflection (AR)/high-reflection (HR) coating on the end facets or favorable mirror to grating phases of an as-cleaved device. The tuning of the Bragg wavelength is done in Fig. 3.25b by a homogeneous longitudinal "stretching" of the DFB laser. Obviously, the stretching factor is equal for both laser length and grating period, which can be expressed mathematically as

$$\frac{L_2}{L_1} = \frac{\Lambda_2}{\Lambda_1}.$$ (3.76)

While the laser wavelength is proportional to the grating period, the number of half wavelengths fitting longitudinally into the laser (or, in other words, the mode number) remains constant during stretching. So a continuous wavelength tuning with the same longitudinal mode is obtained.

Even though a semiconductor cannot be stretched significantly, this operation principle can be applied since the stretching of the optical laser length (*i.e.*, the product of laser length and effective refractive index) is the essential feature, and this parameter can be varied electronically via refractive index changes. Assuming the laser operation at the Bragg wavelength λ_B, we can write for the emission wavelength

$$\lambda = \lambda_\beta = 2\Lambda \mathcal{R}(\bar{n}_{eff}).$$ (3.77)

The total roundtrip phase Φ for the laser light

$$\Phi = 2kL = \frac{4\pi \mathcal{R}(\bar{n}_{eff})}{\lambda} L$$ (3.78)

can be expressed by means of Eq. (3.77) as

$$\Phi = 2\pi \frac{L}{\Lambda}.$$ (3.79)

Accordingly, the total round trip phase and the longitudinal mode order $N = \Phi/2\pi = L/\Lambda$ remain unchanged as the longitudinally homogeneous $\mathcal{R}(\bar{n}_{eff})$ changes induce the laser wavelength tuning corresponding to Eq. (3.77). As a consequence, this kind of wavelength tuning yields an inherently continuous tuning without mode jumps and the corresponding singularities of the laser characteristics.

Practically, significant changes of $\mathcal{R}(\bar{n}_{eff})$ cannot be induced via carrier number changes in the active region of a DFB laser. This is because the gain clamping mechanism (Eq. 3.6) in the laser prevents essential carrier number changes above threshold. Furthermore, wavelength tuning via the active region current is less advantageous because of the simultaneous influence on the laser output power, so that the independent control of power and wavelength would be impossible. Instead, the $\mathcal{R}(\bar{n}_{eff})$ changes in the TTG laser are introduced via an additional tuning region collocated in parallel to the active region along the entire cavity length, as displayed in Fig. 3.26a. The tuning region exhibits a larger band gap energy than

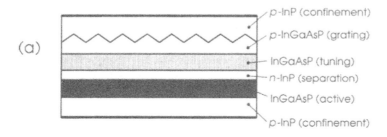

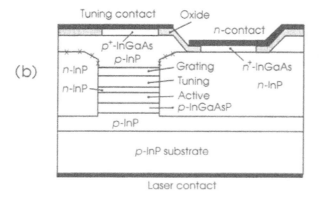

Figure 3.26: Schematic longitudinal (a) and cross-sectional end views (b) of TTG DFB laser structure in InGaAsP/InP for a 1.55 μm wavelength. The tuning diode is designed for forward and backward bias, yielding both electronic and thermal tuning.

the active region, so that the injected carriers undergo no band-to-band stimulated emission by the laser light. This decouples the carriers in the tuning region from the gain clamping mechanism and makes their dynamics independent of the light power. Furthermore, the desired homogeneous longitudinal carrier distribution and index change are achieved in the tuning region because no hole burning effects occur.

For an efficient tuning, it is most important that the tuning region is strongly coupled to the laser mode. With respect to Fig. 3.26a, the tuning region should therefore be placed close to the active region so that the transverse mode confinement is strong in both regions, *i.e.*, the optical power density is large in the active and the tuning region. On the other hand, both regions should clearly be decoupled electronically in order to allow the independent electronic biasing. The latter is achieved by putting an n-doped InP confinement layer between them, while the outer InP confinement layers are doped p-type. Due to the resulting blocking *pnp*-structure, the hole currents from bottom and top into the active and tuning region, respectively, are independent. Additionally, the neutrality condition, which demands that electron and hole injection are each equal in the active and in the tuning region, respectively, provides the decoupling of the electron currents as well. The electrons are supplied laterally from the common n-contact via the burying n-InP confinement layers, as shown in the schematic cross-sectional view of Fig. 3.26b.

The total wavelength change $\Delta\lambda$ caused by tuning reads

$$\Delta\lambda = \Delta\lambda_{el} + \Delta\lambda_{therm} = 2\Lambda\mathscr{R}(\delta\bar{n}_{eff}) + \xi\Delta\vartheta, \tag{3.80}$$

where $\Delta\lambda_{el}$ and $\Delta\lambda_{therm}$ denote the purely electronic and thermal contribution to the total tuning. The latter arises from the temperature change $\Delta\vartheta$ caused indirectly by the electronic tuning. It should be noted that the effect of temperature changes on the wavelength occurs also via changes of the effective refractive index. Since the temperature variations induce threshold changes, the carrier density in the active region changes as well, which finally leads to several contributions to the effective refractive index. These can, in summary, be described most conveniently by the experimentally determined parameter $\xi(\approx +0.1$ nm/K for InGaAsP lasers in the 1.5 μm wavelength regime). Continuous tuning may therefore be performed entirely by thermal tuning. This has recently been demonstrated on a DFB laser with an integrated metal stripe heater on top (Sakano *et al.*, 1992).

The change of $\bar{n}_{eff}$ due to carrier density changes ΔN_a and ΔN_t in the active and tuning region, respectively, can be written as

$$\delta\bar{n}_{eff} = \underbrace{(1 - i/\alpha_t)\Gamma_t\beta_t\Delta N_t}_{\delta\bar{n}_{eff,t}} + \underbrace{(1 - i/\alpha_a)\Gamma_a\beta_a\Delta N_a}_{\delta\bar{n}_{eff,a}}, \qquad (3.81)$$

where $\Gamma_{a,t}$ and $\alpha_{a,t}$ are the optical confinement and linewidth enhancement factors in the active (index "a") and tuning (index "t") region, and $\beta_{a,t}$ measures the refractive index change by carrier injection. The contributions of active and tuning region to $\delta\bar{n}_{eff}$ are labelled by $\delta\bar{n}_{eff,a}$ and $\delta\bar{n}_{eff,t}$, respectively.

In the lasing regime, the total gain/loss changes are cancelled out by the active region, *i.e.*, the mode gain always equals the total loss ("gain clamping mechanism," *cf.* Eq. 3.6) so that the imaginary part of $\delta\bar{n}_{eff}$ equals zero. This yields with the imaginary part of Eq. (3.81)

$$\Gamma_a\beta_a\Delta N_a = -\frac{\alpha_a}{\alpha_t}\Gamma_t\beta_t\Delta N_t. \qquad (3.82)$$

Substituting this into the real part of Eq. (3.81), we obtain for the real part of $\delta\bar{n}_{eff}$ (Amann and Borchert, 1992; Amann and Schimpe, 1990)

$$\mathcal{R}(\delta\bar{n}_{eff}) = \beta_t\,\Gamma_t\,\Delta N_t\left(1 - \frac{a_a}{a_t}\right). \qquad (3.83)$$

Typical values for the material parameters are $\beta_t \approx -7.5\cdot10^{-21}\text{cm}^3$, $\alpha_a \approx 2 - 7$, and $\alpha_t \approx -20$ (Kotaki and Ishikawa, 1989). Since α_t and α_a exhibit different signs, the expression in the parentheses is larger than unity, indicating a constructive interaction between the active and tuning regions with respect to the tuning effect. So the wavelength shift is larger than expected from the index change in the tuning region alone. The total tuning range is now obtained as function of ΔN_t and $\Delta\vartheta$ by inserting Eq. (3.83) into Eq. (3.80):

$$\Delta\lambda = \underbrace{2\Lambda\beta_t\,\Gamma_t\left(1 - \frac{a_a}{a_t}\right)\Delta N_t(I_t)}_{\Delta\lambda_{el}} + \underbrace{\xi\Delta\vartheta(I_t)}_{\Delta\lambda_{therm}}, \qquad (3.84)$$

where both ΔN_t and $\Delta\vartheta$ depend on the tuning current I_t. Neglecting the power escape by light emission from the tuning diode, which particularly applies at high tuning region currents, $\Delta\vartheta$ reads explicitly

$$\Delta\vartheta = R_{th}U_t(I_t)I_t, \qquad (3.85)$$

where R_{th} and U_t stand for the thermal resistance of the laser and the tuning diode voltage. The temperature change $\Delta\vartheta$ is positive for forward and reverse tuning region bias, so that it always leads to a positive $\Delta\lambda$ (red shift). On the other hand, the first expression in Eq. (3.84), which describes the electronic tuning, essentially contributes only under forward bias, yielding a negative $\Delta\lambda$ (blue shift). Accordingly, under forward bias the thermal tuning counteracts the electronic tuning. For a large tuning range, therefore, the thermal resistance and the forward voltage at the tuning diode should be small, which implies low resistive contacts and confinement layers. Exploiting the reverse bias for thermal tuning yields a large wavelength change for a large tuning region breakdown voltage.

In order also to enable the thermal tuning by heating the laser with reverse tuning diode bias, a low breakdown voltage InP homojunction was established in parallel with the tuning diode, as indicated by the crosses in Fig. 3.26b. This was accomplished by a high p-doping of $1.5 \cdot 10^{18}$ cm^{-3} of the upper p-InP confinement layer. Together with the high n-doping ($N_D = 2 - 3 \cdot 10^{18}$ cm^{-3}) of the burying n-InP region, a breakdown voltage near 3 V has been obtained, which is markedly lower than that of the tuning region pn-junction (estimated > 10 V) but larger than the forward voltage ($\simeq 1$ V) of the tuning diode. Accordingly, under reverse bias, the current flows via this shunt path to the tuning diode, yielding a strong heating effect without degrading the tuning region.

The threshold currents of the TTG lasers with as-cleaved end facets range between 5 mA and 15 mA for laser lengths between 200 and 600 μm; and the optical output power per facet at 100 mA current into the laser active region is typically between 3 mW (L = 600 μm) and 10 mW (L = 200 μm). From the viewpoint of output power and threshold current margin, long laser cavities are advantageous. On the other hand, however, the SSR decreases with increasing laser length. So optimum device performance has been achieved for a laser length around 400 μm, where the SSR is typically larger than 40 dB. A representative experimental tuning characteristic is displayed in Fig. 3.27a for a 400-μm-long TTG laser. Using tuning currents between -100 and $+100$ mA, a strictly continuous tuning range of 11 nm or 1.4 THz, respectively, is obtained, keeping the current into the laser active region constant at 100 mA. The maximum tuning range achieved in selected devices is around 13 nm or 1.6 THz, respectively. No degradation of the tuning diode by the reverse biasing and heating has been found. In accordance with previous TTG laser diodes, the largest tuning efficiency $|d\lambda/dI_t|$ is achieved at small (forward) tuning

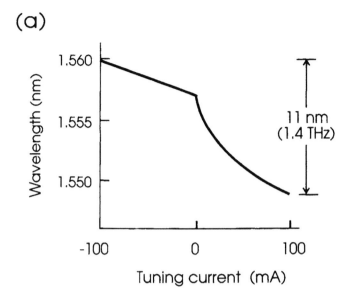

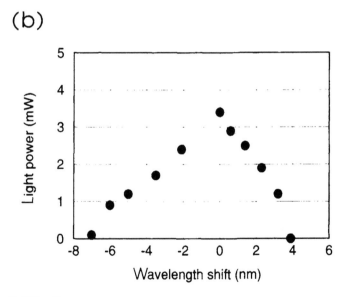

Figure 3.27: Wavelength versus tuning current (a) and light power versus wavelength shift (b) characteristics of an as-cleaved 400-μm-long TTG DFB laser at constant active region current of 100 mA, using forward and backward tuning region bias.

currents. With increasing forward tuning current, the tuning efficiency monotonically decreases while the almost constant breakdown voltage yields an approximately constant tuning efficiency of 0.04 nm/mA under reverse bias. As predicted by the analysis (Amann, 1995), the thermal tuning under reverse bias contributes about 4 nm to the tuning range, corresponding to a temperature rise of 40K and a thermal resistance around 120–130 K/W.

The optical output power at one end facet, measured with a numerical aperture of 0.5 at an active region current of 100 mA, is shown in Fig. 3.27b versus the wavelength shift. Starting with 3.5 mW light output power at zero tuning region bias, the light power almost linearly drops with increasing wavelength shift at both forward and backward bias. An optical output power above 1 mW can be maintained over a tuning range of about 9 nm, keeping the active region current constant at 100 mA. The SSR over this entire tuning range remains above 40 dB. The optical power strongly depends on the wavelength shift in this device, and the tuning range of 11 nm is determined by the extinction of the laser operation at large tuning currents and not by the limitations on the effective refractive index change. This is because the optical absorption losses (IVBA) in the tuning region become the dominant losses during tuning. Accordingly, a further optimization of the tuning and active region with respect to smaller IVBA and larger differential gain might increase the total tuning range.

3.3.2.3   Spectral linewidth

In general, electronic wavelength tuning via carrier injection causes an additional linewidth broadening. It has become clear both experimentally (Kotaki and Ishikawa, 1989; Sundaresan and Fletcher, 1990) and theoretically (Amann and Schimpe, 1990) that for most device types a tradeoff between tuning range and excess linewidth broadening exists, which might limit the applicability of tunable laser diodes in demanding system applications. The carrier number fluctuations in the tuning region(s) due to injection-recombination shot noise (IRSN) has previously been identified as the origin for this broadening (Amann and Schimpe, 1990). In contrast to the active region of the laser, the carrier number fluctuations in the passive tuning region(s) are not damped out by the gain clamping in the lasing regime.

A simplified physical model illustrating this broadening mechanism is shown in Fig. 3.28. Even though a Fabry-Perot cavity is shown here,

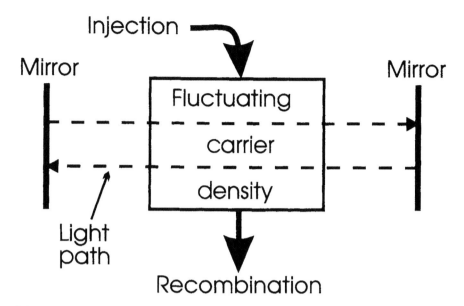

Figure 3.28: Simplified physical model for a laser cavity with linewidth broadening by injection-recombination shot noise (IRSN) due to the carrier injection into the tuning region.

the model equally well applies to more complex resonator structures such as DFB or DBR laser cavities. As can be seen, the IRSN of the carriers injected into the tuning region randomly modulates the refractive index and, hence, the optical length of the laser cavity. As a consequence, the instantaneous laser frequency fluctuates, finally yielding a broadening of the laser line.

The linewidth broadening $\Delta\nu_{IRSN}$ by the IRSN in the tuning regions of current tuned laser diodes (DFB, DBR, and TTG lasers) was analyzed theoretically (Amann and Schimpe, 1990; Hamada *et al.*, 1991), revealing that this linewidth contribution is proportional to the square of the tuning efficiency

$$\Delta\nu_{IRSN} \geq 4\pi e_0 \frac{c^2}{\lambda^4} \left(\frac{d\lambda}{dI_t}\right)^2 I_t. \qquad (3.86)$$

The total laser linewidth is the sum of $\Delta\nu_{IRSN}$ and the Schawlow-Townes-Henry linewidth $\Delta\nu_{STH}$:

$$\Delta\nu = \Delta\nu_{IRSN} + \Delta\nu_{STH}. \qquad (3.87)$$

The plot of the laser linewidth versus the reciprocal optical power according to Eq. (3.87) is shown schematically in Fig. 3.29. As with the single mode lasers (*cf.* Fig. 3.15) straight lines are approximately obtained at small and moderate optical power; however, for $I_t > 0$ these lines do not cross the origin because $\Delta \nu_{IRSN}$ is independent of the optical power. Instead, from the intersection with the vertical axis $\Delta \nu_{IRSN}$ can be estimated.

It is difficult to achieve a narrow spectral linewidth simultaneously with a large continuous tuning range, because the former requires a small tuning efficiency while a large tuning range can be obtained only with a

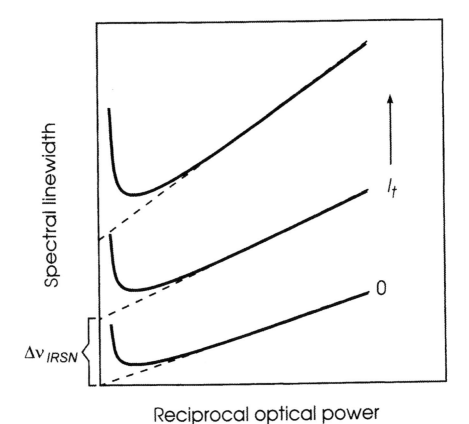

Figure 3.29: Spectral linewidth versus reciprocal optical power of continuously current tuned laser diode, with the tuning current I_t as parameter. The intersection with the vertical axis yields the linewidth contribution by the IRSN.

high tuning efficiency. From quantitative evaluations it turns out that, particularly for multisection DBR devices and TTG lasers with their passive tuning sections, $\Delta \nu_{IRSN}$ represents the essential contribution to the excess linewidth broadening (Amann and Schimpe, 1990; Kotaki and Ishikawa, 1989). Since, for a fixed laser geometry, $\Delta \nu_{IRSN}$ is approximately proportional to the tuning range squared, $\Delta \nu_{IRSN}$ becomes the dominating linewidth contribution in the large range tunable devices, *i.e.*, an electronic tuning range larger than about 5 nm. It should be noted, however, that this broadening effect does not occur by the thermal tuning.

The total laser linewidth of a widely (7 nm) tunable TTG laser with an undoped tuning region is displayed in Fig. 3.30. In accordance with Eq. (3.86), the linewidth reveals a significant enhancement by about an order of magnitude (solid curve) at small tuning currents, where $|d\lambda/dI_t|$ is largest. Using the same laser structure but doping the tuning region with n = $2 \cdot 10^{18}$ cm^{-3} (broken curve) reduces the tuning range by about a factor of two but also reduces the total spectral linewidth in the tuning

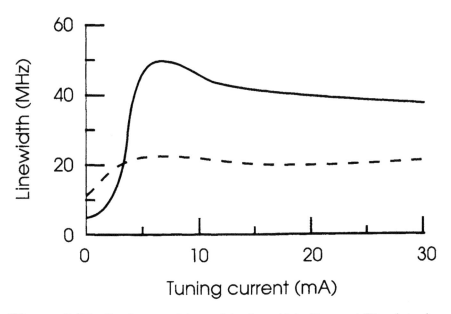

Figure 3.30: Total spectral linewidth of a widely (7 nm at 50 mA tuning current) tunable TTG laser versus the tuning current (solid curve). By heavily doping the tuning region *n*-type ($N = 2 \cdot 10^{18}$ cm^{-3}), the tuning range decreases to 3 nm, which yields a significant reduction of the linewidth (broken curve).

mode. As demonstrated previously (Amann and Borchert, 1992; Amann *et al.*, 1991), the IRSN can effectively be suppressed by applying a low-impedance bias network to the tuning diode without affecting the tuning range. Thereby the tuning diode voltage is kept constant, and the electron and hole quasi-Fermi levels become fixed. Consequently, carrier number fluctuations are strongly reduced in the tuning region, leading to a smaller linewidth broadening. In this way, the spectral linewidth of the TTG DFB laser can be kept below 30 MHz over the 9 nm tuning range. In addition, only slow variations in the linewidth occur by tuning current changes, and the power-linewidth product is almost constant.

Also in the case of the three-section DBR lasers, a large IRSN linewidth broadening occurs (Amann and Schimpe, 1990; dos Santos Ferreira *et al.*, 1992). The resulting linewidth broadening is typically of the order of several MHz (Kotaki and Ishikawa, 1989), which may limit also the applicability of the three-section DBR lasers in linewidth sensitive systems. The IRSN in the tuning mode was recently measured directly in the relative intensity noise (RIN) spectrum of a tunable three-section DBR laser (Sundaresan and Fletcher, 1990).

A further linewidth broadening occurs in tunable DBR and TTG lasers due to the cavity loss fluctuations accompanying the refractive index fluctuations in the tuning layer. Via the gain-clamping mechanism, these loss fluctuations induce immediate carrier density fluctuations in the active region in order to keep the total device gain constant. Depending on the α-factor in the active region, these fluctuations are accompanied by additional index and consequently wavelength fluctuations. This interaction between the active and tuning region has previously been investigated theoretically for the case of the TTG DFB laser, taking into account also the longitudinal inhomogenities of the electromagnetic field in the DFB structure (Amann and Borchert, 1992). A statistical analysis was performed for as-cleaved devices, in order to treat the random grating-facet relationship. It turned out from this calculation that the IRSN linewidth broadening is enhanced in the DFB structure with large coupling-coefficient-length product κL. Compared to the Fabry-Perot limit ($\kappa L \rightarrow 0$), $\Delta \nu_{IRSN}$ is larger by up to about 50% for typical κL-values at equal tuning range.

By exploiting the QCSE for tuning of the TTG laser (Wolf *et al.*, 1992; Yamamoto *et al.*, 1991a), no carriers are injected into the tuning region. Correspondingly, no shot noise broadening occurs, so that the total spectral linewidth can be kept small. In addition, the FM modulation bandwidth can be increased, because the carrier lifetime limitation in the tuning region is removed. On the other hand, however, due to the small optical confinement in the quantum wells, the tuning range is essentially smaller. Improved spectral properties and high-speed FM modulation might also be achieved in future devices using the electron-transfer within an MQW-type tuning region. With the so-called barrier reservoir and quantum-well electron-transfer structure (BRAQWETS) voltage controlled, refractive index changes up to 0.02 (Wegener *et al.*, 1989) have been demonstrated and (parasitic free) switching times well below 100 ps have theoretically been predicted (Wang *et al.*, 1993).

Using a separate confinement heterostructure quantum well (SCH QW) structure in the tuning region that localizes the holes within the wells while the electrons are distributed over the entire tuning region (Sakata *et al.*, 1993) or applying a MQW twin-active-guide (Yamamoto *et al.*, 1991b) might enable a further simultaneous improvement of optical power and spectral linewidth of the TTG laser.

Because of the markedly smaller tuning efficiency and the larger relative contribution of thermal tuning (Tohyama *et al.*, 1993), the IRSN linewidth broading is less important in wavelength tunable multisection DFB lasers. With properly selected devices and by exploiting also the wavelength change by thermal heating, tuning ranges up to 6 nm with spectral linewidth below 2 MHz have been reported for multisection DFB lasers (Kuindersma *et al.*, 1990). Quite recently a spectral linewidth less than 100 kHz has been obtained over a (mostly thermally induced) tuning range of 1.3 nm with a corrugation-pitch-modulated multi-quantum well (MQW) DFB laser (Okai and Tsuchiya, 1993).

Published data on the spectral linewidth for the various continuously tunable laser diodes are compiled in Fig. 3.31 (Kotaki *et al.*, 1989; Kotaki and Ishikawa, 1989; Illek *et al.*, 1990a; Wolf *et al.*, 1994), proving clearly the linewidth broadening with increasing continuous tuning range.

3.3.2.4 Physical limitations on tuning range of continuously tunable laser diodes

The maximum carrier-induced relative refractive index change in InGaAsP at a 1.55 μm wavelength is of the order of 1% for a carrier injection of $N = P = 3 - 4 \cdot 10^{18} \, \text{cm}^{-3}$ (Bennet *et al.*, 1990). With typical optical confinement factors of up to 0.5 in laser diodes, the maximum wavelength change due to the free-carrier plasma effect alone is therefore

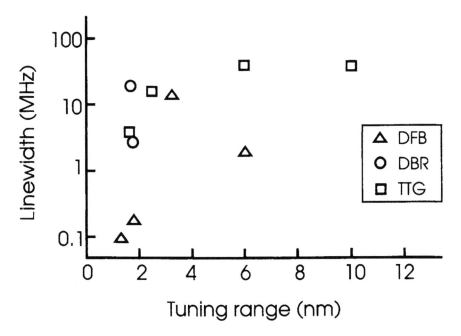

Figure 3.31: Spectral linewidth versus tuning range for various continuously tunable laser diode structures. In spite of the different device parameters, the strong enhancement of the linewidth with increasing tuning range is obvious.

around 0.5%, or 7.5 nm. The temperature span for thermal heating may be extended at most to about 100 K, so that thermal tuning may contribute at most by about 10 nm, yielding a total continuous tuning range of 17.5 nm. It should be noted that this value addresses the limitations of the tuning function by the refractive index tunability. On the other hand, however; the loss and temperature increase by tuning increases the threshold current and may thus limit the tuning range by preventing the laser operation entirely. The power loss change of the laser mode $\Delta\gamma = -2k_0\mathcal{I}(\delta\tilde{n}_{eff,t})$, as caused by the tuning region alone, is obtained from Eq. (3.81) as

$$\Delta\gamma = \frac{4\pi}{\lambda_0}\frac{\Gamma_t\beta_t\Delta N_t}{\alpha_t}. \tag{3.88}$$

Eliminating the product $\Gamma_t\beta_t\Delta N_t$ by means of the purely electronic wavelength shift $\Delta\lambda_{el}$ (cf. Eq. 3.84) and using Eq. (3.77) with the laser wavelength equal to the Bragg wavelength ($\lambda_0 = \lambda_B$) yields finally a relationship between $\Delta\gamma$ and $\Delta\lambda_{el}$:

$$\Delta\gamma = \frac{4\pi\mathscr{R}(\bar{n}_{eff})}{\lambda_0^2(\alpha_t - \alpha_a)}\Delta\lambda_{el}. \qquad (3.89)$$

Taking, for instance, $\mathscr{R}(\bar{n}_{eff}) = 3.3$, $\lambda_0 = 1.55$ μm, $\alpha_t = -20$, and $\alpha_a = 2$ yields $\Delta\gamma = -7.8 \cdot 10^7$ cm$^{-2}\Delta\lambda_{el.}$ For an electronic wavelength shift of -7.5 nm, the loss change is 59 cm$^{-1.}$ Such large additional losses strongly increase the laser threshold and reduce the output power. Consequently, the reduction of the α-factor in the tuning region to larger negative values would be advantageous. However, this requires larger band gap energy of the tuning region material, which, on the other hand, yields smaller refractive index changes (Murata et al., 1988). Recent attempts to improve the relationship between refractive index change and loss change in the tuning region (Sakata et al., 1993; Yamamoto et al., 1991b) have not yet resulted in improved tuning ranges.

Considering that the maximum tuning ranges achieved so far are around 13 nm, it is concluded that only marginal improvements of the continuous tuning range may be feasible with further optimized InGaAsP/ InP laser diodes at 1.55 μm wavelength. The other relevant laser parameters, like output power or spectral linewidth, however, still require an improvement in order to fully exploit the potential of electronically tunable laser diodes in advanced optoelectronics applications and in optical communications.

3.3.3 Discontinuously tunable laser diodes

As mentioned above, a wavelength coverage larger than about 15 nm at 1.55 μm wavelength cannot be achieved continuously, *i.e.*, in a single transverse and axial mode as displayed in Fig. 3.32a. If a larger tuning range is needed, *e.g.*, in wavelength division multiplexing (WDM) systems, one has therefore to accept mode changes and the corresponding inconvenience during tuning. The resulting discontinuous tuning scheme is shown in Fig. 3.32b, exhibiting multiple mode jumps. Since the tuning range in this tuning mode benefits from the mode jumps between different axial modes, a wide tunability can be achieved. On the other hand, however, not all wavelengths within the covered wavelength range can be individually addressed. The maximum tuning range is therefore usually limited only by the gain bandwidth of the active region, which is of the order of 100

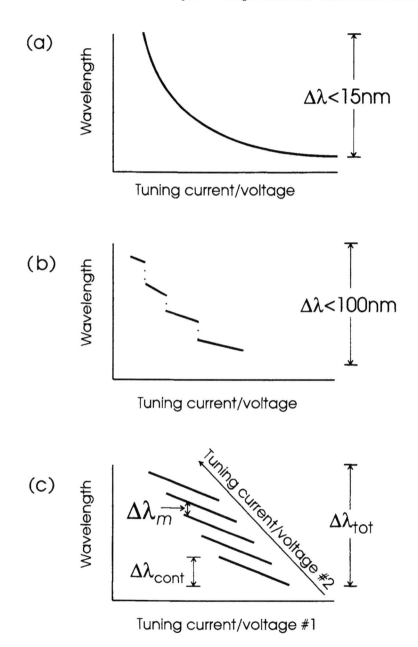

Figure 3.32: Schematic representation of (a) continuous (b) discontinuous and (c) quasi-continuous tuning modes.

nm at 1.5 μm wavelength. Owing to the mode jumps, it is extremely important in practice to ensure that the set of wavelengths needed for the application can be accessed by the laser.

A third principal tuning technique possible with certain devices is the so-called quasi-continuous tuning mode, as displayed in Fig. 3.32c. In these lasers two control currents (or voltages) are used, one of which allows the continuous tuning in each longitudinal mode over a small wavelength range $\Delta\lambda_{cont}$, while the other one enables the discontinuous tuning over the large wavelength range $\Delta\lambda_{tot}$ via longitudinal mode changes. If the continuous tuning range exceeds the longitudinal mode spacing $\Delta\lambda_m$, as shown in Fig. 3.32c, overlapping continuous tuning ranges are obtained, yielding a complete coverage of the entire tuning range $\Delta\lambda_{tot}$. However, an extremely precise and sophisticated device control is required in order to avoid instabilities or mode jumps by operating each near the center of the continuous tuning ranges; this may be done, *e.g.*, by means of a look-ahead table describing the influences of the various control currents and temperature on the laser wavelength. Nevertheless, even a slight degradation of the device parameters can be sufficient to make a readjustment of the wavelength control necessary.

In the discontinuous tuning mode it is of particular importance to characterize the lasers carefully over a wide range of currents and temperatures in order to know exactly under which conditions mode jumps occur and in which operation regimes stable modes exist. Major challenges resulting from the discontinuous tuning are the single mode operation and the wavelength access to as many wavelengths as possible within the tuning range. Accordingly, completely different laser structures have been developed for tuning ranges above 15 nm.

3.3.3.1 DBR-type laser structures

In the previous sections those types of DBR lasers that exploit either only one single grating or two matched gratings with equal Bragg wavelength for the definition of the laser wavelength (*cf.* Figs. 3.5 and 3.20) have been considered. The corresponding reflection spectra essentially exhibit a strongly dominant reflection peak exactly at the Bragg wavelength (Fig. 3.7). On the other hand, a wide discontinuous tuning can be achieved by a DBR-like laser structure as shown schematically in Fig. 3.33a by using two different Bragg-type grating reflectors, R_a and R_b, located at each end of the laser cavity. If both grating reflectors exhibit comb-reflection

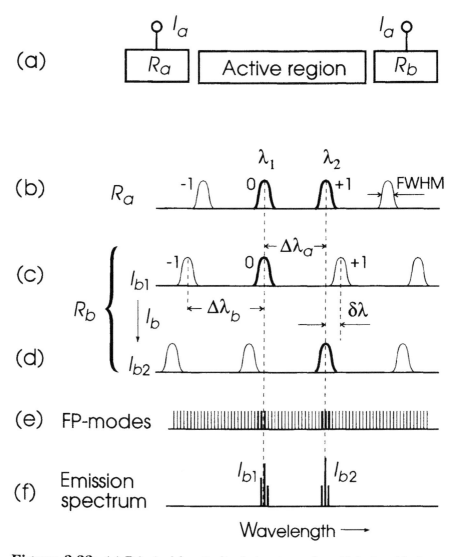

Figure 3.33: (a) Principal longitudinal structure of a widely tunable laser diode using mirrors with comb reflection spectra of different periodicity of the reflection peaks ($\Delta\lambda_a \neq \Delta\lambda_b$); (b) reflection spectrum of R_a; (c) and (d) reflection spectra of R_b for two different values of the bias current I_b; (e) Fabry-Perot comb mode spectrum; and (f) emission spectra for the two different bias currents into reflector R_b.

spectra with different spacings of the reflection peaks, *i.e.*, $\Delta\lambda_a \neq \Delta\lambda_b$ (Fig. 3.33b and c), lasing is possible only around those wavelengths where both spectra simultaneously exhibit a reflection peak, since a closed feedback loop exists only at these "resonance" wavelengths. In close analogy with the (tunable) DBR lasers, lasing occurs exactly at the wavelengths of those Fabry-Perot modes of the total laser cavity, which match closest with the "resonance" wavelengths (Fig. 3.33e and f).

As with the Vernier effect, small shifts of one of the comb-reflection spectra may largely change the "resonance" wavelengths, where both reflectors simultaneously exhibit a reflection peak. In order to avoid lasing at multiple resonances, the spacing $\Delta\lambda_r$ of these multiple resonances must be chosen much larger than the gain bandwidth $\Delta\lambda_{gain}$ of the active region

$$\Delta\lambda_r = \frac{\Delta\lambda_a \cdot \Delta\lambda_b}{|\Delta\lambda_a - \Delta\lambda_b|} >> \Delta\lambda_{gain}. \tag{3.90}$$

To achieve sufficiently small overlap of the reflection peaks of R_a and R_b beneath the resonances, on the other hand, the difference of the periods $\Delta\lambda_a$ and $\Delta\lambda_b$ must be much larger than the full width at half maximum (FWHM) of the reflection peaks

$$|\Delta\lambda_a - \Delta\lambda_b| >> FWHM. \tag{3.91}$$

Considering Fig. 3.33b and c, this means that R_a and R_b are resonant at λ_1 with each of the reflection peaks labelled by 0. The adjacent reflection peaks (labelled ± 1) are mismatched by $|\Delta\lambda_a - \Delta\lambda_b|$, and their overlap should be small enough to prevent laser operation at their wavelengths.

Assuming loss-free Bragg reflectors, the FWHM can be estimated from the DBR theory to be smaller than the spacing between the first zeros of the Bragg reflection r around λ_B. In the example of Fig. 3.7 with $\kappa L_B = 2$, for instance, the zeros are at about ± 0.9 nm, so that the FWHM is less than 1.8 nm. Since these zeros of r correspond to $\Delta\beta = \pm\kappa$, we obtain with Eq. (3.40)

$$FWHM < \frac{\kappa\lambda_B^2}{\pi\bar{n}_{eff,g}}, \tag{3.92}$$

where κ represents the coupling coefficient of reflectors R_a and R_b, which is assumed to be equal for both reflectors and for all reflection peaks. So Eq. (3.91) may be rewritten as

$$|\Delta\lambda_a - \Delta\lambda_b| >> \frac{\kappa\lambda_B^2}{\pi\bar{n}_{eff,g}}. \tag{3.93}$$

For a shift of the reflection spectrum of R_a or R_b by $\delta\lambda = |\Delta\lambda_a - \Delta\lambda_b|$, the resonance wavelengths shift by $\Delta\lambda_b$ or $\Delta\lambda_a$, respectively. If

$$\Delta\lambda_a \approx \Delta\lambda_b >> |\Delta\lambda_a - \Delta\lambda_b|, \tag{3.94}$$

therefore, large changes of the lasing wavelength can be realized with relatively small changes of the reflection spectra. One can define a tuning enhancement factor F that relates the shift of the laser wavelength to the shift of the reflection spectrum. With $\Delta\lambda_a \approx \Delta\lambda_b$, one obtains

$$F \simeq \frac{\Delta\lambda_a}{|\Delta\lambda_a - \Delta\lambda_b|} \qquad \text{(tuning enhancement factor)}. \tag{3.95}$$

Typically, F amounts to about 15 (Jayaraman et al., 1993a).

The spectral shift of the reflection spectra is implemented by carrier injection into the Bragg reflector sections via I_a and/or I_b. Thereby, the reflection spectra can be shifted along the wavelength axis, while the spacing of the comb modes remains almost unaffected. At a certain operation condition, as specified by currents I_a and I_{b1} (Fig. 3.33c), the device lases around wavelength λ_1 (Fig. 3.33f), because only at this wavelength does substantial light reflection occur at both reflectors. Increasing control current I_b slightly to I_{b2} decreases the effective refractive index in R_b, which in turn shifts the reflection peaks of R_b by $\delta\lambda = |\Delta\lambda_a - \Delta\lambda_b|$ toward smaller wavelengths (Fig. 3.33d). As a consequence, the lasing wavelength changes from λ_1 to the longer wavelength λ_2 (Fig. 3.33f), because now a common reflection peak of R_a and R_b appears at λ_2. As can clearly be seen, the change of the lasing wavelength $\Delta\lambda = \Delta\lambda_a$ is much larger than the wavelength shift $\delta\lambda$ induced in R_b. It should be stressed, however, that the tuning behavior is extremely discontinuous and that the single longitudinal mode operation with reasonable SSR is difficult to achieve and must be carefully adjusted. This is because the longitudinal mode spacing of the entire laser cavity is usually much smaller or at most comparable to the bandwidth of the reflector peaks, so that several longitudinal modes may attain equal or similar reflection at the same time.

Comb-reflection spectra can be realized by spatial modulation of a Bragg grating. This modulation may apply either to the amplitude (AM) or to the local spatial frequency (FM) of the gratings. The corresponding devices are denoted as sampled grating (SG) DBR (Jayaraman et al., 1992; Jayaraman et al., 1993b) (AM) and superstructure grating (SSG) DBR (Tohmori et al., 1993; Yoshikuni et al., 1993) (FM) tunable lasers, respectively.

The operation principle of a spatially AM modulated Bragg grating is illustrated in Fig. 3.34. The Bragg grating of period Λ (Fig. 3.34a) is modulated by the sampling function shown in Fig. 3.34b, with sampling period $\Lambda_s = \Lambda_1 + \Lambda_2$ yielding the sampled grating of Fig. 3.34c. The spatial Fourier transform decomposes the sampled grating into a superposition of homogeneous Bragg gratings with different Bragg wavelengths and amplitudes. If the Bragg wavelengths are sufficiently far apart from each other, one can treat each Bragg reflection separately and neglect the influence of the neighboring Fourier components (Jayaraman *et al.*, 1992).

Referring to Eq. (3.25), the z-dependent effective refractive index of the sampled grating in Fig. 3.34c reads

$$n_{eff}(z) = \bar{n}_{eff} + \frac{\Delta n_{eff}}{2} \sin(k_g z) \tag{3.96}$$

$$\cdot \begin{cases} 0 & \text{for } n\Lambda_s + \Lambda_1 \;\; \leq z \leq (n+1)\Lambda_s \\ 1 & \text{for } n\Lambda_s \quad\quad\quad \leq z \leq n\Lambda_s + \Lambda_1 \end{cases},$$

where n is an arbitrary integer. Writing $n_{eff}(z)$ as a Fourier sum,

$$n_{eff}(z) = \bar{n}_{eff} + \sum_{i=0,\pm 1,\pm 2,\dots} \frac{\Delta n_{eff}(i)}{2} \sin\left(k_g + \frac{2\pi i}{\Lambda_s}\right)z, \tag{3.97}$$

yields the Fourier components of Δn_{eff}:

$$\Delta n_{eff}(i) = \Delta n_{eff} \frac{2}{\Lambda_s} \int_0^{\Lambda_1} \sin k_g z \sin\left(k_g + \frac{2\pi i}{\Lambda_s}\right) z \, dz \tag{3.98}$$

$$\tag{3.99}$$

$$= \Delta n_{eff} \frac{\Lambda_1}{\Lambda_s} \left\{ \frac{\sin 2\pi i \frac{\Lambda_1}{\Lambda_s}}{2\pi i \frac{\Lambda_1}{\Lambda_s}} - \frac{\sin(2k_g\Lambda_1 + 2\pi i \frac{\Lambda_1}{\Lambda_s})}{2k_g\Lambda_1 + 2\pi i \frac{\Lambda_1}{\Lambda_s}} \right\}. \tag{3.100}$$

The second expression in the braces can be neglected with respect to the first one if $2k_g >> 2\pi |i|/\Lambda_s$, which with Eq. (3.26) reads

$$|i| << 2\frac{\Lambda_s}{\Lambda}. \tag{3.101}$$

For typical sampling and Bragg grating periods of the order 10 μm and 0.2 μm, respectively, this approximation usually applies well for up to several ten Fourier components.

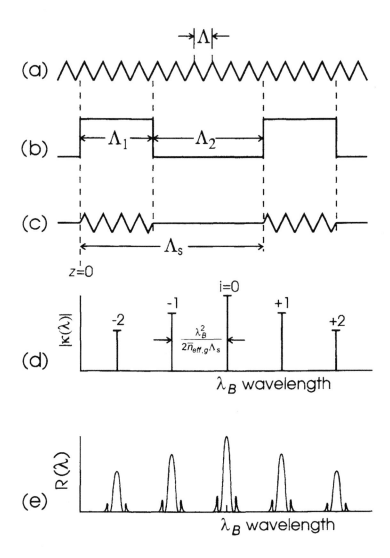

Figure 3.34: Realization of comb reflection spectrum in the sampled grating (SG) laser: The Bragg grating (a) is longitudinally (amplitude) modulated by the sampling function (b) yielding the sampled grating (c). The decomposition of the sampled grating into a Fourier series yields a superposition of a set of simple Bragg gratings with equally spaced Bragg wavelengths and different amplitudes. The resulting coupling coefficients are shown in (d). Each Fourier component can be considered to yield its independent reflection peak, so that a comb reflection spectrum results (e).

With Eqs. (3.100) and (3.101) and the definition of the coupling coefficient (Eq. 3.46), the coupling coefficient of the ith Fourier component can hence be expressed as

$$\kappa_i = \kappa \frac{\Lambda_1}{\Lambda_s} \frac{\sin 2\pi i \frac{\Lambda_1}{\Lambda_s}}{2\pi i \frac{\Lambda_1}{\Lambda_s}}, \qquad (3.102)$$

where κ denotes the coupling coefficient of the unsampled grating. Principally, the κ_is from Eq. (3.102) are positive or negative real numbers. For optimum laser performance, the variations of the κ_is should be as small as possible, so that usually only the components with almost constant positive coupling coefficient placed symmetrically around λ_B are of practical interest.

Referring to the DBR reflection (Eq. 3.59), the peak power reflection of the ith Fourier component of the SG reads

$$r_i = |\tanh(\kappa_i L_{SG})|^2 \qquad (3.103)$$

$$= \tanh^2(|\kappa_i| L_{SG}) \qquad \text{for real } \kappa_i, \qquad (3.104)$$

where L_{SG} is the length of the SG. The Bragg wavelength of the ith component is $\lambda_i = \lambda_B + i\Delta\lambda_s$, and the wavelength spacing between the reflection peaks amounts to $\Delta\lambda_s = \lambda_B^2/2\bar{n}_{eff,g}\Lambda_s$.

The relevant spectrum of $|\kappa_i|$ and the resulting comb-reflection spectrum are shown schematically in Fig. 3.34d and e. Note that the reflection peaks exhibit different magnitudes, with the reflection at λ_B being the largest. According to Eq. (3.102), a small Λ_1/Λ_s ratio reduces the variations of $|\kappa_i|$; however, at the same time all κ_i become smaller by the factor Λ_1/Λ_s. A detailed theoretical treatment and design of the SG tunable laser diodes can be found in Jayaraman *et al.* (1993a).

Likewise a comb-reflection spectrum can be accomplished by an FM-modulated Bragg grating, as shown in Fig. 3.35. Here the instantaneous spatial frequency $2\pi/\Lambda$ is periodically varied; typically the instantaneous spatial frequency is linearly chirped as shown in Fig. 3.35b. The resulting so-called superstructure grating (SSG) (Tohmori *et al.*, 1992) is displayed in Fig. 3.35c. Again, the Fourier analysis yields a decomposition of the SSG as a superposition of unmodulated Bragg gratings, and the $|\kappa|$-spectrum is shown in Fig. 3.35d. The resulting power reflection spectrum is plotted in Fig. 3.35e. As can be seen, the Fourier components of this FM spectrum are significantly different from each other, and the maximum reflection peak may occur at a wavelength different from λ_B, which may degrade the

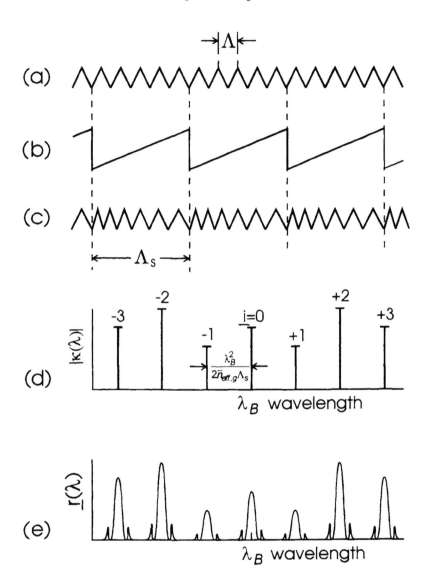

Figure 3.35: Realization of a comb reflection spectrum in the super structure grating (SSG) laser: The period of the Bragg grating (a) is longitudinally (frequency) modulated, *e.g.*, by a ramp function (b), yielding a spatially FM-modulated grating (c). From the spatial Fourier analysis, a set of simple Bragg gratings with equally spaced Bragg wavelengths but different amplitudes results. The corresponding set of coupling coefficients (d) produces the comb reflection spectrum (e).

performance of the corresponding tunable lasers. An improved reflection spectrum with less variation of the reflection peaks was therefore developed (Ishii *et al.*, 1993b; Ishii *et al.*, 1995) using multiple phase shifts within the grating. A theoretical treatment of the SSG reflector was given by Ishii *et al.* (1993a).

While tuning ranges up to 62 nm in cw-operation (Jayaraman *et al.*, 1994) with side-mode suppression above 30 dB were reported with the SG tunable lasers, maximum wavelength coverage has been achieved so far with the SSG lasers. The principal longitudinal section of an SSG laser diode (Tohmori *et al.*, 1993; Ishii *et al.*, 1994b) is shown in Fig. 3.36. Besides the two electronically controllable comb reflectors, a tunable phase shifter has been added in order to enable the quasi-continuous tuning (Ishii *et al.*, 1994a). The typical tuning behavior, as shown in Fig. 3.37, is obtained only by changing either reflector control current I_a or I_b. While the total wavelength coverage of this device is as large as 95 nm, the number of accessible wavelengths is rather small ($\approx$ 11). Maximum wavelength coverage with a comparably small number of accessible wavelengths was as large as 103 nm (Tohmori *et al.*, 1993). On the other hand, in the more useful quasi-continuous tuning mode tuning ranges of 34–62 nm with SSR of 30 dB were reported for SSG lasers from two research groups (Öberg *et al.*, 1995; Ishii *et al.*, 1994a; Ishii *et al.*, 1996).

3.3.3.2 Wavelength tunable Y-laser

Multibranch reflectors are well known as wavelength-selective elements (Miller, 1989). Integrating these structures monolithically into the cavity

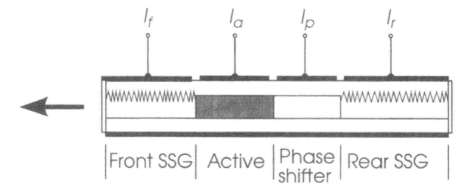

Figure 3.36: Schematic longitudinal section of an experimental superstructure grating (SSG) laser.

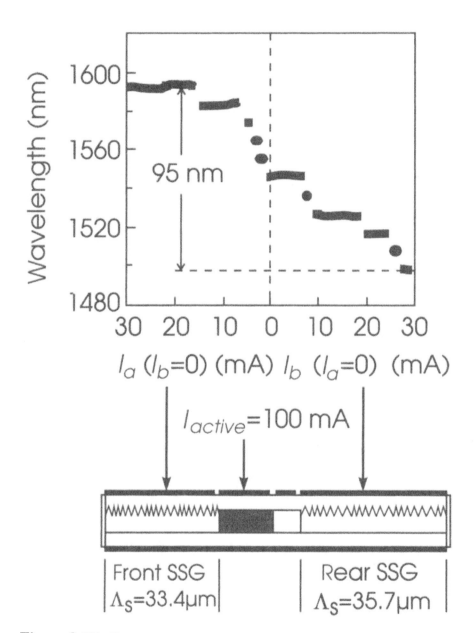

Figure 3.37: Experimental wavelength tuning characteristics of a superstructure (SSG) laser obtained by tuning one of the reflector control currents (I_f or I_r) at a time. (After Ishii *et al.*, 1994b).

of a laser diode and applying an electronic refractive index control, such as the free carrier plasma effect, enables the realization of widely tunable laser diodes. Up to now the most usual structure is the two-branch Y-laser (Schilling *et al.*, 1990; Kuznetsov *et al.*, 1992), particularly the asymmetric Y-laser with two branches of different lengths (Kuznetsov *et al.*, 1992; Dütting *et al.*, 1994). A schematic top view of such a Y-laser is shown in Fig. 3.38. The device operation is illustrated in Fig. 3.39. For the tuning mechanism, the wavelength selectivity of the mode gain characteristic $g(\lambda)$ plays an important role. The principal shape of the mode gain is plotted in Fig. 3.39a. The wavelength selection and tuning essentially

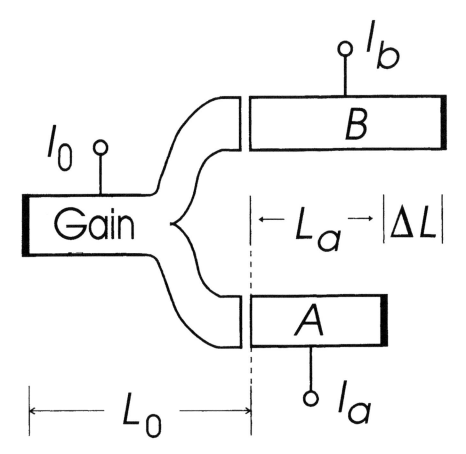

Figure 3.38: Schematic top view of a wavelength tunable Y-laser with asymmetric branches.

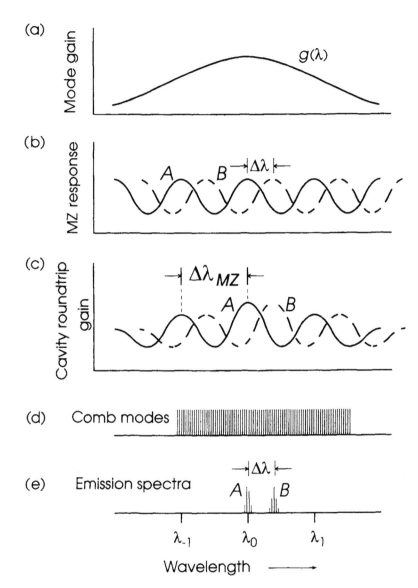

Figure 3.39: Illustration of Y-laser wavelength filtering and tuning: (a) mode gain characteristic; (b) filtering characteristics of Mach-Zehnder interferometer; (c) resulting filtering characteristics of asymmetric Y-laser cavity; (d) Fabry-Perot comb mode spectrum; and (e) resulting emission spectra. The tuning function is indicated by the solid and broken curves A and B, representing two different biasing conditions with variation of ΔL_{eff}.

depend on the Mach-Zehnder (MZ) wavelength filtering caused by the interference of the waves transmitted and reflected in both arms. The filtering effect of the MZ interferometer without tuning is shown in Fig. 3.39b as solid curve A, displaying a $\cos^2$-type function with multiple resonances. The filtering peaks appear at a set of wavelengths λ_i satisfying the relation

$$k_0(\lambda_i)n_{eff}\Delta L = i\pi \quad i \in N,\tag{3.105}$$

yielding

$$\lambda_i = \frac{2n_{eff}\Delta L}{i},\tag{3.106}$$

and the wavelength spacing $\Delta\lambda_{MZ}$ of the filtering peaks is inversely proportional to the length difference ΔL (cf. Fig. 3.38) of arms A and B:

$$\Delta\lambda_{MZ} = \frac{\lambda_0^2}{2n_{eff,g}\Delta L}.\tag{3.107}$$

Equations (3.106) and (3.107) apply for the case of equal effective refractive index n_{eff} in both arms. In case of different effective refractive indexes n_a (in branch A) and n_b (in branch B), the product $n_{eff}\Delta L$ must be replaced by

$$n_{eff}\Delta L \rightarrow n_bL_b - n_aL_a,\tag{3.108}$$

where an effective optical length difference between the two arms can be defined as

$$\Delta L_{eff} = n_bL_b - n_aL_a.\tag{3.109}$$

The combination of the good far selection of the mode gain and the good near selection (but ambiguity) of the MZ filtering results in the composite filtering characteristic shown in Fig. 3.39c as solid curve A. As can be seen, an unambiguous wavelength selection occurs, which leads to laser operation around λ_0. The number of lasing modes, the SSR, and the exact laser wavelength are determined by the relative position of the filtering curve with respect to the Fabry-Perot comb-mode spectrum, which is shown in Fig. 3.39d. It should be stressed that no relationship or even synchronization exists between the comb-mode spectrum and the filtering characteristic. The resulting emission spectrum of the Y-laser peaks around λ_0, as shown in Fig. 3.39e (mode group A). Obviously, the single longitudinal mode operation is not easily achieved and requires the most accurate control of all relevant laser parameters, including temperature.

The tuning of the laser wavelength is done by changing the control currents I_a and I_b, which changes the effective refractive indexes in both arms. With Eqs. (3.106), (3.108), and (3.109), the wavelength shift of the filtering peaks and, consequently, the laser wavelength tuning is obtained as

$$\Delta \lambda = \lambda_0 \frac{\delta L_{eff}}{\Delta L_{eff}} = \lambda_0 \frac{\delta n_a L_a - \delta n_b L_b}{n_a L_a - n_b L_b} , \qquad (3.110)$$

where δL_{eff} is the change of ΔL_{eff} caused by the effective refractive index changes δn_a and δn_b of n_a and n_b, respectively.

The shifted Mach-Zehnder filtering characteristic and cavity round trip gain curve are indicated schematically as broken curves B in Fig. 3.39b and c. Assuming the effective length changes to be large enough, the tuning range $\Delta \lambda$ is limited by the spacing of the filtering peaks, *i.e.*, $\Delta \lambda \leq \Delta \lambda_{MZ}$. The resulting laser emission occurs now in mode group B of Fig. 3.39e, which is shifted by (approximately) $\Delta \lambda$.

Because not the absolute but the relative changes of the effective length difference are essential for the tuning effect, optimum tuning range can be obtained by using almost symmetrical Y-lasers with equally long branches (*cf.* Eq. 3.110). In this design, however, the near selection is degraded, because the larger spacing of the filtering peaks inevitably implies a smaller curvature of the $\cos^2$-function around λ_0. Thus the single-mode operation and tuning range cannot be optimized independently. As with the SSG and SG, the tuning mechanism of the Y-laser is based on the Vernier effect, exploiting the resonance condition of the two coupled reflector arms.

Practical Y-lasers usually consist of an all-active waveguide structure with a typical length around 1 mm. By the separation of the top p-contact into three to four sections, the two interferometer arms A and B can be biased independently. Experimentally, up to 51 nm discontinuous tuning was reported for InGaAsP/InP Y-lasers at 1.55 μm wavelength (Kuznetsov *et al.*, 1992; Schilling *et al.*, 1992). An experimental tuning characteristic of an asymmetric Y-laser with $\Delta L = 62$ μm is shown in Fig. 3.40 (Kuznetsov *et al.*, 1992). The section lengths were $L_0 = 300$ μm and $L_a = 1201$ μm, so that the Fabry-Perot mode spacing is about 0.2 nm and $\Delta \lambda_{MZ} \approx 5$ nm. It should be noted that in the figure the wavelength versus current relation

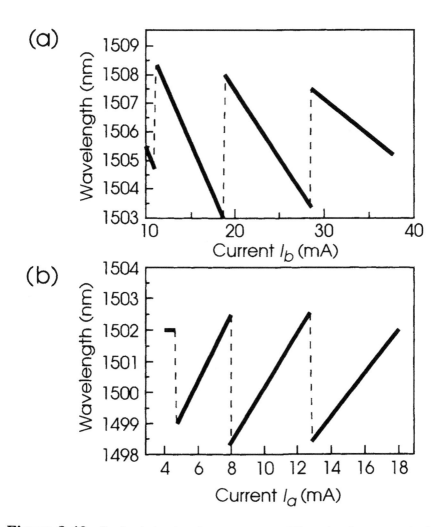

Figure 3.40: Tuning behavior of an asymmetric Y-laser by changing only the current in the longer (a) and shorter (b) branch. Tuning occurs in numerous short wavelength jumps along the solid lines. (After Kuznetsov *et al.*, © 1992 IEEE).

is drawn as solid line, while in practice it consists of numerous small wavelength jumps (approximately 18–20 jumps between 1508 and 1503 nm). The free spectral range $\Delta\lambda_{MZ}$ of 5 nm corresponds well to the estimation, and the tuning behavior appears rather regular. Nevertheless, the 5-nm wavelength interval comprises 25 Fabry-Perot modes, which obviously were not completely addressable.

Using Y-lasers as tunable wavelength converters, a conversion of 2.5 Gb/s data streams has been demonstrated (Schilling *et al.*, 1992).

Differing from the DBR-type tunable SSG and SG lasers, the Y-laser exhibits a not unambiguous internal wavelength reference such as a Bragg grating but relies on the multiple resonances of the (closely spaced) Fabry-Perot modes. Accordingly, the device characterization as well as the wavelength control effort appear more critical.

Improved performance is theoretically expected from multiple branch lasers exhibiting three or more branches (Miller, 1989). The first experimental approaches with three-branch lasers consisting of two cascaded Y-junctions (Y3-laser) showed promising performance with respect to the number of accessible wavelength channels (Kuznetsov *et al.*, 1994; Kuznetsov *et al.*, 1993); however, further progress is to be expected from these approaches.

3.3.3.3   Codirectionally mode-coupled lasers

A further principal approach for a wide wavelength tuning is the application of laser cavities in which the lasing mode consists of a superposition of codirectionally (or forward) coupled waveguide modes. In contrast to the well-established DFB and DBR lasers that exploit the contradirectional coupling between the forward and backward propagating wave of a single-mode waveguide, the codirectionally coupled devices require a twin-waveguide structure with two guided waveguide modes.

The phase-matching condition for the codirectional coupling can be understood from the wavevector diagram shown in Fig. 3.41a. As shown schematically in Fig. 3.41b, the two codirectionally propagating waves R and S, with effective refractive indexes n_R and n_S, respectively, are coupled together by a periodic longitudinal perturbation of the waveguide. As a consequence of the same propagation direction, wavevectors $k_R = k_0 n_R$ and $k_S = k_0 n_S$ exhibit equal signs, and the size of the grating vector $|k_g|$ $= |k_R - k_S|$ is much smaller than for the contradirectional coupling, where $k_g = 2k_R$. Typically k_R and k_S differ by only about 3%, so that k_g is 1–2 orders of magnitude smaller than in case of the DFB/DBR lasers. We designate the modes such that mode R is the fundamental one, so that it always exhibits the largest effective refractive index ($n_R > n_S$). Solving the phase-matching condition $k_R = k_S + k_g$ for the phase matching wavelength λ_0, one obtains

$$\lambda_0 = \Lambda \cdot (n_R - n_S) = \Lambda \cdot \Delta n_{RS}, \qquad (3.111)$$

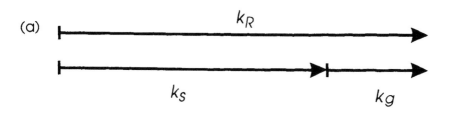

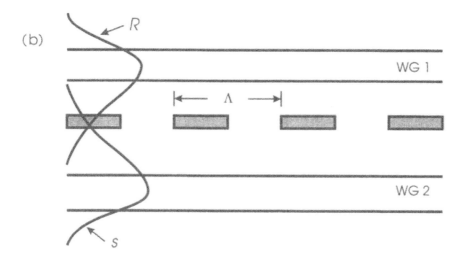

Figure 3.41: (a) Wavevector diagram for a codirectional coupling grating. k_R and k_S are the wavevectors of the coupled modes R and S, respectively, and k_g is the grating wavevector. (b) Schematic longitudinal section of a twin-waveguide with periodic perturbation for codirectional coupling between modes R and S.

where Λ is the grating pitch and Δn_{RS} is the effective refractive index difference between modes R and S. Tuning is done by changing Δn_{RS} via changes of n_R (δn_R) or n_S (δn_S). Denoting the change of Δn_{RS} by $\delta n_{RS} = \delta n_R - \delta n_S$, the tuning range reads

$$\frac{\Delta\lambda}{\lambda_0} = \left|\frac{\delta n_{RS}}{\Delta n_{RS}}\right|. \tag{3.112}$$

At equal index changes, the tuning range can be appreciably larger than for the tunable DBR laser (Eq. 3.75), because the denominator of the right expression in Eq. (3.112) (Δn_{RS}) is a much smaller number as compared with both effective indexes n_R and n_S.

The wavelength tuning method proposed above requires the ability to selectively change the effective refractive index of one single waveguide mode. Therefore a waveguide structure must be chosen in which the two codirectionally travelling modes exhibit a clear spatial separation, enabling the selective effective refractive index change by local index variations. Taking into account the dispersion of Δn_{RS}, the tuning range reads

$$\Delta\lambda = \left|\frac{\delta n \Delta\Gamma_t \lambda_0}{\Delta n_{RS}\beta}\right| = \left|\frac{\delta n \Delta\Gamma_t \Lambda}{\beta}\right|, \tag{3.113}$$

where δn is the refractive index change in the tuning region, $\Delta\Gamma_t$ is the difference of the optical confinement factors of modes R and S in the tuning region so that

$$\delta n_{RS} = \delta n \Delta\Gamma_t/\beta, \tag{3.114}$$

and

$$\beta = 1 - \Lambda\frac{\partial\Delta n_{RS}}{\partial\lambda}\bigg|_{\lambda=\lambda_0} \tag{3.115}$$

measures the dispersion of Δn_{RS}. The most common waveguide configuration used so far for the codirectionally coupled filter devices is the twin waveguide with periodic perturbations (Alferness *et al.*, 1992; Alferness *et al.*, 1989; Sakata *et al.*, 1992; Illek *et al.*, 1991b), the longitudinal section of which is shown schematically in Fig. 3.41b. Due to the vertical integration, this structure may be designated as *Vertical Coupler Filter* (VCF) (Alferness *et al.*, 1992).

One clearly sees in the figure that the two modes are confined preferentially either in WG 1 (mode R) or WG 2 (mode S), and the mode coupling is accomplished by the periodic index perturbations of period Λ. At the perturbations, both modes should exhibit significant amplitude because the coupling coefficient is proportional to the geometrical average of the confinement factors of both modes within the perturbation region (Amann and Illek, 1993). As with the previous TTG lasers (Illek *et al.*, 1990b), the tuning function can be established, for instance, by carrier injection into WG 2. Thereby the refractive index of WG 2 and n_S are reduced while n_R

is kept almost constant. As a consequence, both Δn_{RS} and the coupling wavelength λ_0 increase.

A large-range tunable laser diode is achieved by monolithic integration of the VCF with an active gain section and a mode filtering mechanism. The mode filtering function is required to force the laser operation at the coupling wavelength defined by the VCF and to prevent solitary laser operation of mode R or S, respectively (Chuang and Coldren, 1991). The simplest approach for the mode filter represents an optical absorber placed on two opposite arms of the VCF. The corresponding composite laser structure, which essentially consists of a monolithic integration of an Amplifier, a Coupler, and an Absorber (ACA-laser), is shown schematically in Fig. 3.42. Here the optical gain is induced by the gain control current I_a into the amplifying gain section, and the coupling wavelength λ_0 is tuned by the wavelength control current I_t into the VCF.

The mode selection under tuning in this composite cavity structure is illustrated in Fig. 3.43. The phase condition requires that an integer multiple of half waves fits into the laser resonator. A corresponding comb-mode spectrum is plotted schematically in Fig. 3.43a. The wavelength spacing of the modes is inversely proportional to the composite cavity

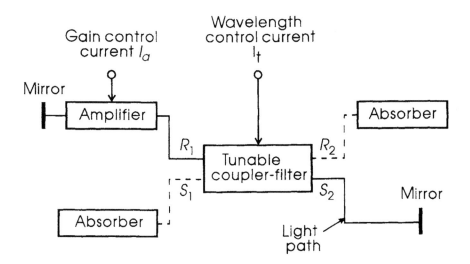

Figure 3.42: Wavelength tunable ACA laser with a tunable vertical codirectional filter (VCF). The optical feedback path is indicated. No feedback occurs along the dashed path.

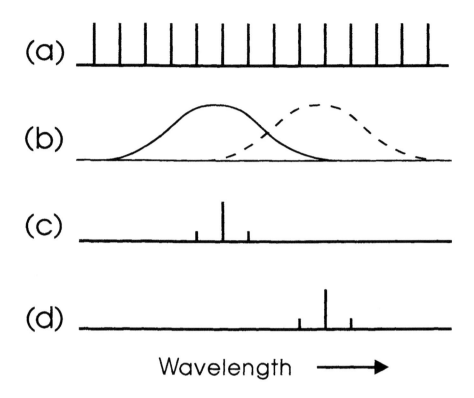

Figure 3.43: Illustration of the tuning mechanism of ACA laser: (a) comb-mode spectrum of the composite laser cavity; (b) filter curves for two different tuning currents and resulting mode spectra (c and d).

length. Besides the phase condition that determines the comb-mode spectrum (Eq. 3.4), the effective optical gain also determines the lasing wavelength. The wavelength dependence of the effective optical gain of the composite structure comprises the wavelength-dependent mode gain of mode R in the amplifier as well as the filter characteristic of the VCF. Resulting effective cavity gain curves for two different tuning currents are displayed schematically in Fig. 3.43b. The resulting emission spectra for the two I_t-values in Fig. 3.43b are displayed in Figs. 3.43c and d. Evidently, the wavelength tuning manifests itself as successive jumps from one cavity mode to the neighboring one as determined by the maximum of the effective cavity gain, and a large SSR requires a careful adjustment of effective gain peak and comb-mode spectrum.

For a VCF length L_c equal to one coupling length, which is typically of the order of 200–800 μm, the fractional filter bandwidth (FWHM) is given approximately as (Alferness *et al.*, 1989)

$$\Delta\lambda_{VCF}(FWHM) \simeq 0.8\lambda_0 \frac{\Lambda}{\beta L_c}. \tag{3.116}$$

The mode selectivity can be estimated by relating the filter bandwidth $\Delta\lambda_{VCF}(FWHM)$ to the mode spacing $\Delta\lambda_m$ (*cf.* Eq. 3.7) as

$$\frac{\Delta\lambda_{VCF}(FWHM)}{\Delta\lambda_m} \simeq 1.6 \frac{n_g^{av}}{\Delta n_{RS}} \frac{L}{\beta L_c}, \tag{3.117}$$

where L stands for the total laser length and the average group effective refractive index n_g^{av} is obtained as

$$n_g^{av} = \frac{1}{2}\{n_{gR}(L_a + L_c/2) + n_{gS}(L_c/2 + L_w)\}, \tag{3.118}$$

with L_a and L_w denoting the lengths of the gain and the window regions (*cf.* Fig. 3.44b), respectively, and n_{gR} and n_{gS} are the group effective refractive indexes of modes R and S.

For very large values of this expression, the spectral output consists essentially of a multimode comb spectrum, whose spectral position can be varied by the tuning. For smaller $\Delta\lambda_{VCF}(FWHM)/\Delta\lambda_m$ ratios, instead, an almost single-mode spectrum may be achieved with distinct mode hops during tuning. Since it seems difficult in practice to adjust λ_0 precisely to a given comb mode, a stable and reproducible single-mode operation will not be easily attained with these devices.

As can be seen from Eq. (3.116), the mode selectivity mainly depends on Δn_{RS} and on the ratio of coupler length L_c and total laser length L. On the other hand, however, the tuning range is also inversely proportional to Δn_{RS} (Eq. 3.112), revealing that mode selectivity and tuning range cannot be optimized independently. Detailed theoretical treatments of the codirectionally mode-coupled tunable lasers can be found in the literature (Heismann and Alferness, 1988; Amann and Illek, 1993; Amann *et al.*, 1995).

Owing to the close relationship to the previous TTG laser, the first codirectionally coupled tunable lasers were based on the $\lambda = 1.5$ μm InGaAsP/InP TTG laser structure (Illek *et al.*, 1991b; Illek *et al.*, 1991a). The device concept of Fig. 3.42 has therefore been realized in the Amplifier Coupler Absorber (ACA) structure, the longitudinal section of which is

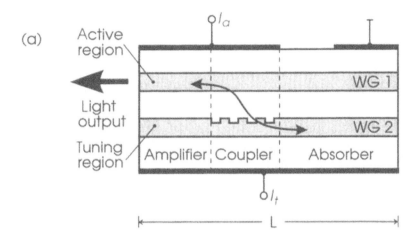

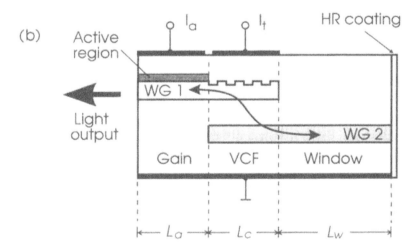

Figure 3.44: Schematic longitudinal view of widely tunable ACA lasers using an integrated vertical coupler filter (VCF). The absorber function is done either by an unbiased (or grounded) active layer (a) or by the window technique (b).

shown in Fig. 3.44a. This structure consists of the TTG modified by applying a two-mode waveguide, as achieved by a thick n-InP separation layer, and by introducing a longitudinal sectioning. The two waveguide layers are coupled vertically, providing the advantage of larger coupling strength than in a laterally integrated configuration, thus making the lasers significantly shorter. The device structure was simplified further by using the active layer (λ_g = 1.57 μm) simultaneously as one of the waveguide layers (WG 1, *cf.* Fig. 3.41b) in the coupler region and as absorbing layer in the absorber region. The latter is accomplished by grounding the corresponding p-electrode. The presence of an amplifying waveguide in the coupler section, on the other hand, may deteriorate the tuning effect at larger optical power because spatial hole burning leads to a power-dependent longitudinally inhomogeneous effective refractive index profile in the coupler. The bottom waveguide layer (WG 2) consists of a transparent λ_g = 1.3 μm layer, which is low n-doped.

The amplifier gain is controlled by current I_a fed into the active layer in both the amplifier and coupler sections. Wavelength tuning is accomplished by current I_t fed into the tuning layer (WG 2) along the entire cavity length. The lengths of the amplifier (L_a), coupler (L_c), and absorber (L_{abs}) were chosen each to be 200 μm so that an overall device length L of only 600 μm results. With this ACA laser structure, a tuning range of around 30 nm was demonstrated, however, with a rather irregular tuning behavior showing a poor wavelength access (Illek *et al.*, 1991b; Illek *et al.*, 1991a).

Another approach to the idealized ACA laser principle, which was presented by Alferness *et al.* (1992), is shown schematically in Fig. 3.44b. In this particular structure, both absorbing regions are replaced by window regions in which the respective waveguide is cut so that the modal power is radiated away so that practically no reflections occur into the waveguides via the end facets.

It should be noted, however, that with the window technique a waveguide discontinuity is introduced at the interfaces to the VCF. As a consequence, both eigenmodes R and S of the VCF become excited by the incoming light, so that a complete coupling from R to S and vice versa can be achieved only with a weakly coupled twin-guide (Huang and Haus, 1989). This in turn implies that relatively long couplers are required, yielding a correspondingly large device length.

Differing from the first ACA laser (Fig. 3.44a), the coupler is comprised of purely passive optical waveguides, providing a better control and stability of the tunable filter function. Improved device performance is also realized by using a quantum well active region. The coupling strength is significantly smaller than with the previous structure, so that an increased device length of 1.5–2.5 mm results (Alferness, 1993; Kim

et al., 1993; Kim *et al.*, 1994). Typical laser lengths range between 1.5 mm
(Kim *et al.*, 1994) and 2.5 mm (Alferness, 1993). High-power conversion
efficiency is achieved by applying a high-reflection coating onto the rear
mirror. Since wavelength tuning can be induced by changing either n_R
or n_S, respectively, the wavelength tuning is conveniently done by carrier
injection into the upper waveguide layer, which changes n_R.

This improved ACA-type laser showed maximum tuning ranges up
to 74 nm (Kim *et al.*, 1994) with an impressive side-mode suppression
ratio between 25 and 34 dB; however, only about 25% of the comb modes
were accessible. Optimizing the laser with respect to wavelength access,
on the other hand, resulted in slightly reduced tuning ranges, as shown
in Fig. 3.45 (Kim *et al.*, 1994). As can be seen, the variation of the tuning
current I_t into the VCF section from 0 to 500 mA yields a wavelength
tuning of about 72 nm. It should be noted that the wavelength versus
tuning current characteristic is not a smooth function (as plotted here for
convenience); instead it consists of numerous small mode and wavelength
jumps. From the 360 comb modes of this 1.5-mm-long laser, about 105

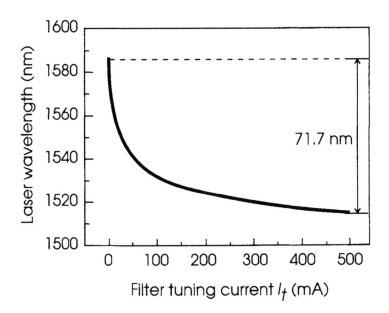

Figure 3.45:   Laser wavelength versus VCF control current I_t for experimental
VCF-type tunable lasers. Tuning along the solid curve is discontinuous with about
100 discrete wavelengths. (After Kim *et al.*, 1994).

were individually addressable within the tuning range of 72 nm. On average, therefore, almost every third comb mode could be selected. Nevertheless, even this percentage represents a performance too poor for many applications. The physical origin for the incomplete wavelength access in this laser type may be the small backward reflections at the perturbations and at the discontinuities at the sections interfaces. These reflections result in multiple mirror effects that irregularly disturb the filter function due to the random mirror phases.

It was pointed out quite recently that the tapering of the waveguide inhomogeneities might effectively deteriorate the backward reflections (by 2–4 orders of magnitude) by simultaneously retaining the forward coupling (Amann *et al.*, 1996).

Various approaches have been presented in the past to optimize the wavelength selectivity of the codirectionally coupled tunable lasers. Among those are the distributed forward-coupled (DFC) laser (Amann *et al.*, 1993; Amann *et al.*, 1995) and the vertical Mach-Zehnder interferometer (VMZ) laser (Amann *et al.*, 1994; Borchert *et al.*, 1994). In both of these devices, the forward coupling is extended over the entire laser axis ($L_c \simeq L$), so that the ratio of VCF filter bandwidth and comb-mode spacing (Eq. 3.117) is reduced. This results in an improved mode selection by the VCF and a larger SSR. As a matter of fact, the VMZ laser achieved an access to about 80% of the comb modes with an SSR>20 dB over a tuning range of 30 nm (Amann *et al.*, 1994).

Improved tuning performance can be expected also by a combination of the SG or SSG structure with the codirectional coupling (Öberg *et al.*, 1993; Rigole *et al.*, 1995). The schematic longitudinal structure of these devices is shown in Fig. 3.46. In this case, the wide tuning of the VCF

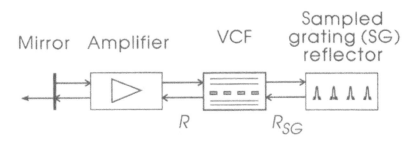

Figure 3.46: Schematic longitudinal section of a widely tunable laser combining a VCF for wide tuning and a sampled grating reflector for high selectivity.

fits well with the strong near selection of the SG (SSG). The laser operation principle is illustrated in Fig. 3.47. Here the top plot shows the VCF filter curves for two tuning currents (curve a and b, respectively). Together with the sharp but multiple reflection peaks R_{SG} of the SG (or SSG)

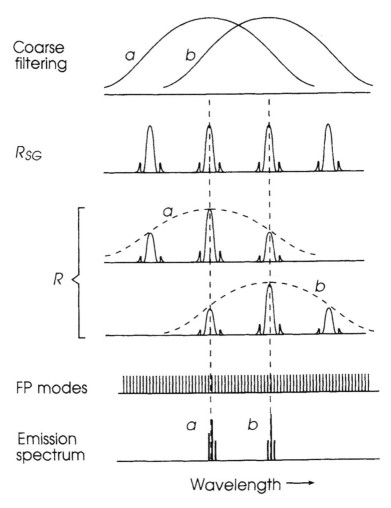

Figure 3.47: Operation principle of the VCF-type laser with sampled grating (SG) reflector: The coarse filtering is done with the VCF, while the fine filtering is due to the SG reflector. Compared with the simple VCF-type laser, the SSR may be improved, but the number of accessible laser modes is reduced because of the large spacing of the comb reflection peaks.

reflector, the effective right-hand side reflection R as seen by the amplifying region (*cf.* Fig. 3.46) clearly selects one wavelength channel. Depending on the comb-mode positions, the lasing spectra sketched in the bottom plot of Fig. 3.47 are observed. Apparently, the improved near selection via the SG (or SSG) leads to a reduced number of well-defined wavelength channels. A representative experimental tuning characteristic of a VCF-SG integrated tunable laser is plotted in Fig. 3.48 (Öberg *et al.*, 1993). As can be seen, a tuning range of 74 nm is achieved with about 16 individually addressable wavelengths. Quite recently, the integration of the VCF with an SSG resulted in a record tuning range of 114 nm (Rigole *et al.*, 1995).

3.3.3.4 Outlook

The present status of electronically tunable monolithic laser diodes strongly differs if narrow (< 15 nm at 1.5 μm wavelength) or widely ($>$

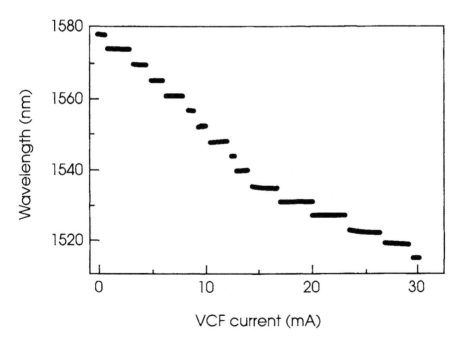

Figure 3.48: Wavelength tuning characteristic of VCF-type laser with integrated SG reflector. (After Öberg *et al.*, © 1993 IEEE).

15 nm at 1.5 μm wavelength) tunable devices are considered. While the development of the former yielded continuous tuning and almost ultimate performance with respect to the tuning range, the widely tunable laser diodes show a discontinuous or at most a quasi-continuous tuning behavior, the performance of which appears not yet sufficient for most applications. This is because the underlying operation principles of the widely tunable laser diodes exploit the slight differences of mode numbers or reflection maxima (Vernier effect); therefore even rather small variations of these parameters occurring, *e.g.*, during the device fabrication, yield large relative changes of the device characteristics, particularly of the laser wavelength. Thus the device characteristics show large scattering, even for devices from the same wafer, making the laser control extremely difficult with respect to the unevitable mode jumps. The successful further development of these novel components and the implementation of a reasonable handling convenience therefore basically require a highly precise and homogeneous fabrication technique with reproducible and stable device parameters; this essentially concerns the crystal growth, for which, *e.g.*, a layer thickness control on the 0.01 μm scale is required.

A common deficiency of all these approaches for an extended tuning range is the lack of a continuous tunability hindering the access to any wavelength within the tuning range. This still limits the practical applicability of these lasers, particularly as the number of longitudinal modes covered by the wavelength tuning is typically of the order of 100. A challenge for the future development is therefore the improvement of the wavelength access, *e.g.*, with laser structures exhibiting an enhanced wavelength selectivity or with quasi-continuously tunable devices.

3.4   Conclusion

The state of the art of single-mode and electronically tunable laser diodes as required in numerous applications in optical fiber communications, sensing, and measurement was reviewed. For most applications, the continuous tuning mode and the simple and unambiguous wavelength control are most essential, so that the recent development has focused mainly on the achievement of large continuous tuning range and simple device handling, *i.e.*, only one control current for the wavelength setting. For tuning ranges below about 10 nm, various types of high-performance (single-mode, continuous tuning, narrow linewidth, easy to handle) devices already exist, based on the technologically well-developed DFB and DBR laser structures. For physical reasons, the maximum continuous tuning range of electronically tunable monolithic diode lasers is limited to about 15 nm, corresponding to a relative range of 1% at 1.55 μm wavelength. Experimentally up to 13 nm continuous tuning ranges have been achieved so far with the TTG laser.

In applications where one can accept the discontinuous tuning mode, novel device concepts have been presented offering discontinuous tuning up to about 100 nm, which almost equals the gain bandwidth of InGaAsP at 1.55 μm. However, the wavelength access is still an issue for these new structures, because typically only about 10 to 30 channels with different spacing can individually be addressed. The essential device parameter influencing the wavelength access is the bandwidth of the filtering curve, so that the present development concentrates particularly on the improvement of the wavelength selectivity.

References

Agrawal, G. P., and Dutta, N. K. (1986) *Long-wavelength semiconductor lasers* New York: Van Nostrand Reinhold.

Alferness, R. C. (1993). *Proc. Opt. Fiber Conf.* San Jose, CA, USA, 11.

Alferness, R. C.; Koch, T. L.; Buhl, L. L.; Storz, F.; Heismann, F.; and Martyak, M. J. R. (1989). *Appl. Phys. Lett.* **55**, 2011.

Alferness, R. C.; Koren, U.; Buhl, L. L.; Miller, B. I.; Young, M. G.; Koch, T. L.; Raybon, G.; and Burrus, C. A. (1992). *Proc. Opt. Fiber Conf.* San Jose, CA, USA, 321.

Amann, M.-C. (1990). *Electron. Lett.* **26**, 569.

Amann, M.-C. (1994). *Techn. Dig. 5ᵗʰ Optoelectron. Conf.*, Makuhari Messe, Japan, 208.

Amann, M.-C. (1995). *Optoelectronics—Dev. Technol.* **10**, 27.

Amann, M.-C., and Borchert, B. (1992). *Archiv für Elektronik und Übertragungstechnik—Int. J. Electron. Commun.* **46**, 63.

Amann, M.-C.; Borchert, B.; Illek, S.; and Steffens, W. (1996). *Electron. Lett.* **32**, 221.

Amann, M.-C.; Borchert, B.; Illek, S.; and Veuhoff, E. (1994). *Conf. Dig. 14ᵗʰ IEEE Semicond. Laser Conf.* Maui, Hawaii, USA, PD12.

Amann, M.-C.; Borchert, B.; Illek, S.; and Wolf, T. (1993). *Electron. Lett.* **29**, 793.

Amann, M.-C.; Borchert, B.; Illek, S.; and Wolf, T. (1995). *IEEE J. Sel. Topics Quantum Electron.* **1**, 387.

Amann, M.-C., and Illek, S. (1993). *IEEE J. Lightwave Technol.* **11**, 1168.

Amann, M.-C.; Illek, S.; and Lang, H. (1991). *Electron. Lett.* **27**, 531.

Amann, M.-C.; Illek, S.; Schanen, C.; and Thulke, W. (1989). *Appl. Phys. Lett.* **54**, 2532.

Amann, M.-C., and Schimpe, R. (1990). *Electron. Lett.* **26**, 279.

Amann, M.-C., and Thulke, W. (1990). *IEEE J. Sel. Areas Commun.* **8**, 1169.

Asada, M.; Kameyama, A.; and Suematsu, Y. (1984). *IEEE J. Quantum Electron.* **20**, 745.

Bandelow, U.; Wünsche, H. J.; and Wenzel, H. (1993). *IEEE Photon. Technol. Lett.* **5**, 1176.

Baumann, G. G., Stegmüller, B., and Amann, M.-C. (1988). *Siemens Forsch.-und Entwickl.- Berichte* **17**, 264.

Bennett, B. R.; Soref, R. A.; and Del Alamo, J. A. (1990). *IEEE J. Quantum Electron.* **26**, 113.

Borchert, B.; David, K.; Stegmüller, B.; Gessner, R.; Beschorner, M.; Sacher, D.; and Franz, G. (1991). *IEEE Photon. Technol. Lett* **3**, 955.

Borchert, B.; Illek, S.; Wolf, T.; Rieger, J.; and Amann, M.-C. (1994). *Electron. Lett.* **30**, 2047.

Braagaard, C.; Mikkelsen, B.; Durhuus, T.; and Stubkjaer, K. E. (1994). *IEEE Photon. Technol. Lett.* **6**, 694.

Broberg, B., Koyama, F., Tohmori, Y., and Suematsu, Y. (1984). *Electron. Lett.* **20**, 692.

Buus, J. (1986). *IEE Proc., Pt. J* **133**, 163.

Buus, J. (1991) *Single frequency semiconductor lasers.* Bellingham: SPIE.

Casey, H. C. Jr., and Carter, P. L. (1984). *Appl. Phys. Lett.* **44**, 82.

Chawki, M. J.; Auffret, A.; and Berthou, L. (1990). *J. Optics. Commun.* **11**, 98.

Childs, G. N.; Brans, S.; and Adams, R. A. (1986). *Semicond. Sci. Technol.* **1**, 116.

Chuang, Z. M., and Coldren, L. A. (1991). *Techn. Dig. LEOS Ann. Meet.* San Jose, CA, USA, 77.

Coldren, L. A., and Corzine, S. W. (1987). *IEEE J. Quantum Electron.* **23**, 903.

Delorme, F.; Gambini, G.; Puleo, M.; and Slempkes, S. (1993). *Electron. Lett.* **29**, 41.

Dieckmann, A., and Amann, M.-C. (1995). *Opt. Engineering* **34**, 896.

Dopheide, D.; Faber, M.; Reim, G.; and Taux, G. (1988). *Experiments in Fluids* **6**, 289.

dos Santos Ferreira, M. F.; Ferreira da Rocha, J. R.; and de Lemos Pinto, J. (1992). *IEEE J. Quantum Electron.* **28**, 833.

Durhuus, T.; Pedersen, R. J. S.; Mikkelsen, B.; Stubkjaer, K. E.; Öberg, M.; and Nilsson, S. (1993). *IEEE Photon. Technol. Lett.* **5**, 86.

Dutta, N. K., Napholtz, S. G., Piccirilli, A. B., and Przybylek, G. (1986). *Appl. Phys. Lett.* **48**, 1419.

Dutta, N. K.; Temkin, H.; Tanbun-Ek, T.; and Logan, R. (1990). *Appl. Phys. Lett.* **57**, 1390.

Dütting, K.; Hildebrand, O.; Baums, D.; Idler, W.; Schilling, M.; and Wünstel, K. (1994). *IEEE J. Quantum Electron.* **30**, 654.

Ebberg, A., and Noé, R. (1990). *Electron. Lett.* **26**, 2009.

Economou, G.; Youngquist, R. C.; and Davies, D. E. N. (1986). *IEEE J. Lightwave Technol.* **4**, 1601.

Favre, F., and Le Guen, D. (1991). *Electron. Lett.* **27**, 183.

Furuya, F. (1985). *Electron. Lett.* **21**, 200.

Hamada, M.; Yamamoto, E.; Suda, K.; Nogiwa, S.; and Oki, T. (1991). *Jpn. J. Appl. Phys.* **31**, L1552.

Heismann, F., and Alferness, R. C. (1988). *IEEE J. Quantum Electron.* **24**, 83.

Henry, C. H. (1982). *IEEE J. Quantum Electron.* **18**, 259.

Huang, K.-Y., and Carter, G. M. (1994). *IEEE Photon. Technol. Lett.* **6**, 1466.

Huang, W.-P., and Haus, H. A. (1989). *IEEE J. Lightwave Technol.* **7**, 920.

Illek, S.; Thulke, W.; and Amann, M.-C. (1991a). *Electron. Lett.* **27**, 2207.

Illek, S.; Thulke, W.; Borchert, B.; and Amann, M.-C. (1991b). *Proc. 17th Europ. Conf. Opt. Commun. (Part 3: Post-Deadline Papers)* Paris, France, 21.

Illek, S.; Thulke, W.; Schanen, C.; Drögemüller, K.; and Amann, M.-C. (1990a). *Conf. Dig. 12th Internat. Semicond. Laser Conf.* Davos, Switzerland, 60.

Illek, S.; Thulke, W.; Schanen, C.; Lang, H.; and Amann, M.-C. (1990b). *Electron. Lett.* **26**, 46.

Ishida, O.; Tada, Y.; and Ishii, H. (1994). *Electron. Lett.* **30**, 241.

Ishida, O.; Toba, H.; Tohmori, Y.; and Oe, K. (1989). *Electron. Lett.* **25**, 703.

Ishii, H.; Kano, F.; Tohmori, Y.; Kondo, Y.; Tamamura, T.; and Yoshikuni, Y. (1994a). *Electron. Lett.* **30**, 1134.

Ishii, H.; Kano, F.; Tohmori, Y.; Kondo, Y.; Tamamura, T.; and Yoshikuni, Y. (1995). *IEEE J. Sel. Topics Quantum Electron.* **1**, 401.

Ishii, H.; Tanobe, H.; Kano, F.; Tohmori, Y.; Kondo, Y.; and Yoshikuni, Y. (1996). *Electron. Lett.* **32**, 454.

Ishii, H.; Tohmori, Y.; Tamamura, T.; and Yoshikuni, Y. (1993a). *IEEE Photon. Technol. Lett.* **5**, 393.

Ishii, H.; Tohmori, Y.; Yamamoto, M.; Tamamura, T.; and Yoshikuni, Y. (1994b). *Electron. Lett.* **30**, 1141.

Ishii, H.; Tohmori, Y.; Yoshikuni, Y.; Tamamura, T.; and Kondo, Y. (1993b). *IEEE Photon. Technol. Lett.* **5**, 613.

Jayaraman, V.; Chuang, Z.-M.; and Coldren, L. A. (1993a). *IEEE J. Quantum Electron.* **29**, 1824.

Jayaraman, V.; Cohen, D. A.; and Coldren, L. A. (1992). *Appl. Phys. Lett.* **60**, 2321.

Jayaraman, V., Heimbuch, M. E., Coldren, L. A., and DenBaars, S. P. (1994). *Electron. Lett.* **30**, 1492.

Jayaraman, V.; Mathur, A.; Coldren, L. A.; and Dapkus, P. D. (1993b). *IEEE Photon. Technol. Lett.* **5**, 489.

Kano, F., Tohmori, Y., Kondo, Y., Nakao, M., Fukuda, M., and Oe, K. (1989). *Electron. Lett.* **25**, 709.

Kano, F.; Yamanaka, T.; Yamamoto, N.; Yoshikuni, Y.; Mawatari, H.; Tohmori, Y.; Yamamoto, M.; and Yokoyama, K. (1993). *IEEE J. Quantum Electron.* **29**, 1553.

Kim, I.; Alferness, R. C.; Buhl, L. L.; Koren, U.; Miller, B. I.; Newkirk, M. A.; Young, M. G.; Koch, T. L.; Raybon, G.; and Burrus, C. A. (1993). *Electron. Lett.* **29**, 664.

Kim, I.; Alferness, R. C.; Koren, U.; Buhl, L. L.; Miller, B. I.; Young, M. G.; Chien, M. D.; Koch, T. L.; Presby, H. M.; Raybon, G.; and Burrus, C. A. (1994). *Appl. Phys. Lett.* **64**, 2764.

Kobayashi, K. and Mito, I. (1988). *IEEE J. Lightwave Technol.* **6**, 1623.

Koch, T. L., and Koren, U. (1990). *IEEE J. Lightwave Technol.* **8**, 274.

Koch, T. L.; Koren, U.; Gnall, R. P.; Burrus, C. A.; and Miller, B. I. (1988a). *Electron. Lett.* **24**, 1431.

Koch, T. L.; Koren, U.; and Miller, B. I. (1988b). *Appl. Phys. Lett.* **53**, 1036.

Koelink, M. H.; Slot, M.; de Mul, F. F. M.; Greve, J.; Graaff, R.; Dassel, A. C. M.; and Aarnoudse, J. G. (1992). *Appl. Optics* **31**, 3401.

Kogelnik, H., and Shank, C. V. (1972). *J. Appl. Phys.* **43**, 2327.

Kojima, K., and Kyuma, K. (1990). *Semicond. Sci. Technol.* **5**, 481.

Kojima, K.; Kyuma, K.; and Nakayama, T. (1985). *IEEE J. Lightwave Technol.* **3**, 1048.

Kotaki, Y., and Ishikawa, H. (1989). *IEEE J. Quantum Electron.* **25**, 1340.

Kotaki, Y., and Ishikawa, H. (1991). *IEE Proc. Pt. J.* **138**, 171.

Kotaki, Y.; Matsuda, M.; Ishikawa, H.; and Imai, H. (1988). *Electron. Lett.* **24**, 503.

Kotaki, Y.; Ogita, S.; Matsuda, M.; Kuwahara, Y.; and Ishikawa, H. (1989). *Electron. Lett.* **25**, 990.

Kuindersma, P. I.; Scheepers, W.; Cnoops, J. M. H.; Thijs, P. J. A.; v. d. Hofstad, G. L. A.; v. Dongen, T.; and Binsma, J. J. M. (1990). *Conf. Dig. 12th IEEE Semicond. Laser Conf.* Davos, Switzerland, 248.

Kuznetsov, M.; Verlangieri, P.; and Dentai, A. G. (1994). *IEEE Photon. Technol. Lett.* **6**, 157.

Kuznetsov, M.; Verlangieri, P.; Dentai, A. G.; Joyner, C. H.; and Burrus, C. A. (1993). *IEEE Photon. Technol. Lett.* **5**, 879.

Kuznetsov, M. (1988). *IEEE J. Quantum Electron.* **24**, 1837.

Kuznetsov, M.; Verlangieri, P.; Dentai, A. G.; Joyner, C. H.; and Burrus, C. A. (1992). *IEEE Photon. Technol. Lett.* **4**, 1093.

Lau, K. Y. (1990). *Appl. Phys. Lett.* **57**, 2632.

Lee, C. W.; Peng, E. T.; and Su, C. B. (1995). *IEEE Photon. Technol. Lett.* **7**, 664.

Luo, Y.; Nakano, Y.; Tada, K.; Inoue, T.; Hosomatsu, H.; and Iwaoka, H. (1990). *Appl. Phys. Lett.* **56**, 1620.

Mehuys, D.; Mittelstein, M.; Yariv, A.; Sarfaty, R.; and Ungar, J. E. (1989). *Electron. Lett.* **25**, 143.

Meyer, R.; Grützmacher, D.; Jürgensen, H.; and Balk, P. (1988). *J. Crystal Growth* **93**, 285.

Miller, D. A. B.; Chemla, D. S.; Damen, T. C.; Gossard, A. C.; Wiegmann, W.; Wood, T. H.; and Burrus, C. A. (1984). *Phys. Rev. Lett.* **53**, 2173.

Miller, S. E. (1989). *IEEE J. Lightwave Technol.* **7**, 666.

Murata, S.; Mito, I.; and Kobayashi, K. (1987). *Electron. Lett.* **23**, 403.

Murata, S.; Mito, I.; and Kobayashi, K. (1988). *Electron. Lett.* **24**, 577.

Noé, R.; Rodler, H.; Ebberg, A.; Meißner, E.; Bodlaj, V.; Drögemüller, K.; and Wittmann, J. (1992). *Electron. Lett.* **28**, 14.

Numai, T.; Murata, S.; and Mito, I. (1988). *Electron. Lett.* **24**, 1526.

Öberg, M.; Nilsson, S.; Streubel, K.; Wallin, J.; Bäckbom, L.; and Klinga, T. (1993). *IEEE Photon. Technol. Lett.* **5**, 735.

Öberg, M.; Rigole, P.-J.; Nilsson, S.; Klinga, T.; Bäckbom, L.; Streubel, K.; Wallin, J.; and Kjellberg, T. (1995). *IEEE J. Lightwave Technol.* **13**, 1892.

Öberg, M.; Nilsson, S.; Klinga, T.; and Ojala, P. (1991). *IEEE Photon. Technol. Lett.* **3**, 299.

Ogita, S.; Kotaki, Y.; Kihara, K.; Matsuda, M.; Ishikawa, H.; and Imai, H. (1988). *Electron. Lett.* **24**, 613.

Ohtoshi, T., and Chinone, N. (1989). *IEEE Photon. Technol. Lett.* **1**, 117.

Okai, M. (1994). *J. Appl. Phys.* **75**, 1.

Okai, M.; Sakano, S.; and Chinone, N. (1989). *Proc. 15th Europ. Conf. Opt. Commun.* Gothenburg, Sweden, 122.

Okai, M.; Suzuki, M.; Taniwatari, T.; and Chinone, N. (1994). *Jap. J. Appl. Phys.* **33**, 2563.

Okai, M., and Tsuchiya, T. (1993). *Electron. Lett.* **29**, 349.

Okai, M.; Tsuchiya, T.; Takai, A.; and Uomi, K. (1992). *Proc. Conf. Lasers and Electro-Optics.* Baltimore, MD, USA, 66.

Okuda, M., and Onaka, K. (1977). *Jap. J. Appl. Phys.* **16**, 1501.

Oshiba, S.; Nagai, K.; Kawahara, M.; Watanabe, A.; and Kawai, Y. (1989). *Appl. Phys. Lett.* **55**, 2383.

Pan, X.; Olesen, H.; and Tromborg, B. (1988). *IEEE J. Quantum Electron.* **24**, 2423.

Pan, X.; Olesen, H.; and Tromborg, B. (1990). *IEEE Photon. Technol. Lett.* **2**, 312.

Petermann, K. (1979). *IEEE J. Quantum Electron.* **15**, 566.

Reichmann, K. C.; Magill, P. D.; Koren, U.; Miller, B. I.; Young, M.; Newkirk, M.; and Chien, M. D. (1993). *IEEE Photon. Technol. Lett.* **5**, 1098.

Rigole, P.-J.; Nilsson, S.; Bäckbom, L.; Klinga, T.; Wallin, J.; Stålnacke, B.; Berglind, E.; and Stoltz, B. (1995). *IEEE Photon. Technol. Lett.* **7**, 697.

Rosenberger, M.; Köck, A.; Gmachl, C.; and Gornik, E. (1993). *Proc. Internat. Symp. Phys. Concepts and Materials for Novel Optoelectron. Device Applic.* Trieste, Italy, 1985.

Sakano, S.; Tsuchiya, T.; Suzuki, M.; Kitajima, S.; and Chinone, N. (1992). *IEEE Photon. Technol. Lett.* **4**, 321.

Sakata, H.; Takeuchi, S.; and Hiroki, T. (1992). *Electron. Lett.* **28**, 749.

Sakata, Y.; Yamaguchi, M.; Takano, S.; Shim, J.-I.; Sasaki, T.; Kitamura, M.; and Mito, I. (1993). *Proc. Opt. Fiber Conf.* San Jose, CA, USA, 9.

Sasaki, T.; Takano, S.; Henmi, N.; Yamada, H.; Kitamura, M.; Hasumi, H.; and Mito, I. (1988). *Electron. Lett.* **24**, 1408.

Schell, M.; Huhse, D.; and Bimberg, D. (1993). *Techn. Dig. 19th Europ. Conf. Opt. Commun.* Montreux, Switzerland, 229.

Schilling, M.; Idler, W.; Baums, D.; Dütting, K.; Laube, G.; Wünstel, K.; and Hildebrand, O. (1992). *Conf. Dig. 13th IEEE Semicond. Laser Conf.* Takamatsu, Japan, 272.

Schilling, M.; Schweitzer, H.; Dütting, K.; Idler, W.; Kühn, E.; Nowitzki, A.; and Wünstel, K. (1990). *Electron. Lett.* **26**, 243.

Schneider, I.; Nau, G.; King, T. V. V.; and Aggarwal, I. (1995). *IEEE Photon. Technol. Lett.* **7**, 87.

Slotwinski, A. R.; Goodwin, F. E.; and Simonson, D. L. (1989). *SPIE Proc.-Laser Diode Technol. and Applic.* **1043**, 245.

Soref, R. A., and Lorenzo, J. P. (1986). *IEEE J. Quantum Electron.* **22**, 873.

Stoltz, B.; Dasler, M.; and Sahlen, O. (1993). *Electron. Lett.* **29**, 700.

Strzelecki, E. M.; Cohen, D. A.; and Coldren, L. A. (1988). *IEEE J. Lightwave Technol.* **6**, 1610.

Suematsu, Y.; Arai, S.; and Kishino, K. (1983). *IEEE J. Lightwave Technol.* **1**, 161.

Sundaresan, H., and Fletcher, N. C. (1990). *Electron. Lett.* **26**, 2004.

Susa, N., and Nakahara, T. (1992). *Appl. Phys. Lett.* **60**, 2457.

Takemoto, A., Ohkura, Y., Kawama, Y., Nakajima, Y., Kimura, T., Yoshida, N., Kakimoto, S., and Susaki, W. (1989). *IEEE J. Lightwave Technol.* **7**, 2072.

Tanbun-Ek, T.; Logan, R. A.; Temkin, H.; Olsson, N. A.; Sergent, A. M.; and Wecht, K. W. (1990). *IEEE Photon. Technol. Lett.* **2**, 453.

Tiemeijer, L. F.; Thijs, P. J. A.; de Waard, P. J.; Binsma, J. J. M.; and v. Dongen, T. (1991). *Appl. Phys. Lett.* **58**, 2738.

Tohmori, Y.; Jiang, X.; Arai, S.; Koyama, F.; and Suematsu, Y. (1985). *Jap. J. Appl. Phys.* **24**, L399.

Tohmori, Y. and Oishi, M. (1988). *Jpn. J. Appl. Phys.* **27**, L693.

Tohmori, Y.; Yoshikuni, Y.; Tamamura, T.; Yamamoto, M.; Kondo, Y.; and Ishii, H. (1992). *Techn. Dig. 13ᵗʰ IEEE Intern. Semicond. Laser Conf.* Takamatsu, Japan, 268.

Tohmori, Y.; Yoshikuni, Y.; Kano, F.; Ishii, H. Tamamura, T.; and Kondo, Y. (1993). *Electron. Lett.* **29**, 352.

Tohyama, M.; Onomura, M.; Funemizu, M.; and Suzuki, N. (1993). *IEEE Photon. Technol. Lett.* **5**, 616.

Tromborg, B.; Olesen, H.; and Pan, X. (1991). *IEEE J. Quantum Electron.* **27**, 178.

Tsang, W. T. (1990). *J. Crystal Growth* **105**, 1.

Utaka, K.; Akiba, S.; Sakai, K.; and Matsushima, Y. (1986). *IEEE J. Quantum Electron.* **22**, 1042.

Uttam, D, and Culshaw, B. (1985). *IEEE J. Lightwave Technol.* **3**, 971.

Wagner, R. E., and Linke, R. A. (1990). *IEEE LCS*, **28**.

Wang, J.; Leburton, J. P.; and Educato, J. L. (1993). *J. Appl. Phys.* **73**, 4669.

Wang, J.; Schunk, N.; and Petermann, K. (1987). *Electron. Lett.* **23**, 715.

Weber, J.-P. (1994). *IEEE J. Quantum Electron.* **30**, 1801.

Wegener, M.; Chang, T. Y.; Bar-Joseph, I.; Kuo, J. M.; and Chemla, D. S. (1989). *Appl. Phys. Lett.* **55**, 583.

Westbrook, L. D. (1986). *IEE Proc., Pt. J* **133**, 135.

Willner, A. E.; Chapuran, T. E. Wullert, Jr. II, Meyer, J.; and Lee, T. P. (1992). *Electron. Lett.* **28**, 1526.

Wolf, T., Borchert, B., Drögemüller, K., and Amann, M.-C. (1991). *Jpn. J. Appl. Phys.* **30**, L 745.

Wolf, T.; Drögemüller, K.; Borchert, B.; Westermeier, H.; Veuhoff, E.; and Baumeister, H. (1992). *Appl. Phys. Lett.* **60**, 2472.

Wolf, T.; Illek, S.; Rieger, J.; Borchert, B.; and Amann, M.-C. (1994). *Techn. Dig. Conf. Lasers and Electro-Optics.* Anaheim, CA, USA, CWB.

Wu, M. C.; Chen, Y. K.; Tanbun-Ek, T.; Logan, R. A.; and Sergent, A. M. (1990). *Proc. Conf. Lasers and Electro-Optics.* Anaheim, CA, USA, 667.

Wu, M. S.; Vail, E. C.; Li, G. S.; Yuen, W.; and Chang-Hasnain, C. J. (1995). *Electron. Lett.* **31**, 1671.

Yamada, H., Sasaki, T., Takano, S., Numai, T., Kitamura, M., and Mito, I. (1988). *Electron. Lett.* **24**, 147.

Yamamoto, E.; Hamada, M.; Suda, K.; Nogiwa, S.; and Oki, T. (1991a). *Appl. Phys. Lett.* **59**, 2721.

Yamamoto, E.; Suda, K.; Hamada, M.; Nogiwa, S.; and Oki, T. (1991b). *Jpn. J. Appl. Phys.* **30**, L1884.

Yamamoto, H.; Asada, M.; and Suematsu, Y. (1985). *Electron. Lett.* **21**, 579.

Yasaka, H.; Ishii, H.; Yoshikuni, Y.; and Oe, K. (1995). *IEEE Photon. Technol. Lett.* **7**, 161.

Yoshikuni, Y.; Tohmori, Y.; Tamamura, T.; Ishii, H.; Kondo, Y.; Yamamoto, M.; and Kano, F. (1993). *Proc. Opt. Fiber Conf.* San Jose, CA, USA, 8.

Zatni, A., and Le Bihan, J. (1995). *IEEE J. Quantum Electron.* **31**, 1009.

Zhang, L., and Cartledge, J. C. (1993). *IEEE Photon. Technol. Lett.* **5**, 1143.

Zhang, L., and Cartledge, J. C. (1995). *IEEE J. Quantum Electron.* **31**, 75.

Zucker, J. E.; Bar-Joseph, I.; Miller, B. I.; Koren, U.; and Chemla, D. S. (1988). *Appl. Phys. Lett.* **54**, 10.

Chapter 4

High-Power Semiconductor Lasers

David G. Mehuys

*SDL Inc., 80 Rose Orchard Way, San Jose CA 95134, U.S.A.,
Tel: (408) 943-9411 FAX: (408) 943-1070*

4.1   Introduction

In the 35 years since the first demonstration of the semiconductor injection laser, much progress has been made toward increasing power output, improving reliability, and broadening the range of wavelengths available. By virtue of their small size, high efficiency, and large lifetime, semiconductor lasers have the potential to supplant other bulkier lasers in scientific applications and become integrated within electronic and telecommunications subsystems. Most applications demand stable power output and beam quality over long time periods. In this chapter, the issues that determine the beam quality and power output of high-power semiconductor lasers are discussed. To illustrate the basic concepts, examples are chosen from state-of-the-art research results and commercial lasers.

259

4.1.1 Why high power?

Semiconductor lasers play a key role in many applications ranging from consumer products to research studies to military applications. Much of their versatility stems from their operation at wavelengths from the visible to the far infrared. For many applications, the power requirements are quite low. For example, compact disk players employ 3 mW GaAs lasers emitting at $\lambda = 780$ nm wavelength. Supermarket scanners use visible ($\lambda = 670$ nm) 3 mW AlGaInP lasers, and for fiber optic telecommunications 5 mW InGaAsP lasers are used, emitting single-mode, single-frequency ($\lambda = 1.3$ μm or $\lambda = 1.55$ μm) in the near-infrared.

However, an important class of applications requires much higher powers either in pulsed or continuous-wave (cw) operation. Operation in a single spatial mode or on a single frequency line is often desirable, but not necessarily a requirement. One example where raw power output is most important is diode-pumped YAG lasers, where the pumping laser array fabricated from GaAlAs material emits kilowatts of incoherent quasi-cw pulses (100 μs–1 ms pulse length). The wavelength of the GaAlAs laser array is well-tuned to the absorption spectrum of the YAG crystal (806–810 nm), providing much higher overall efficiency than flashlamp pumping.

Applications where both high-power and near-diffraction-limited beam quality (*i.e.*, coherent emission) are required include many communication and most scientific applications, where transmission over long distances and/or the ability to focus the light to a small spot of high power density is desired. In general, the power available from incoherent sources exceeds that available from coherent sources by more than an order of magnitude. However, the brightness or radiance (power/beam divergence) of coherent, diffraction-limited sources is significantly higher.

The distinction between low-power and high-power semiconductor lasers is not very well defined. It may depend on the type of laser and the application for which it is designed. Generally speaking, 50 mW (or more) *cw* for single-mode, single-frequency lasers and 500 mW (or more) *cw* for broad-area multimode lasers and laser arrays are considered to be high power. Although diode lasers are capable of emitting kWatts of peak *power* under pulsed conditions, pulse *energies* are limited to much less than other solid-state lasers due to the diode short carrier lifetime (a few ns).

4.1.2 Power limitations

The semiconductor laser differs from other laser systems in that the population inversion required to reach lasing threshold results from electrical charge injection into the active region, contrary to other solid-state, liquid dye, and gas lasers, which are typically optically pumped using flashlamps or other lasers. The resulting laser output power is characterized by the light-output versus current-input (L–I) curve, as plotted in Fig. 4.1a. Ideally, a straight line of slope corresponding to one photon emission per electron injection results.

In practice, there are four key factors that limit the useful output power of a semiconductor laser:

1. Multimode operation.

2. Catastrophic optical degradation.

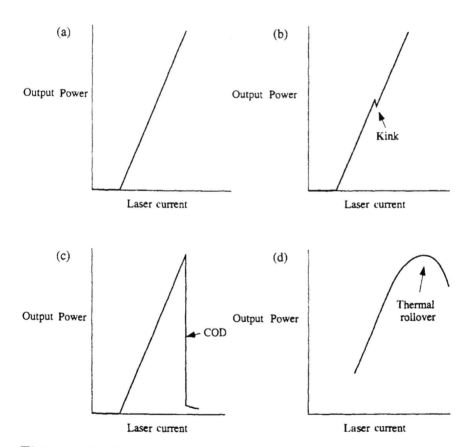

Figure 4.1: Conceptual laser diode light-current (L-I) curves illustrating (a) ideal output, (b) kink, (c) catastrophic optical damage (COD), and (d) thermal rollover.

3. Thermal rollover.

4. Age.

The first three of these conditions can usually be diagnosed by nonideal features that appear in the L–I curve (Thompson, 1980), as illustrated in Figs. 4.1b, c, and d. Multimode operation, which is the onset of higher-order spatial modes that causes a distortion of the far-field radiation pattern of the laser, is usually accompanied by a discontinuity in output power or slope (a "kink") in the cw light-current curve as illustrated in Fig. 4.1b. This kink is the most common power-limiting mechanism in single-mode lasers. The physical phenomenon that gives rise to the simultaneous oscillation of two or more spatial modes is the perturbation of the waveguide refractive index or gain profile caused by inhomogeneities in the local carrier population or temperature (see Sects. 4.2.4 and 4.3.2). For most lasers, operating through these modal kinks is a noncatastrophic event (changes are reversible, albeit with a possible hysteresis).

Catastrophic optical damage, commonly referred to as COD, refers to irreversible damage at the laser facet caused by heating due to high optical power density. Surface recombination centers at the cleaved crystalline facet, and the associated depletion of charge carriers causes the active region near the facet to become absorbing at the lasing wavelength. As a result, heat is generated, and the local temperature is raised. At some critical optical flux density, the heating causes sufficient shrinkage in the local band gap resulting in increased optical absorption. As the temperature reaches the material melting point, the facet is mechanically damaged (usually a blister is formed), its reflectivity drops, and the laser output is diminished irreversibly. Figure 4.1c shows the large decrease in laser output that typically accompanies COD. In AlGaAs lasers with uncoated facets, the critical optical power density is approximately 1–5 MW/cm^2. Facet coatings and passivation techniques can increase this damage level to approximately 10–20 MW/cm^2.

The third power-limiting mechanism, thermal rollover, is illustrated in Fig. 4.1d. Its characteristic slow decrease in laser efficiency with increased current injection is caused by the increased fraction of ohmic losses incurred at high drive levels. This increased ohmic loss results in increased heat dissipated near the *pn* junction, raising the temperature of the active region and lowering its conversion efficiency. Eventually the optical output saturates and even decreases with increasing current. A principal reason for the decreased optical output after saturation is the increased spillage of charge carriers out of the active region (see Sect. 4.3.3) into the carrier confinement or cladding layers, where they cannot contribute to radiative recombination. As the temperature rise becomes dramatic, the laser gain decreases at fixed current density, eventually causing the optical output to diminish to near zero. Thermal rollover, like multimode operation, is usually not a catastrophic event but is reversible. It is often the power-limiting mechanism on lower-efficiency long-wave-length lasers ($\lambda > 1.3 - 1.5\ \mu$m) or on lasers with poor thermal resistance, such as those mounted junction-side up.

One final limitation to the output power of a semiconductor laser is its age. All semiconductor lasers exhibit some degradation in optical output if operated at constant current and temperature for extended periods of time. Aging behavior and the associated reliability of a given semiconductor laser are highly dependent upon its structure and operating conditions. The topic of reliability merits an extended discussion of its own, and as such, it will be discussed only briefly here (see Sect. 4.3.4). We comment in passing that semiconductor laser reliability is usually limited by either thermal dissipation, high current density, or COD. Thermal and current-density induced degradation mechanisms are associated with the forma-tion of point defects or dark-line defects, which can migrate into and along the active region, degrading the laser output over time. Often, the lifetime of a laser is specified as the number of hours at which the operating current increases by some percentage, for example 20%, to maintain a specified output power.

4.1.3 Case study: index-guided single-mode lasers

The most widely produced semiconductor laser in the world today is the index-guided single-mode laser. Commonly called "single-stripe" lasers, they are designed to propagate the lowest order ("fundamental") optical mode that has a *near*-Gaussian shape in both lateral and transverse dimensions. The mode shape is not strictly Gaussian but typically decays exponentially, giving M^2 values between 1.0 and 1.3 (see Sect. 4.2.5.3). Single-mode lasers are widely used in compact disc players, supermarket scanners, telecommunciations, and laser pointers in the low-power regime $P < 5$ mW. At higher output powers, $P > 30$–200 mW, single-stripe lasers have applications in optical recording, printing, frequency-doubling, com-munications, and laser and fiber amplifier pumping. A key application is 980 nm wavelength fiber-coupled lasers used as pumps for Er-doped optical fiber amplifiers. In this section, we examine the characteristics of high-power, single-stripe semiconductor lasers.

A popular single-stripe laser structure is the ridge-waveguide laser, shown in Fig. 4.2 (Jaeckel *et al.*, 1991). The active region of the example laser is a single 70-Angstrom-thick InGaAs quantum well, which is centered within a 0.2-μm-thick (*i.e.*, 0.1 μm each side of quantum well) graded-index separate-confinement AlGaAs heterostructure, which provides the high-refractive-index core of the optical waveguide perpendicular to the pn junction. The low-refractive-index cladding regions are 1.8-μm-thick $Al_{0.4}Ga_{0.6}As$ on both p and n sides to provide a symmetrical, positive refractive index waveguide perpendicular to the pn junction. The waveguide in the plane of the pn junction is provided by etching almost all of the p-side cladding region away from a narrow ridge approximately 3 μm in width. In the etched regions outside the ridge, the waveguide perpendicular to the pn junction is asymmetric, displacing the light toward the n-type cladding layer, thereby lowering its effective refractive index (see Sect. 4.2.2). Thus, as shown in Fig. 4.3a, the two-dimensional optical mode is largely confined to the graded-index AlGaAs layers beneath the

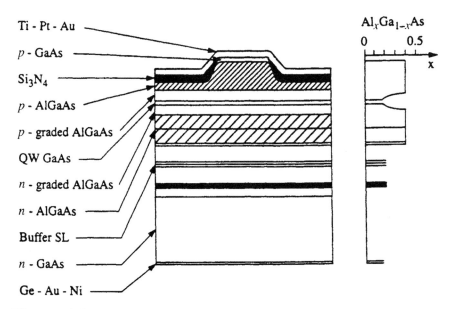

Figure 4.2: Schematic diagram of an index-guided ridge-waveguide laser, showing the epitaxial layer structure (from Jaeckel *et al.*, 1991).

(a)

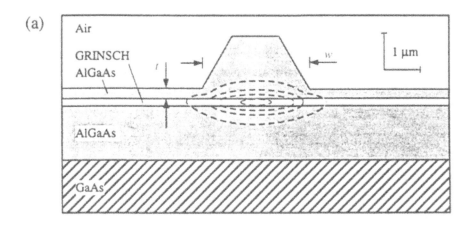

(b)

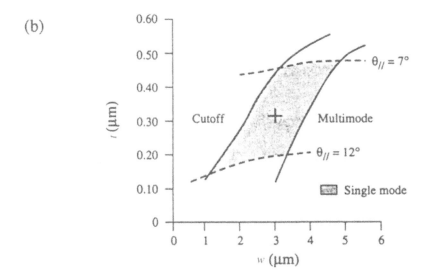

Figure 4.3: Ridge waveguide (a) schematic diagram, showing definition of ridge width and depth, and (b) calculated results showing single-mode regime (shaded) and tolerances for w and t (from Jaeckel *et al.*, 1991).

ridge. Contact metallizations are fabricated on top of the ridge to channel light through the ridge waveguide and enable lasing.

Maintaining fundamental mode operation to high-power output requires careful optimization of the ridge waveguide geometry, in particular

the ridge width at its base, w, and the separation t between the bottom
of the ridge and the quantum-well active region (where the optical mode
is centered). Figure 3b plots the area within (w,t) space for which the
ridge waveguide supports only a single spatial mode (shaded). Based on
this plot, target values of $w = 3.1$ μm and $t = 0.3$ μm have been established,
and tolerances of $\Delta w = +/- 0.5$ μm and $\Delta t = 0.1$ μm have been identified
(Jaeckel *et al.*, 1991).

For high-power testing, the ridge waveguide lasers were cleaved into
750 μm lengths, after which the facets were passivated and deposited
with dielectric layers to provide $R > 90\%$ reflectivity on the rear facet
and $R = 10\%$ reflectivity on the front facet. The lasers were mounted p-
side up onto heatsinks for cw testing. The resulting cw L–I curve is plotted
in Fig. 4.4. The output power increases nearly linearly up to 380 mW at
a corresponding laser current of 400 mA. A further increase in the laser
current leads to thermal saturation of the laser output at a maximum
power of 425 mW, then thermal rollover, and finally the device stops lasing
due to excessive heating at 780 mA.

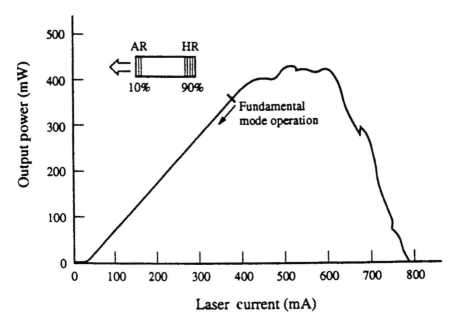

Figure 4.4: Measured cw L–I curve for the ridge waveguide lasers of Figs.
4.2 and 4.3, showing kinks and thermal rollover (from Jaeckel *et al.*, 1991). Fundamental spatial mode operation is obtained up to approximately 350 mW.

As discussed briefly above, the passivation of surface states associated with laser facets can greatly increase the threshold for COD (Susaki *et al.*, 1977; Yonezu *et al.*, 1979). Figures 4.5a and b plot the L–I curves of 4-μm-wide index-guided single-stripe lasers with and without facet passivation. The laser without facet passivated facets exhibits COD at approximately 300 mW, corresponding to an optical flux density of 7.5 MW/cm^2. The laser with passivated facets does not exhibit COD but exhibits thermal saturation at 800 mA injection current, corresponding

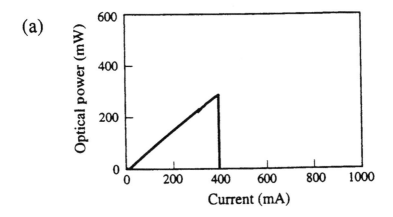

(a)

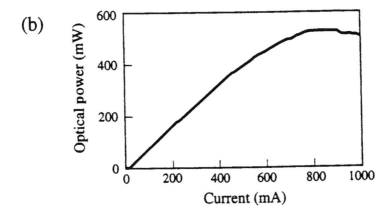

(b)

Figure 4.5: Measured L–I curves of index-guided lasers (a) without facet passivation, showing COD, and (b) with facet passivation.

to approximately 520 mW of output power. The useful output of the passivated laser is limited by multimode operation at the kink in its L–I curve at approximately 180 mW, as witnessed by the broadening and steering of its lateral radiation pattern shown in Fig. 4.6b, compared to that of a laser without a kink in Fig. 4.6a.

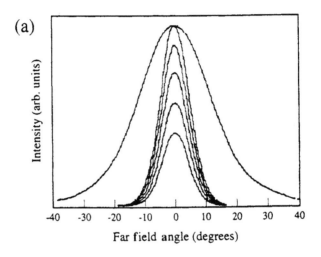

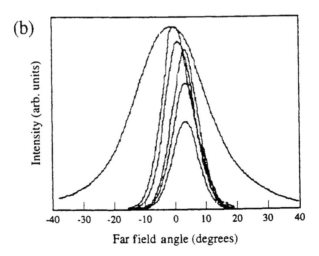

Figure 4.6: Measured lateral far field patterns of ridge waveguide lasers at output powers of 40, 80, 120, 160, and 200 mW cw output for (a) a laser without an L–I kink, and (b) a laser with an L–I kink at approximately 150 mW.

In order to ensure reliable single-mode operation over extended periods of time, commercially available single-mode lasers are rated at power levels well *below* the thresholds for kinks, beam-steering, COD, and thermal rollover. In recent years, the state of the art has been steadily advancing, enabling the commercialization of highly reliable 50 mW, 100 mW, 150 mW, and 200 mW single-stripe lasers (Scifres, 1993).

4.2 Beam quality

The ability to transmit light over long distances or to focus light to a small, diffraction-limited spot relies upon having a high-quality, spatially coherent beam. In this section the groundwork will be laid for the analysis and characterization of semiconductor laser beam quality. Concepts of gain and loss, index and gain waveguiding, gain saturation, and filamentation are reviewed because they determine the beam quality of high-power, near-diffraction-limited semiconductor lasers. Then, concepts of brightness (or radiance), Strehl ratio, and M^2 will be discussed because of their utility in characterizing laser beam quality. This section will conclude with a case study of a high-power, clean-beam master oscillator power amplifier diode laser.

4.2.1 Carrier-induced gain and refractive index

Fundamental to the operation of a semiconductor laser is the way that light and matter interact within the active medium. This interaction is more complex in semiconductor lasers than in gas or other solid-state lasers and results in unique issues and challenges to be addressed. The interaction of light and matter comprises not only stimulated and spontaneous emission, whereby electronic charge carriers are converted into photons, but also the carrier-density-dependent gain and refractive index experienced by the electromagnetic light wave. In turn, the spatially varying gain and refractive index within the semiconductor laser determine its modal profiles. In this section, we derive expressions for the carrier-dependent gain and refractive index profile. This treatment is necessarily mathematical, and the reader wishing only an overview of the subject may omit detailed study on first reading.

Mathematically, the interaction between radiation and matter is manifest in the complex-valued dielectric constant $\varepsilon(\omega)$ seen by a traveling wave of radian frequency ω. However, the response of a material to an electric field is embodied in its polarization, and so we write the dielectric constant as a function of the material susceptibility $\chi(\omega)$,

$$\varepsilon(\omega) = \varepsilon_0 (1 + \chi(\omega)), \qquad (4.1)$$

where ε_0 is the permittivity of free space. One distinguishes between the resonant component $X(\omega)$ and the nonresonant component X_0 of the semiconductor susceptibility: $\chi(\omega) = X_0 + X(\omega)$. The commonly used (nonresonant) refractive index n_0 is defined by

$$n_0^2 = 1 + X_0. \qquad (4.2)$$

However, it is the resonant component of the susceptibility that embodies the interaction between light and matter: It is complex-valued (Yariv, 1989),

$$X(\omega) = X_r(\omega) + i X_i(\omega), \qquad (4.3)$$

to allow for changes in the phase or speed (X_r—dispersion) and in the amplitude (X_i—gain or loss) of electromagnetic waves propagating in such a medium.

In order to see how X_r and X_i are related to the refractive index and gain experienced by an electromagnetic wave, consider a plane wave of frequency ω_0,

$$E(z,t) = \hat{x} E_0 \exp i\{k(\omega_0)z - \omega_0 t\}, \qquad (4.4)$$

where $\hat{x}$ is a unit vector in the x-direction and E_0 is a constant. The propagation constant of the wave is defined by

$$k(\omega_0) = \omega_0 [\mu_0 \varepsilon(\omega)]^{1/2}, \qquad (4.5)$$

where μ_0 is the permeability of free space, and by substitution of $\varepsilon(\omega)$ from equation (4.1) we obtain for the propagation constant

$$k(\omega_0) = (n_0 \omega_0/c)\{1 + X(\omega_0)/2n_0^2\}. \qquad (4.6)$$

Since X is complex-valued, we obtain the following correspondences between the resonant susceptibility $X(\omega)$ and the gain $g(\omega_0)$ and the index of refraction

$$n(\omega_0) = n_0 \{1 + X_r(\omega_0)/2n_0^2\} \qquad (4.7a)$$

$$g(\omega_0) = -k_0 X_i(\omega_0)/n_0, \qquad (4.7b)$$

where $k_o = 2\pi/\lambda$ and λ is the wavelength in air. Alternatively, gain is often expressed as the imaginary component of the refractive index:

$$\text{Im}\{n\} = X_i(\omega_o)/2n_o = -g(\omega_o)/2k_o . \tag{4.8}$$

The final stone in the foundation is laid when we relate the resonant susceptibility X to the carrier population inversion. This relationship is complicated by the semiconductor band structure, density of states in the conduction and valence bands, and lineshape broadening functions. Its explicit calculation is beyond the scope of this chapter. However, as an example, consider the *calculated* curves of X_r and $-X_i$ as a function of energy above the band gap, $E-Eg$, for a single quantum well (Vahala *et al.*, 1983) presented in Fig. 4.7. X_i is directly proportional to the laser gain (note that $X_i > 0$ represents loss, and $X_i < 0$ represents gain), while X_r gives the change in refractive index from its nonresonant value. The different curves in Fig. 4.7 illustrate the qualitative changes in X_r and X_i as a function of carrier density, N. As the carrier density increases, the peak quantum-well gain increases $(d\{-X_i\}/dN > 0)$, while the corresponding index decreases $(dX_r/dN < 0)$. In addition, the wavelength of peak gain changes, as indicated by the location of the dots in Fig. 4.7. The consequences of this carrier-dependent gain and refractive index are of considerable importance in determining the waveguiding properties of semiconductor lasers.

4.2.2 The effective index method for finding spatial modes

Having determined expressions for gain and refractive index, the next step is to determine the complex-valued electromagnetic mode profile. To solve this full two-dimensional or three-dimensional problem generally requires extensive numerical modeling involving finite-element calculations (Hadley *et al.*, 1991). For many applications, however, it is sufficient to solve simpler one-dimensional problems, and that is the approach that we outline here.

In a semiconductor laser, the electric field $E_x(x, y, z, t)$ is a complicated superposition of many lateral (x-dependent), transverse (y-dependent), and longitudinal (z-dependent) modes, all of which oscillate at different frequencies. For simplicity, we consider a single longitudinal mode and thereby select a single frequency ω_o. As is customary, transverse electric (TE) polarization is considered exclusively here due to the nominally

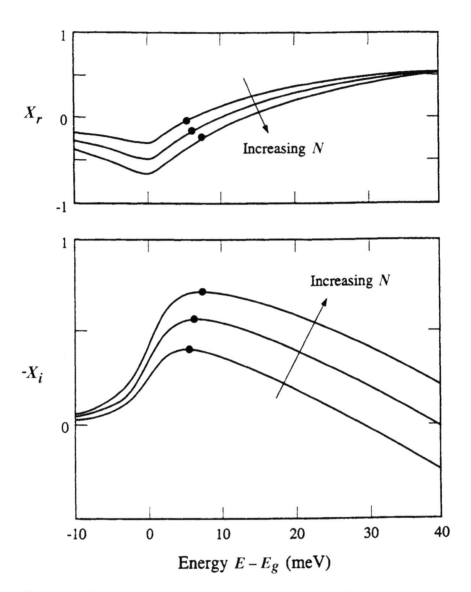

Figure 4.7: (a) Real and (b) imaginary parts of the complex-valued susceptibility function X as a function of photon energy above the band gap energy, for a
quantum-well laser diode. The different curves represent increasing values of the
carrier density, N, and the dots represent the photon energy of peak gain.

higher gain and increased reflection coefficients of TE modes (Ikegami *et al.*, 1972). The *x*-polarized electric field is a traveling wave along the *z*-direction of the form

$$E_x(x,y,z,t) = E(x) F(x,y) \exp i \{\beta z - \omega_0 t\}, \qquad (4.9a)$$

satisfying the scalar wave equation

$$[\nabla^2 + k_o^2 n^2(x,y)] E_x(x,y,z,t) = 0, \qquad (4.9b)$$

where $\beta = k_o \eta$ is the complex modal propagation constant and η is the complex modal "effective index." In Eq. (4.9), $E(x)$ is the lateral mode we wish to determine, and $F(x,y)$ is the transverse mode, which is usually a slowly varying function of the lateral coordinate, x. These functions are illustrated for a stripe geometry laser in Fig. 4.8a.

The procedure for determining $F(x, y)$, $E(x)$, and η in the case of a stripe-geometry laser is illustrated graphically in Fig. 4.8. We will outline the corresponding mathematical procedure shortly. Briefly, cross-sections of the two-dimensional complex index profile are constructed in the transverse (y) direction at fixed x, as in Fig. 4.8c. The transverse mode $F(x, y)$ at fixed x and its effective index $n_{eff}(x)$ are the eigensolutions of these waveguides. The lateral mode $E(x)$ and its modal propagation constant β are the eigensolutions of the complex index profile subsequently constructed from $n_{eff}(x)$, as shown in Fig. 4.8d. This technique, known as the effective index approximation (Streifer and Kapon, 1979), is valid when the lateral refractive index and electric field variations are much slower than the transverse ones. A common type of laser where the effective index method has been applied successfully is the single-stripe index-guided laser of Sect. 4.1.3.

Mathematically, the effective index approximation proceeds as follows: Substitution of equation (4.9a) into (4.9b) gives the following partial differential equation:

$$E \, d^2F/dy^2 + d^2(EF)/dx^2 + k_o^2 n^2(x,y) \, EF - \beta^2 \, EF = 0. \qquad (4.10)$$

In order to decouple equations for E and F, we hypothesize that F is the solution of a transverse waveguide problem that can be posed at fixed x and can subsequently be computed as a slowly varying function of x. Mathematically, in the application of the effective index method, we therefore neglect the x-derivatives of F. Thus F is proposed to be the eigensolution of

$$d^2F(x,y)/dy^2 + k_o^2 \{n^2(x,y) - n^2_{eff}(x)\} F(x,y) = 0, \qquad (4.11a)$$

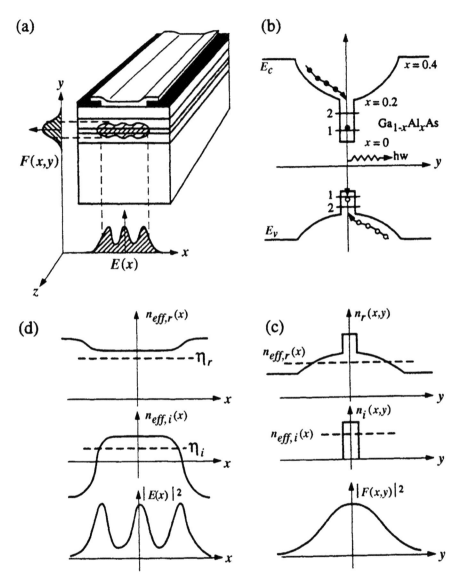

Figure 4.8: The effective index model for a broad area laser. (a) Coordinate definitions showing transverse and lateral optical modes. (b) Band diagram in the transverse dimension. (c) Transverse gain $n_i(x, y)$ and index $n_r(x, y)$ profiles at a particular value of x, with the transverse mode $F(x, y)$. (d) Lateral gain $n_{eff,i}(x)$ and index $n_{eff,r}(x)$ profiles, with the lateral mode $E(x)$.

where the term $k_o^2 n_{eff}^2(x)$ acts as a mathematical "separation constant." Under these conditions, Eq. (4.10) becomes a decoupled equation for the lateral mode E:

$$d^2E(x)/dx^2 + k_o^2\{n^2_{eff}(x) - \eta^2\}E(x) = 0. \qquad (4.11b)$$

Equation (4.11b) is known as the complex-valued Helmholtz equation (*i.e.*, one-dimensional, reduced-wave equation). The solutions of the Helmholtz equation are in general not unique, since it is an eigenvalue equation. Each eigensolution $E(x)$ is referred to as a "lateral mode," and η is its effective modal index, that is, $\beta = k_o\eta$ is its propagation constant. The real part of β corresponds to the modal propagation constant, and the imaginary part of β corresponds to the modal gain (or loss), as discussed below. Waveguides that support more than one confined optical mode are termed "multimode." In particular, semiconductor lasers are often classified as single-mode or multimode structures.

In practice, the intermediate step of finding the eigensolutions $F(x, y)$ and $n_{eff}(x)$ is often omitted, especially in the case of gain-guided lasers (see Sect. 4.2.3). Instead approximate shapes for $F(x, y)$ and values for $n_{eff}(x)$ are estimated or assumed. For example, using Eq. (4.8), the modal gain of the transverse mode is proportional to the imaginary component $n_{eff,i}(x)$ and can be estimated by reducing the gain of the active region by the transverse optical confinement factor:

$$n_{eff,i}(x) = (d/W_{eff})g(x)/2k_0 = \Gamma g(x)/2k_0, \qquad (4.12a)$$

where $g(x)$ is the peak gain within the active region as appearing in Eq. (4.8), d is the active region thickness, and W_{eff} is the effective height of the transverse optical mode $F(x, y)$. The ratio d/W_{eff} is defined to be the transverse optical confinement factor Γ. Similarly, a value for $n_{eff,r}(x)$ is often estimated using tabulated values for bulk materials. While the absolute value of $n_{eff,r}(x)$ is not critical, the change $\Delta n_{eff,r}(x)$ in $n_{eff}(x)$ that is due to the presence of lateral variations in transverse structure or in injected carrier density determines $E(x)$ and η. In the case of laser beams guided only by injected carrier density, $\Delta n_{eff,r}(x)$ is often assumed to be simply related to the change $\Delta n_{eff,i}(x)$ in the gain that is due to injected carriers. Figure 4.7 indicates that at the wavelength of peak gain, the change in real part of the refractive index is proportional to the change in imaginary part of refractive index, that is:

$$\Delta n_{eff,r}(x) = -b\Delta n_{eff,i}(x), \qquad (4.12b)$$

where b is referred to as the antiguiding parameter (Thompson, 1980). In most cases, b is taken to be a constant. In general, however, b is a function of frequency ω_0 and carrier density $N(x)$. In the limit of small changes in carrier density, b is simply given by

$$b = -dN_{eff,r}/dN \bigg/ dN_{eff,i}/dN = -(dX_r/dN)/(dX_i/dN) = \alpha \quad (4.13)$$

and is referred to as the α-parameter or linewidth enhancement factor because of the role it plays in broadening a laser's linewidth beyond the Schawlow-Townes formula (Henry, 1982). Figure 4.9 plots α as a function of current density for a QW laser, calculated from gain and index spectra similar to Fig. 4.7 (Arakawa and Yariv, 1985). For this plot, α is calculated at the wavelength of peak gain at each current density, for active regions comprised of one, three, and five quantum wells. The figure illustrates that the magnitude of α generally decreases with increasing carrier density.

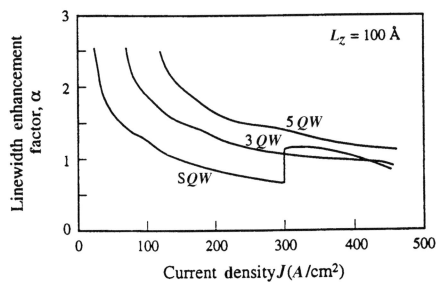

Figure 4.9: The linewidth enhancement factor α as a function of the injected current at photon energy, at which the gain is maximum with various carrier densities N. The curves are parameterized by number of quantum wells, each of thickness 100 Angstroms (from Arakawa and Yariv, 1985).

Using Eqs. (4.8) and (4.12b), the complex refractive index (at the frequency ω_0) is taken to be

$$n_{eff}(\omega_0) = n_{eff,0}(\omega_0) - (\alpha + i)\,\Gamma g(\omega_0)/2k_0\,, \qquad (4.14)$$

where $n_{eff,0}$ is the complex index of refraction at the carrier density corresponding to optical transparency.

The model of the effective index described above is often used in postulating Eq. (4.11b) at a particular fixed frequency ω_0 corresponding to the frequency of peak gain. Solutions for $E(x)$ and η are then determined by solution of the one-dimensional Helmholtz equation (4.11b). Alternatively, numerical techniques can be applied to solve the full two-dimensional problem of Eq. (4.9b). In the 1–d case, the allowed lateral modes $E(x)$ may be ordered according to real or imaginary parts of their effective modal index (Nash, 1973). For semiconductor lasers, it is most useful to order the modes according to imaginary effective index: According to Eq. (4.8), the mode with the lowest value of $Im\{\eta\}$ has the highest modal gain and will be the mode to lase at lowest pump current. In single-stripe lasers, for example, this mode has a near-Gaussian intensity profile. Other (higher-order) modes typically have multiple spots or lobes in their near- and far-field patterns.

4.2.3 Gain and index waveguiding

Laser diode modes can be classified as either gain-guided or index-guided (Thompson, 1980; Lang, 1986), depending on whether the imaginary or real part of the refractive index provides the dominant guiding mechanism. Gain-guided lasers (that is, those operating in a gain-guided mode) rely only on current confinement to channel the charge carriers to a narrow stripe of the active region. The optical mode that develops is a result of its preferential amplification along the stripe, as discussed below. Index-guided lasers, on the other hand, augment the current confinement of gain-guided lasers with a real-refractive-index profile that is produced by varying the structure or composition of the material that bounds the active region; an example is the ridge waveguide laser discussed in Sect. 4.1.3.

The stripe geometry laser of Fig. 4.8 illustrates the difference between gain and index waveguiding: The gain-guided lateral mode is guided by a significantly different mechanism from the index-guided transverse mode. Light guided in the transverse direction, perpendicular to the epitaxial layers, is channeled by total internal reflection, confining most of the energy within its core heterostructure. Outside the positive index waveguide within the cladding, the electromagnetic field is evanescent and decays exponentially. Energy is stored in the electromagnetic tails, but there is no power flow there. The electromagnetic wavefront is flat, indicating power flow in the axial direction. Such light, which is guided chiefly by total internal reflection, is termed "index-guided."

The lateral waveguide, however, is defined only by the current confinement to a striped area. Thus it exhibits a negative refractive index step that is due to the carrier-induced refractive index coupling of the gain. Thus, for angled incidence, light is not totally internally reflected but is refracted out of the waveguide. The Helmholtz equation (4.11b) yields a solution $E(x) = \cos(kx)$ that, like the index-guided case, consists of two plane waves propagating at equal but opposite angles to the longitudinal axis. However, the constituent plane waves experience gain in traversing the waveguide (note that k is, in general, complex-valued) and hence do not have equal amplitudes everywhere across the width. Since the core experiences higher gain than does the cladding, the modal wavefront is advanced there. This directs power into the cladding (an additional loss) in order to equalize the energy derived from stimulated emission everywhere in the mode. Such a field propagates self-consistently along the longitudinal axis and thereby satisfies the definition of a lateral mode. Such modes are called "gain-guided," and, rather than rely on total internal reflection at the core-cladding interfaces to guide the light, these modes rely on a continuous generation of photons within the guide itself to compensate for those lost at those interfaces. Importantly, since the "waveguide" of gain-guided lasers depends upon the local carrier density, gain-guided modes are in general a function of the laser current level and output power. This dependence has a direct impact on the beam quality, as discussed below.

4.2.4   Filamentation and the α-parameter

One goal of semiconductor laser researchers has been to develop a laser with a broad-aperture, high-power beam that is spatially coherent. This goal has been frustrated by the spontaneous deterioration of broad-aperture beams into narrow filaments, a consequence of the coupling of gain and refractive index within a semiconductor (Eq. 4.12b). This inherent nonlinear interaction with traveling laser beams has enormous implications on laser beam quality, especially at high power. The complex refractive index is dependent on the electromagnetic field intensity through the depletion of the carrier population that accompanies stimulated emission. As laser threshold is surpassed, the electromagnetic modal gain becomes clamped at the level of the resonator losses, as is required for steady-state operation. Charge carriers introduced into the active region by additional pumping are depleted by stimulated emission in proportion to the increasing intensity of the optical field. However, the optical field is not of spatially uniform intensity and therefore neither is the steady-state carrier population. The result is a spatially varying gain and refractive index. In particular, where the optical intensity exhibits a local maximum, the carrier density has a local minimum, implying locally decreased gain and increased refractive index. This is the essence of the nonlinear interaction between the laser beam and the semiconductor material that determines its beam quality (Thompson, 1972; Lang, 1979).

This intensity dependence is incorporated easily into the carrier-dependent dielectric constant because the population inversion may be written as a function of the field intensity via the charge carrier and photon rate equations (Lau and Yariv, 1985). The optical nonlinearity arises by introducing gain saturation, namely $g = g(\omega_o, |E|^2)$, into the effective refractive index $n_{eff}(\omega_o)$. Gain saturation is the process whereby stimulated emission decreases the carrier population, and in turn the laser gain, from its unsaturated value. The unsaturated modal gain is the gain when no photons are present: It is approximated near threshold by the following simple linearization:

$$g = \Gamma\beta(J - J_o), \tag{4.15}$$

where β is the differential gain (not to be confused with the propagation constant), J is the current density, and J_o is the current density required for optical transparency at frequency ω_o. For quantum-well lasers at high pump currents, this equation should be modified to include band-filling effects, which cause a sublinear dependence of modal gain on current density (Arakawa and Yariv, 1985). However, the values $\Gamma\beta$ and J_o can be determined experimentally from measuring the threshold currents of broad-area lasers of different lengths and fitting the results to the linearized gain-current relation of Eq. (4.15) (Welch *et al.*, 1990a). The dependence of modal gain (and carrier density) on field intensity is manifest in the reduction of modal gain by gain saturation when photons are present (Thompson, 1972; Lang *et al.*, 1993):

$$g(x) = \Gamma\beta(J - J_0)/(1 + P(x)/P_{sat}), \qquad (4.16a)$$

where $P(x)$ is the optical power $|E(x)|^2$, and P_{sat} is the saturation power density, given by

$$P_{sat} = hc/\lambda q \cdot \eta_{int}/\Gamma\beta, \qquad (4.16b)$$

where η_{int} is the internal quantum efficiency and q is the electronic charge.

Figure 4.10 depicts the effect of gain saturation in the case of fundamental mode propagation within a stripe-geometry laser. The lateral gain and refractive index profiles, plus the self-consistent mode intensity and wavefront, are shown below laser threshold (dashed) and above laser threshold (solid). Because the refractive index above threshold is increased where the optical intensity is highest, thereby creating a local refractive index waveguide, the mode is further narrowed and its peak intensity increases. This "self-focusing" mechanism and its accompanying local decrease in gain ("spatial hole-burning") are collectively termed "filamentation" because of their tendency to focus broader beams down into narrow filamentlike propagating streams of light (Thompson, 1972).

Theory and experiments have shown that incident beams broader than approximately 10 μm are likely to break apart into many filaments, each 5–10 μm in width, in the presence of gain saturation. The resulting intensity modulation is accompanied by wavefront modulation as well, causing large aberrations in the optical beam (Goldberg *et al.*, 1993). While gain saturation, namely the depletion of carriers via stimulated emission, is the root of filamentation, its effect on beam quality is more directly caused by the coupling of gain saturation into the local refractive index via the α-parameter.

4.2.5   Brightness, Strehl ratio, and M^2

In the preceeding sections we have outlined the dominant mechanisms that determine the shape of the lasing mode. How best then to measure the "quality" of the emitted beam that results? There are many ways of characterizing a laser output beam. The three most common are simply its power, its spectrum, and its radiation pattern (or "far field"). In most cases, however, these measures define only in a qualitative way the usefulness of the laser in a particular application. A more quantitative mea-

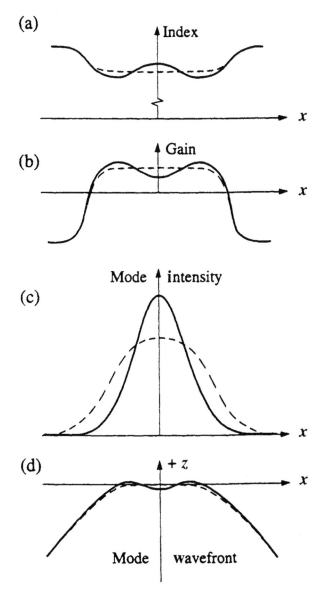

Figure 4.10: Self-focusing in a broad-area laser diode, resulting from gain saturation above threshold. (a) Refractive index, (b) gain, (c) mode intensity, and (d) mode wavefront, all shown below threshold (dashed lines) and above threshold (solid lines) (from Thompson, 1972).

sure is required for determining how much power can be focussed to a small spot or how much power can be propagated a large distance from the laser and recovered in a detector, for example. For this reason, in this section we review three quantitative measures of laser beam quality: brightness, Strehl ratio, and M^2.

4.2.5.1 Brightness

The brightness (or radiance) of a source is defined as the emitted power per unit solid angle per unit area (Leger, 1993). The brighter the source, the more intense the spot to which it can be focussed, and the further the distance it can be propagated. Light from an aberrated source diffracts over a much larger radiation angle than does light from a nonaberrated source, and likewise light from a source of small size diffracts over a much larger radiation angle than does light from a large source (given that both have nonaberrated wavefronts). The utility of the definition of brightness lies in the fact that it is independent of the optical system that follows the source and is used to collimate or focus the light, assuming that such an optical system is lossless. This fact, sometimes called the "radiance theorem" or "conservation of brightness," implies that brightness is solely a function of the laser source (Leger, 1993).

For example, consider the freely diffracting output of a laser diode, as shown in Fig. 4.11. Assume that the optical mode height perpendicular to the junction is d, that the optical mode width in the plane of the pn junction is D, and that its divergence at full-width-half-maximum (FWHM) points is θ_x by θ_y radians. For an output power of P Watts, the brightness of the source is

$$I = P/d\theta_y D\theta_x. \qquad (4.17)$$

If the laser output is diffraction-limited, then $\theta_x = \lambda/D$, $\theta_y = \lambda/d$, and the brightness reduces to

$$I = P/\lambda^2. \qquad (4.18)$$

The brightness is therefore proportional to the total power divided by the square of the wavelength for a diffraction-limited source. Furthermore, at a given power output, brightness is maximized for a diffraction-limited source, because for spatially incoherent beams $\theta_x > \lambda/D$ and $\theta_y > \lambda/d$.

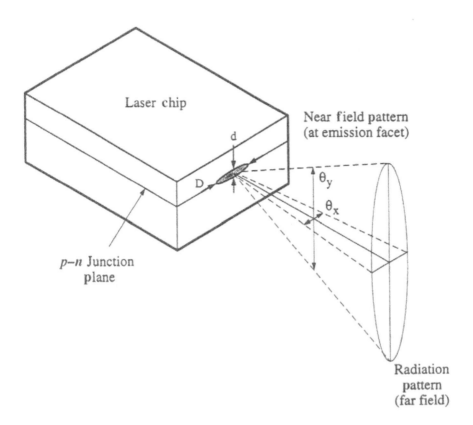

Figure 4.11: Schematic diagram of a laser chip showing definition of near field width D, height d, and far-field divergences θ_x and θ_y.

4.2.5.2 Strehl ratio

Many coherent laser applications require maximum brightness from a laser source. As discussed above, maximum brightness from a given source is attained when the radiation pattern is diffraction-limited, implying negligible amplitude and phase distortion in the laser beam. When a source is diffraction-limited, its energy can be propagated over long distances maintaining maximum on-axis intensity. This quality is desirable in applications such as free-space communication, where satellite receivers must maximize detected signal in order to maintain acceptable bit-error-rates. Furthermore, diffraction-limited sources can be focussed to a small spot on the order of a wavelength in dimension and provide maximum intensity at its focal spot. The latter quality is important for fiber-coupling, optical recording, frequency-doubling, and other scientific applications including laser tweezers and scissors. For many such applications, the far-field pattern of a source can be examined to predict behavior within an optical system. In this section, the Strehl ratio is defined as a quantitative measure of performance.

The far-field intensity pattern $I(x, y, z)$ of a laser is proportional to the square of the Fourier transform of the near-field complex amplitude $E(x, y)$ (Leger, 1993):

$$I(x,y,z) = (1/\lambda z)^2 \left| \int \int E(x',y') \exp\{-i2\pi/\lambda z (xx' + yy')\} \, dx' dy' \right|^2. \qquad (4.19)$$

The Strehl ratio of a laser beam is proportional to the on-axis intensity $I(0,0, z)$, as this quantity is desired to be maximized in many applications. It can be shown (Leger, 1993) that the Strehl ratio of a beam is maximized when the near field $E(x, y)$ of the source exhibits constant intensity and phase across its emitting aperture. As a result, the Strehl ratio S of an arbitrary beam with nonconstant near field intensity and phase is defined as the ratio between the on-axis far-field intensity of the test laser beam and the on-axis far-field intensity of a uniformly illuminated (*i.e.*, flat intensity and phase) aperture of the same size as the laser source under test. Therefore, ideal beams have $S = 1$, and all real beams have $S < 1$. The on-axis intensity of the uniformly illuminated aperture can be evaluated from Eq. (4.19):

$$I_u(0,0, z) = (1/\lambda z)^2 K^2 A^2, \qquad (4.20)$$

where K is the field amplitude (which is constant) over the uniformly illuminated aperture and A is its area. Note that K must be defined such that $K^2 A$ is the total power output P. Therefore the Strehl ratio is given by

$$S = I(0,0, z)/I_u(0,0, z) = (1/PA) \left| \int \int E(x',y') \, dx' dy' \right|^2. \qquad (4.21)$$

In a real source, $E(x, y)$ is not a constant over the emitting aperture, and therefore the Strehl ratio is reduced from unity by nonuniformities in the near-field amplitude and phase distribution.

In many systems, the complex-valued amplitude of the electric field at the source-emitting aperture is not readily measured, and thus the Strehl ratio must be estimated from the laser far-field pattern. Several common methods used to estimate the Strehl ratio are (1) power through a diffraction-limited slit, (2) integration of the far field pattern over its diffraction-limited angle, and (3) power coupled into a single-mode fiber.

4.2.5.3 M^2-factor

Another quantitative measure of beam quality that has been recently introduced is the M^2 factor (Siegmann, 1990). This factor relates the divergence of a real laser beam in the far field to its near-field waist size and has a limiting value of unity for the ideal case of a Gaussian beam. In a sense, it complements characterization of a beam via its Strehl ratio, because it is based on beamwidth rather than on on-axis intensity. Unlike the Strehl ratio, for which real beams have values of $S<1$ and the "ideal beam" is of constant intensity over a finite near-field aperture, M^2 values of real beams are >1 and the "ideal beam" is the smoothly varying fundamental Gaussian distribution ("TEM$_{oo}$").

The M^2 definition of beam quality is based on determination of the space-beamwidth product for a given laser beam, where the spatial and angular widths (or near-field and far-field beamwidths) for the laser beam are defined as the variances (or second moments) of the beam intensity profile in the spatial and spatial-frequency domains. As this definition is completely analogous to the Uncertainty Principle, it is not surprising that the TEM$_{oo}$ Gaussian beam should be the standard by which all other beams are measured.

For any real beams, the variance σ_x of its normalized intensity profile $I(x, y, z)$ is defined as

$$\sigma_x^2(z) = \iint (x - \bar{x})^2 I(x,y,z)dx\,dy \,, \tag{4.22}$$

where $\bar{x}$ is the mean position in the x-direction. If $I(x, y, z)$ is a Gaussian beam, then the Gaussian spot size $w(z)$ can be defined in terms of its variance as (strictly true in the paraxial approximation)

$$w(z) = 2\,\sigma_x(z) \,. \tag{4.23}$$

In analogous fashion, we define the "spot size" $W(z)$ of an arbitrary *real beam* at any z-plane as

$$W(z) = 2\sigma_x(z) \,. \tag{4.24}$$

If the beam waist (minimum spot size) is located at $z = z_0$, then by definition $W_0 = 2\,\sigma_x(z_0)$. For simplicity, we write $w_0 = 2\sigma_0$, and $W_0 = 2\,\sigma_{xo}$.

In similar fashion to Eq. (4.23), the angular variance $\sigma_s(z)$ can be defined in terms of the Fourier transform of the spatial beam profile. The main result of the theory is then expressed in the following two equations generalized from Gaussian beam propagation:

$$W^2(z) = W_0^2 + M^4 \lambda^2 / \pi^2 W_0^2 (z - z_0)^2 , \qquad (4.25)$$

where $M^2 = 4\pi\sigma_{xo}\sigma_{so}$, $M^2 > 1$.

For a Gaussian beam, $\sigma_{so} = 1/2\pi w_0$, and therefore $\sigma_{xo}\sigma_{so} = 1/4\pi$. Therefore $M^2 = 1$ for a Gaussian beam, and (it can be shown that) $M^2 > 1$ for all other beams.

To get a feeling for the meaning of M^2, we consider the approximation of Eq. (4.25) in the far field $(z-z_0) \gg \pi W_0^2/\lambda$:

$$W(z) = M^2 (\lambda / \pi W_0) (z - z_0) . \qquad (4.26)$$

Because $\lambda/\pi W_0$ represents the "diffraction-limited" angular divergence of a Gaussian beam, we can interpret M^2 as the "number of times" diffraction limit of the beam divergence. Thus the M^2 formalism provides a rigorous definition of beam quality that can be used to characterize distorted, multimode, and partially coherent laser beams in terms of one simple number.

Conveniently, in the early 1990s, several companies introduced commercial products that can measure M^2 (in both x- and y-directions) in near real time. These instruments can be used in a variety of applications, such as optimization of external cavity laser alignment to minimize M^2, identification of multimode operation (M^2 discontinuous with injection current), and the measurement of beam astigmatism.

4.2.6   Case study: monolithic flared amplifier master oscillator power amplifier

As described in Sect. 4.1.3, single-mode index-guided lasers are capable of emitting up to 200 mW in a diffraction-limited radiation pattern and offer reliable operation projected to beyond 10^5 hours (Scifres, 1993). It has been the goal of laser diode researchers for many years to increase beyond 1 Watt the coherent, diffraction-limited power available from a semiconductor laser. Because of the optical power density limitations discussed in Sect. 4.1.2, much research has been focussed on broadening the laser aperture over which single-mode, diffraction-limited operation can be maintained. The various approaches include (but are not limited to) unstable resonators, phase-locked laser arrays, master oscillator power amplifiers (MOPAs), external-cavity lasers/laser arrays, and antiguide arrays (Chinn, 1994; Botez, 1993). All of these approaches have met with some success. Widespread use of these devices has been hindered by the tight restrictions and tolerances placed upon the uniformity of broad-area and multielement laser architectures.

Recently, the MOPA laser architecture has been demonstrated to successfully boost the clean-beam output power available from single-stripe lasers to beyond the 1 Watt level. Consider the monolithic flared-amplifier master oscillator power amplifier (MFA-MOPA) presented in Fig. 4.12 (Welch *et al.*, 1992). In this section, we discuss the design and operating characteristics of the MFA-MOPA in order to illustrate some of the concepts outlined in Sect. 4.2.1–4.2.5. This laser has demonstrated, as of this writing, the highest diffraction-limited, single-mode output of any monolithic semiconductor laser or laser array, at 3.0 W cw.

In the MFA-MOPA architecture, a high-power flared-contact amplifier is monolithically integrated onto the same chip as a high-power single-stripe laser. The single-stripe laser, termed the master oscillator, is a 4-μm-wide index-guided laser whose reflectors are not cleaved crystal facets

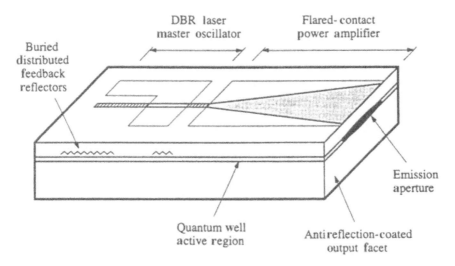

Figure 4.12: Schematic diagram of monolithically integrated master oscillator power amplifier (MOPA) utilizing a DBR master oscillator and a flared-contact power amplifier (from Welch *et al.*, 1992).

but integrated gratings—distributed Bragg reflectors, or DBRs for short. The DBRs provide frequency-selective feedback at a wavelength proportional to the grating period (Suematsu *et al.*, 1985), such that the master oscillator lases in a single spectral mode. The laser waveguide is index-guided, similar to those discussed in Sect. 4.1.3, and therefore operates in the fundamental lateral mode, which has a near-Gaussian intensity profile. Such DBR lasers, operated without an adjacent amplifier, have recently demonstrated diffraction-limited performance in excess of 100 mW cw (Major and Welch, 1993). As indicated in Fig. 4.12, the front grating is fabricated significantly shorter than the rear grating in order to lower its reflectivity and thereby transmit the majority of the DBR laser output into the adjacent amplifier.

The DBR laser output is amplified by the flared-contact power amplifier placed adjacent to the master oscillator. Unlike the master oscillator, whose index-guided waveguide confines the light to a narrow stripe along its length, the amplifier is a gain-guided device (Walpole *et al.*, 1992). Current to the amplifier is restricted to a tapered region whose width varies linearly up to 250 μm at the end of its 2-mm length. Because the amplifier lacks lateral index confinement, the injected beam from the master oscillator spreads laterally by diffraction and is simultaneously amplified. The taper angle, at ~5°, is wider than the beam diffraction angle whose full width at half maximum (FWHM) is approximately $\theta_x = \lambda/Dn_{eff} = 0.98/4*3.3 = 0.07$ radians or 4.25°, where $n_{eff} = 3.3$ is the effective refractive index of the DBR lateral mode. The output facet of the MOPA is antireflection-coated to achieve a residual reflectivity $R < 0.1\%$, thereby minimizing feedback from the amplifier to the master oscillator and also suppressing parasitic lasing of the amplifier alone.

The L–I curve of the MFA-MOPA is presented in Fig. 4.13. For this measurement, the DBR master oscillator was biased at 100 mA so as to inject approximately 50 mW into the amplifier, and the amplifier current was varied up to 6 A. A maximum cw output power of 3.6 W is measured. When the DBR master oscillator is unbiased, laser oscillation is quenched, and only low-power amplified spontaneous emission (ASE) noise exits the amplifier. The MFA-MOPA spectrum under master oscillator injection is shown in Fig. 4.14. The spectrum displays a single longitudinal mode near 990 nm wavelength; the exact frequency tunes slightly to longer wavelength with increased amplifier current due to the temperature increase of the DBR laser.

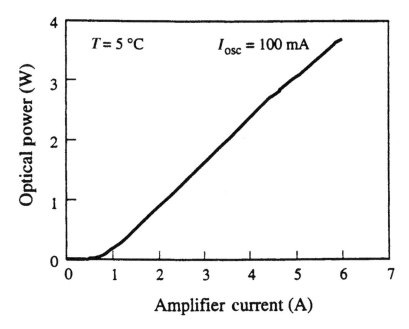

Figure 4.13: Light-current curve of the flared-contact MOPA of Fig. 4.12.

The beam quality of the MFA-MOPA emission is determined by its near-field and far-field patterns. The near field at the output of the flared amplifier can be estimated theoretically by applying numerical beam propagation methods to the output of the DBR laser (Lang *et al.*, 1993). Fig 4.15 plots the expected near-field intensity pattern as a function of the amplifier current. At low currents, the near field is near-Gaussian, replicating the output of the master oscillator. This replication is expected because, at low output powers, the effect of gain saturation (Eq. 4.15) is minimal. At higher power, gain saturation within the central part of the amplifier reduces the gain of high-intensity on-axis rays, relative to lower-intensity off-axis rays, resulting in a flattened intensity profile at the amplifier output. The phase distribution, or wavefront, of the amplified beam is near-spherical, corresponding to diffraction from a narrow (4-μm-wide) aperture at the amplifier input.

The directly measured radiation pattern from the flared output aperture of the MFA-MOPA is a broadly diverging beam characteristic of emission from a $d = 1$-μm-high by $D = 4$-μm-wide source. The waist of

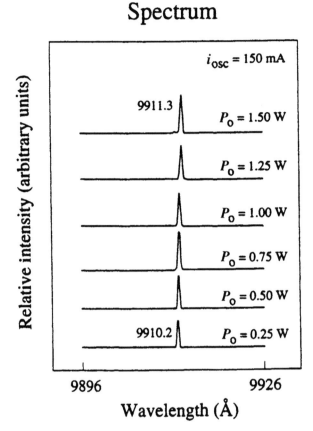

Figure 4.14: Spectrum as a function of output power of the flared-contact MOPA of Fig. 4.12.

emission perpendicular to the *pn* junction is located at the output facet; however, the waist of the emission in the plane of the *pn* junction is located near the amplifier input. Consequently, the output beam is highly astigmatic, as indicated in Fig. 4.16a. The astigmatism *a* can be corrected using conventional bulk optics. In order to characterize the corrected, or stigmated, MFA-MOPA output, the quadratic component of the emitted wavefront can be removed by placing an ideal thin lens at the output facet and forming a far-field image in the Fourier plane of a subsequent imaging lens, as shown in Fig. 4.16a. In practice, a single lens is used to make the measurement, as shown in Fig. 4.16b. The so-measured far-

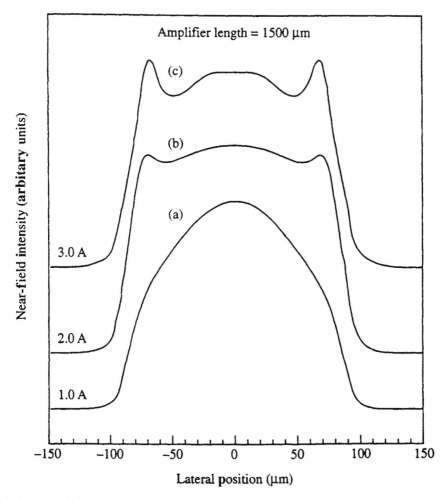

Figure 4.15: Calculated near-field intensity pattern of the flared amplifier output as a function of amplifier current (from Lang *et al.*, 1993).

field pattern is plotted as a function of output power in Fig. 4.17. The far-field displays a single, near-Gaussian lobe with a narrow FWHM = 0.25° indicative of diffraction-limited emission (see Sect. 2.5.1) from the 250-μm-wide output aperture of the flared amplifier. In the direction transverse to the *pn* junction, a single-lobed far field of FWHM = 28° is measured.

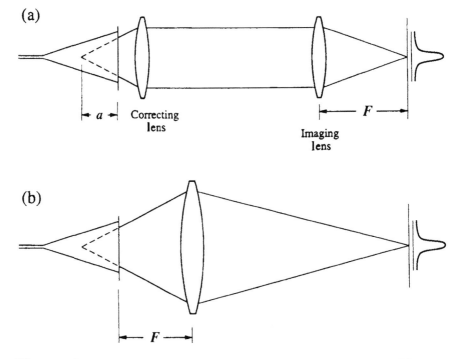

Figure 4.16: Schematic diagram showing (a) conceptual method of forming a lateral far-field image from a flared amplifier, with astigmatism a, using two lenses, and (b) practical method using a single lens.

The brightness of the MOPA source, at 1 W cw output power, can be calculated using Eq. (4.18) to be 100 MW/cm^2-ster. This value may be compared to that of the single-stripe laser of Sect. 4.1.3, whose brightness at 200 mW output is 20 MW/cm^2-ster. Since both lasers are diffraction-limited, the brightness scales simply with the output power. At the time of this writing, the MFA-MOPA is the brightest monolithic semiconductor source to be demonstrated.

The Strehl ratio and M^2 factor of the MFA-MOPA have also been characterized. As discussed in Sect. 4.2.5.2, the Strehl ratio is often estimated by measurement of the fractional power delivered through a diffraction-limited slit placed at the far-field image plane. Figure 4.18 plots the slit transmission versus output power of an MFA-MOPA laser. The Strehl ratio is observed to decrease gradually from ~100% at low power to 77%

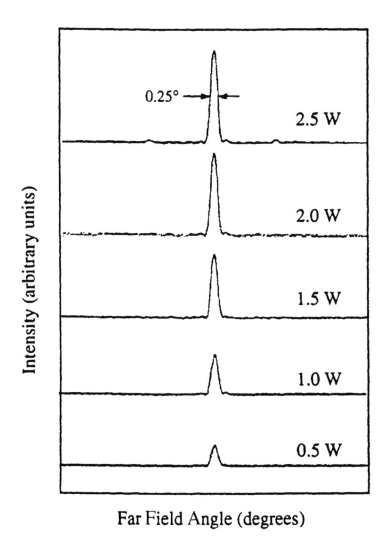

Figure 4.17: Lateral far-field pattern of the flared-amplifier MOPA of Fig. 4.12, measured according to the scheme of Fig. 4.16b.

at 1 W cw output. This decrease is attributed to the formation of filaments in the output as discussed below.

The M^2-value of the MFA-MOPA at 1 W cw output was measured according to its definition in Eq. (4.25). As shown in Fig. 4.19, an f = 8-mm-lens was placed in the MOPA output beam to form a waist in the

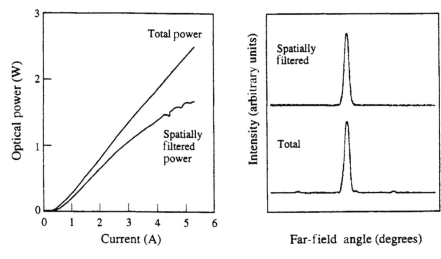

Figure 4.18: Total power and spatially filtered power, measured through a slit placed at the far-field image plane of the MOPA emission of Fig. 4.17. The ratio is an approximation to the Strehl ratio of the beam.

plane of the *pn* junction. The measured convergence and divergence of the beam in the vicinity of the waist is compared to an ideal Gaussian (TEM$_{oo}$), from which a value of $M^2 = 1.25$ is extracted. The measured deviation of M^2 from its ideal value of unity does not indicate a poor-quality beam but merely reflects the expectation according to the near field of Fig. 4.15 that the MFA-MOPA output is flattened by gain saturation at high power to a non-Gaussian profile.

Finally, it is instructive to consider one power limitation to the MFA-MOPA. At higher power levels, slight inhomogeneities of gain or refractive index within the amplifier impress small intensity ripples and phase modulation upon the diffracting beam, and filamentary output can result. Figure 4.20 presents the near-field pattern at the output of an MFA-MOPA emitting near 860 nm wavelength. The near field clearly displays intensity modulation over the central portion of the beam and two large intensity spikes near the stripe edges. The corresponding far-field pattern (not shown) is dominated by a central diffraction-limited lobe; however, its peak height increases sublinearly with increasing output power. Above a few Watts, the intensity modulation of the output grows, causing an increasing fraction of the total power to be emitted in non–diffraction-limited components. In any case, the MFA-MOPA therefore represents a

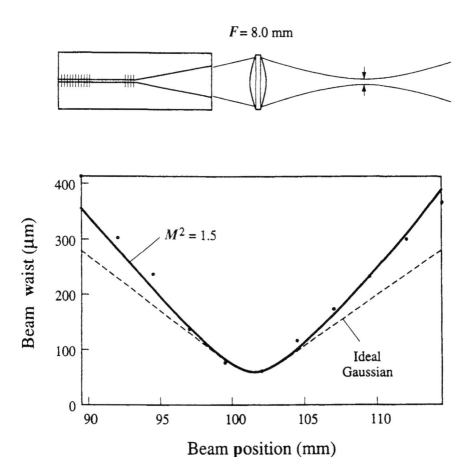

Figure 4.19: Measurement of the M^2 parameter of the flared-amplifier MOPA emission at 1.0 Watt cw output power.

significant increase in the power output and brightness of the single-stripe lasers described in Sect. 4.1.3. The extent to which these high-power lasers become utilized in industrial and scientific applications in the future will hinge upon their demonstration of long-term reliability.

4.3   Thermal management

Many high-power semiconductor lasers provide diffraction-limited, high-brightness output to powers of 1 Watt and beyond before beam degradation

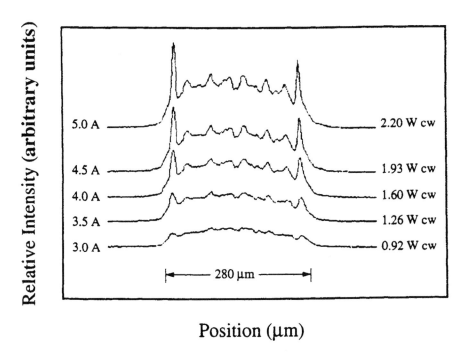

Position (μm)

Figure 4.20: Measured near-field intensity profile of a flared-amplifier MOPA emitting near 860 nm wavelength, as a function of output power.

occurs. Arrays or bars of lasers can provide up to hundreds of Watts of raw power. Almost all semiconductor lasers operate to higher peak output powers under pulsed operation than under CW operation, regardless of beam quality specification. This fact reveals the necessity of proper heat-sinking and temperature control in order to maximize laser performance at high drive levels. In this section, we discuss the important issues related to such thermal management.

4.3.1   Operating parameters

In this section, we review the important electrical and optical properties that characterize diode laser operation and that have an impact on laser output power, efficiency, and reliability. The semiconductor laser is unique in that the population inversion required to reach lasing threshold results from electrical charge injection into the active region. The accumulating electrical charge provides gain for the spontaneously emitted photons recirculating within the laser resonator. As the injection current is increased above its threshold value, the oscillation intensity builds up, and lasing results. Lasing is evidenced both by a sharp increase in output power and a narrowing of the broad spontaneous emission spectrum into distinct lasing lines.

Figure 4.21 presents a schematic diagram of the semiconductor laser light-current (*L–I*) and current-voltage (*I–V*) relations. The (*L–I*) curve is principally characterized by its threshold current I_{th} and slope efficiency η_d, while the *I–V* curve is principally characterized by its junction voltage V_j and series resistance R_s. All of these parameters have a strong impact on overall laser efficiency, which in turn has an impact on laser operating temperature and reliability as discussed in Sects. 4.3.2, 4.3.3, and 4.3.4.

The laser threshold current is defined as that current required to increase the unsaturated gain (see Eqs. 4.7b and 4.16a) of the optical mode to the level of the resonator losses and thereby initiate lasing. The

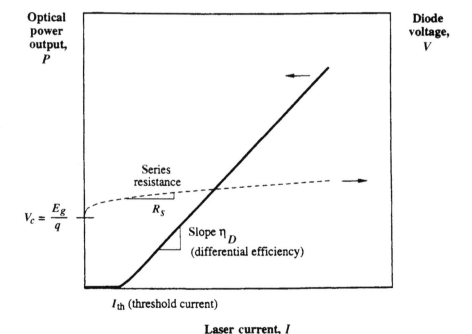

Figure 4.21: Schematic diagram of the light output vs. current plot and the voltage-current relation of a laser diode.

resonator losses are commonly lumped into two terms: the distributed optical loss α present throughout the resonator and the output coupling losses resulting from the use of cavity reflectors of finite reflectivity R_1 and R_2 (<1). At lasing threshold, the requirement of "gain equals loss" implies

$$\Gamma g_{th} = \alpha + (1/2L)\ln(1/R_1R_2),\tag{4.27}$$

where the second term is the transmissive loss (*i.e.*, laser output) normalized to unit length by the active region length, L. Lasing threshold is evidenced by a knee in the L–I curve as indicated in Fig. 4.21. Usually the threshold current is measured to be the intersection of the extrapolation of the linear portion of the L–I curve with the current axis. Since threshold current scales as active region area and is therefore device-dependent, a commonly used figure-of-merit is the threshold current density, J_{th}, defined as the threshold current divided by the active region area:

$$J_{th} = I_{th}/\text{Area}.\tag{4.28}$$

The threshold current density scales in proportion to the resonator loss, with low-loss resonators characterized by low values of J_{th}. Commonly, broad area lasers of width 100–200 μm and length 1–2 mm are fabricated as standards to evaluate J_{th}: In unstrained GaAs/AlGaAs lasers values of J_{th} less than 100 A/cm^2 (Chen *et al.*, 1987) have been measured, and in compressively strained InGaAs/AlGaAs lasers values of J_{th} as low as 50 A/cm^2 (Chand *et al.*, 1991) have been reported.

By using Eqs. (4.15) and (4.27), the threshold current density may be re-expressed as

$$J_{th} = J_o + (1/\Gamma\beta)(\alpha + [1/2L]\ln(1/R_1R_2)).\tag{4.29}$$

As described earlier (see Eq. 4.15), this equation assumes a linear gain-current relation, which for quantum-well lasers must be modified to an approximately logarithmic relation for application in resonators with moderate to high losses (Arakawa and Yariv, 1985). According to Eq. (4.29), the threshold current density is comprised of two components: that required to pump the active region to optical transparency (J_o) and that required to overcome the resonator losses.

Above threshold, the resulting stimulated emission shortens the lifetime of the inverted charge carriers such that the inversion level remains essentially clamped at its threshold value. Ideally, all additional electrons injected into the laser recombine radiatively, each producing a photon of energy $h\upsilon$, so that laser light is produced at the rate of one photon per injected electron. In practice, not all injected electrons recombine radiatively; taking the probability that an injected carrier recombines radiatively within the active region as η_i, the rate of photons produced within the laser resonator by stimulated emission is

$$P_s/h\upsilon = \eta_i(I - I_{\text{th}})/q , \qquad (4.30)$$

where q is the electronic charge. η_i is commonly referred to as the internal quantum efficiency and is typically as high as 0.95 for GaAs/AlGaAs lasers. Part of the stimulated power produced is dissipated inside the laser resonator via the distributed loss α, and the remainder is coupled out through the cavity feedback reflectors. The laser power output through the cavity mirrors as a function of current above threshold can thus be written as (Yariv, 1985):

$$P = \eta_i(I - I_{\text{th}})(h\upsilon/q)\{\ln(1/R_1R_2)/[2\alpha L + \ln(1/R_1R_2)]\} . \quad (4.31)$$

The external differential quantum efficiency, η_d, is defined as the ratio of photon output rate to electron input rate and can be measured from the L–I curve of Fig. 4.21 according to

$$\eta_d = d(P/h\upsilon)/d((I - I_{\text{th}})/q) = \lambda(\mu m)/1.24P(\text{Watts})/(I - I_{\text{th}})(\text{Amps}) , \tag{4.32}$$

where $h\upsilon = hc/\lambda$ has been used, and the constant hc/q is taken as 1.24 W-μm/A. While the differential quantum efficiency is limited by energy conservation to $\eta_d < 1$, the slope of the L–I curve measured in W/A depends on the lasing wavelength λ. At best, when $\eta_d = 1$, the slope efficiency $P/(I - I_{\text{th}}) = 1.24/\lambda(\mu m)$ W/A, where the wavelength λ is measured in units of microns. The measured value of differential efficiency can therefore be used to extract some of the parameters of Eq. (4.31), because according to Eqs. (4.31) and (4.32), η_d satisfies

$$\eta_d^{-1} = \eta_i^{-1}\{1 + \alpha 2L/\ln(1/R_1R_2)\} . \tag{4.33}$$

Commonly, the values of η_i and α are extracted from experimental data by plotting measured values of η_d^{-1} as a function of L for lasers cleaved to different lengths, where the reflectivities R_1 and R_2 are known (for example, $R_1 = 0.31 = R_2$ for GaAs/AlGaAs lasers with uncoated facets). In such a plot, η_i^{-1} is the intercept of the plotted data with the

η_d^{-1} axis, and $2\alpha/\ln(1/R_1R_2)$ is the slope of the plotted data. Figure 4.22 plots a typical data set for broad area lasers lasing at a wavelength $\lambda = 980$ nm, for which an internal quantum efficiency of $\eta_i = 0.90$ and distributed optical loss of $\alpha = 1.5$ cm^{-1} are extracted.

The advent of quantum well lasers in the past 10 years has dramatically improved the optical properties of semiconductor lasers. The optical losses, α, and transparency current density, J_o, in quantum well lasers are much less than those of double heterostructure lasers due to the reduced overlap of the optical mode with the active region. Distributed losses as low as 1–2 cm^{-1} and transparency current densities less than 50 A/cm^2 have been measured, enabling the fabrication of longer cavities without paying a substantial penalty in increased threshold current or decreased differential efficiency, while reducing the thermal resistance of

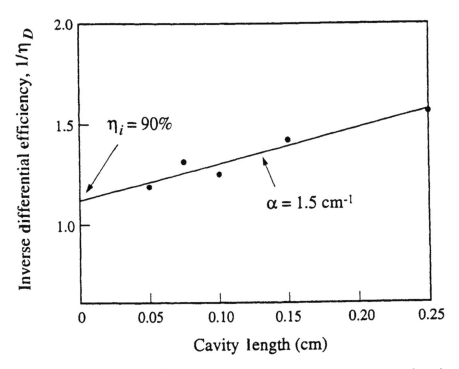

Figure 4.22: Inverse differential quantum efficiency plotted vs. cavity length for a strained-layer InGaAs/AlGaAs laser emitting near 980 nm wavelength. The internal quantum efficiency and unsaturable loss are extracted according to Eq. (4.33).

the laser and increasing the reliability. Higher catastrophic power limitations are also a result of the lower optical overlap.

Figure 4.21 also plots the current-voltage (*I–V*) characteristic of the semiconductor laser. The various heterostructure layers that make up the laser are comprised of a core *pn* junction diode surrounded by cladding layers to provide charge carrier confinement, optical waveguiding, and thermal conductivity. The laser chip *I–V* characteristic is therefore dominated by a *pn* junction diode characteristic in series with a resistor that models the electrical conductivities of the heterostructure and substrate layers. Of course, to complete a useful device, contact metallizations are deposited, the lasers are soldered to heatsinks, and wire bonds are affixed to provide electrical connections. Real laser diodes therefore have complex equivalent circuits comprising junction capacitances and resistances, parasitic capacitances and inductances, and non-ohmic *p*-side and *n*-side contact metallizations. However, in practice, the voltage across high-power lasers is often simplified to the form

$$V = V_j + R_s I, \tag{4.34}$$

where V_j is the junction voltage and R_s is the series resistance. The junction voltage is usually approximated by a constant slightly larger than the energy gap of the active region, $V_c = E_g/q$ (see Fig. 4.21), and the series resistance R_s can be extracted from the slope of the *I–V* curve above laser threshold. The power consumption of the laser diode is calculated using $P = VI$; of interest is the dissipated electrical power:

$$P_{\text{diss}} = V_j I + R_s I^2, \qquad\qquad I < I_{\text{th}}; \tag{3.35a}$$

$$P_{\text{diss}} = V_j I + R_s I^2 - \eta_d (h\nu/q)(I - I_{\text{th}}), \qquad I > I_{\text{th}}. \tag{3.35b}$$

The dissipated electrical power is important because it governs the temperature rise of the active region during laser operation, which in turn has an impact on device performance and reliability (see Sects. 4.3.2, 4.3.3, and 4.3.4). It is desirable to minimize the dissipated electrical power and thereby maximize the electrical-to-optical conversion efficiency of the laser. The electrical-to-optical conversion efficiency, η_c, is defined as the ratio of optical output power to electrical input power and calculated as

$$\eta_c = \eta_d (I - I_{\text{th}})(h\nu/q)/I(V_j + R_s I). \tag{4.36}$$

In recent years, the conversion efficiency of laser diodes has been increased to the 50–60% level (Welch *et al.*, 1990b), and it is this high value of "wallplug" efficiency that makes semiconductor lasers so attractive for a variety of applications.

4.3.2   Thermal resistance

The temperature dependence of the laser diode characteristics are critical to device performance at high output powers, where heat generation is highest. In addition, some applications require laser operation over a wide temperature range. The thermal characteristic is a result of the temperature dependence of the lasing threshold current (its T_o, defined in Sect. 4.3.3), the thermal resistance of the laser and heatsink, and the conversion efficiency η_c of the laser. In this section, we discuss thermal resistance.

The temperature rise ΔT of the active region of a diode is governed by its thermal resistance, R_{th}, according to the equation

$$\Delta T = R_{th} P_{diss}, \tag{4.37}$$

where P_{diss} is the dissipated electrical power defined by Eq. (4.35). The temperature rise of the active region ΔT is usually defined with respect to a controlled temperature reference, such as recirculating cooling water or a thermistor placed within the laser heatsink to provide feedback to an active cooler such as a Peltier element. The finite thermal resistance of a laser diode results from the finite thermal conductivity of the epitaxial layers, solders, submounts, and heatsinks that are part of the laser diode assembly. Heat is dissipated within these layers and components en route to removal by active or passive coolers and contributes to temperature drops across each material, which cumulatively result in a temperature rise ΔT of the active region with respect to the reference temperature. The temperature rise ΔT of the active region is commonly measured by determining the wavelength shift of the laser diode as a function of output power. The band gap of the active region decreases as a function of increasing temperature, causing an emission wavelength shift to longer wavelength. For GaAs, this wavelength shift is approximately 0.3 nm/°C of temperature rise. If such a measurement is conducted as a function of laser current I, and the dissipated electrical power can be extracted from Eq. (4.35), then the thermal resistance of the laser can be estimated empirically.

The laser diode thermal resistance can also be predicted and optimized using numerical models. Commonly, finite-element solutions to

Laplace's equation (*i.e.*, heat diffusion) are computed using a structured model of the laser epitaxial layers, metallizations, and subassemblies. For example, a simple model of a stripe-geometry semiconductor laser is shown in Fig. 4.23 (Joyce and Dixon, 1975). The laser is shown mounted junction-side down using indium solder to a copper heatsink. The thermal conductivity and thickness of the heterostructure layers, substrate, metallizations, and copper submount are shown in the figure. The active region length and width are 375 μm and 12 μm, respectively. By imposing boundary conditions on the temperature of the bottom edge of the copper heatsink, and simulating heat generation due to a specified power dissipation in the active region, Laplace's equation can be solved numerically for the

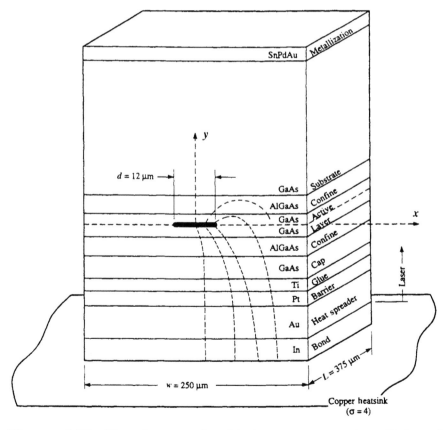

Figure 4.23: Three-dimensional model of a stripe-geometry laser diode used for thermal analysis (from Joyce and Dixon, 1975).

resulting temperature rise of the active region. In the case of this example, the calculated thermal resistance is R_{th} = 20.6 °C/W.

It is important to note some of the relative values of thermal resistivity W of common materials, shown in Table 4.1.

Because of the relatively low thermal conductivity of GaAs, and especially AlGaAs (Adachi, 1985), overall laser thermal resistance is lowered by bonding devices in a junction-side-down configuration. Use of high-thermal-conductivity submounts such as Cu or diamond will also lower R_{th}. In addition, thermal resistance is lowered by optimizing device geometry to encourage two-dimensional heat flow wherever possible (analagous to electrical resistance, thermal resistance is lowered by parallel heat flow). For this reason, thermal resistance is lowered almost in direct proportion to the active region length: Increasing the laser length by a factor of two will halve the thermal resistance. Of course, the laser length must be chosen also with regard to threshold current and differential quantum efficiency.

4.3.3   Characteristic temperature, T_o

An important effect of the temperature increase within the laser diode active region is its impact on laser threshold current. In general, threshold current increases with increasing active region temperature, which significantly increases the dissipated power and the heat generated. The principal mechanism for the threshold increase is a reduction in the gain available at any frequency ω_o at a fixed current density. Charge carriers

Material	W (°C-cm/W)
GaAs	2.27
$Al_xGa_{1-x}As$	$2.27 + 28.83x - 30x^2$
AlAs	1.10
Au	0.314
In	1.15
Cu	0.25
Diamond (CVD)	0.067–0.10

Table 4.1: Thermal resistivity W of some optoelectronic materials.

populate the conduction and valence bands according to Fermi-Dirac statistics: At any finite temperature, the carrier distribution extends into the upper levels of the bands. As the temperature increases, more carriers populate the higher energy levels, reducing the gain available at the lasing wavelength. In addition, carrier leakage over the heterobarriers of the active region increases, which reduces the internal quantum efficiency.

The following empirical relation has been developed to model the increase of laser diode threshold current with increasing temperature of the laser:

$$I_{th} (T + \Delta T) = I_{th} (T) \exp\{\Delta T/T_o\}, \qquad (4.38)$$

where $I_{th}(T)$ is the threshold current at temperature T, $I_{th} (T + \Delta T)$ is the threshold current at an elevated temperature $T + \Delta T$, and T_o ("tee-zero") is the characteristic temperature of the laser (Thompson, 1980). As T_o decreases toward zero, the laser threshold current becomes more and more sensitive to temperature increases. As examples, the measured T_o for (InGa)As/AlGaAs quantum-well lasers (λ = 780–980) range from $100 < T_o < 200$ °C, while those of InGaAsP/InP lasers (λ = 1.3–1.5 μm) range from $50 < T_o < 90$ °C [Agrawal and Dutta, 1986].

As expected from the preceding discussion, T_o is a strong function of the semiconductor laser active region and heterostructure design. A well-designed carrier-confinement heterostructure should provide several $k_B T$ (k_B = Boltzmann's constant) of carrier confinement within the active region to prevent carrier leakage over the heterobarriers. Recently, graded-index separate confinement heterostructures (GRINSCH) have been incorporated to grade the confinement layers bordering the heterobarriers away from the quantum-well active region in order to more effectively funnel carriers into the active region and to reduce the density of states at the energy levels most likely to be populated by "hot" carriers at elevated temperatures. Such structures can be optimized for high-temperature performance, allowing CW lasing at temperatures approaching T = 200 °C.

4.3.4 Reliability and activation energy

The reliability or lifetime of a semiconductor laser is an important factor to consider in the initial design. The reliability of a laser is determined by dissipated heat, thermal stresses, current density, handling, and power density at the facet. All of these parameters are dependent on the waveguide structure, resonator configuration, thermal resistance, and packaging. The specific parameters that drive a design are dependent on the particular application. For example, applications such as pyrotechnics or laser fuses require high output powers but need not exhibit thousands of hours of reliability. For such a laser, optical overlap of the active region can be reduced to minimize the optical density at the laser facets, thus increasing the catastrophic power limit. The cost to device performance is an increased threshold current and a lower conversion efficiency.

In this section, the major factors that determine laser reliability will be briefly reviewed (Thompson, 1980; Welch *et al.*, 1990c), including (1) thermally assisted degradation due to defect migration into the active region; (2) current-density enhanced degradation, which also results in active region defects; (3) catastrophic optical degradation (COD) resulting in facet damage; and (4) electrostatic discharge (ESD). In general, increased reliability resulting from lower degradation rates occurs at lower temperature, lower output power, and lower duty-cycle (pulsed or quasi-cw) operation.

Thermal dissipation is considered the dominant degradation mechanism of many semiconductor lasers. Thermal dissipation within the crystal can result in migration of crystal defects from the cladding regions, or substrate, into the active region. The crystal defects are often nonradiative centers that emit no light and propagate assisted by the electron-hole recombination energy (Thompson, 1980). The migration of such nonradiative centers are manifest as "dark-line defects" (DLDs) that no longer emit light and can be imaged using electroluminescence or cathodoluminescence techniques (Yellen *et al.*, 1993). The increased number of nonradiative recombination sites increases the threshold current and operating current of the laser over time.

Mechanisms that are thermally activated are associated with an activation energy, E_a. The activation energy governs the rate at which degradation is increased at elevated temperatures. An empirical relationship developed for the semiconductor laser lifetime (when determined by thermal dissipation) is (Thompson, 1980):

$$\tau = K \, \exp\{E_a/k_B T\}, \text{ or} \tag{4.39a}$$

$$\tau(T_1) = \tau(T_2)\exp\{E_a(1/k_B T_1 - 1/k_B T_2)\}, \; T_1 > T_2, \tag{4.39b}$$

where K is a constant dependent upon the laser structure. The activation energy E_a is determined by accelerated aging of diode lasers at different elevated temperatures. Laser lifetimes at elevated temperatures, often at 50–70 °C, are extrapolated from measured degradation rates. The laser degradation rate is typically the rate increase of operating current, in mA/ hr, measured during constant-power operation at elevated temperature. Commonly, the laser lifetime is defined as the number of hours required to increase the laser operating current by 20%, for example. By comparing extrapolated lifetimes at elevated temperatures, the activation energy E_a can be calculated, which in turn allows estimation of the laser lifetime at room temperature.

Depending on the type of laser, activation energies ranging from 0.7 eV to 0.2 eV have been reported, where higher values of E_a are associated with long laser lifetimes and high reliability. The variation is indicative of the many different types of nonradiative centers that can migrate to the active region. Because of the direct impact of the laser operating temperature on reliability, it is important to minimize the thermal resistance R_{th} and to maximize T_o.

Current density in the active region may have similar effects on device performance as thermal dissipation. The current density creates an electric field associated with the quasi-Fermi levels that lowers the effective binding energy of many impurities. The result is a migration of nonradiative traps into the active region similar to those associated with thermally activated degradation. Migration caused by current density is not temperature-dependent, however, and cannot be assigned a comparable activation energy. Degradation rates due to high current density are observed to vary only slightly as a function of temperature.

Laser failures associated with power density at the output facet have an entirely different characteristic than either thermally activated degradation or current-density-activated degradation. Failures due to optical power density are a result of the decrease of the catastrophic power limit with operating time. This COD limit decreases as a function of time under normal operating conditions to a saturated level. The device will continue to operate with a characteristic gradual degradation rate resulting from either thermal or current-density mechanisms until the COD limit drops below the operating condition. If the device is operated in a "constant power" mode, the laser will die suddenly as the feedback increases laser current to maintain constant power output. If the laser is operated under constant current feedback, then the laser output will decrease at a high rate until the entire facet has degraded. The occurrence of COD can be suppressed greatly by proper passivation of the facet (Yonezu, 1979;

Epperlein *et al.*, 1993), thus "enabling" these lasers to be limited at higher output powers by thermal dissipation.

If the laser is operated well beneath its COD threshold, then its lifetime scales as a function of power output approximately as:

$$\text{Life}(P_1) = \text{Life}(P_2)\{P_2/P_1\}^n, \qquad P_2 > P_1, \qquad (4.40)$$

where n is a constant that depends upon device design and is typically near 2.

Finally, proper handling of laser diodes is extremely important to maintain long lifetimes. Electrostatic discharge (ESD) caused by inadvertent human contact or voltage and current spikes from nonregulated power supplies can result in *pn* junction breakdown anywhere on the chip. Junction breakdown results in high leakage currents channeled into the damaged regions, quenching laser output.

For highly reliable operation of high-power lasers several trade-offs must be considered, which depend upon the laser application. If the reliability is limited by the power density at the facet, then a reduction in the optical overlap can be employed to increase the COD threshold. If the reliability is limited by thermal dissipation, then an increase in optical overlap may be warranted in order to lower the threshold current density and increase the conversion efficiency. Finally, if the reliability is limited by the current density in the crystal, then alternative solutions may be to (1) increase device size to lower current density at fixed operating current, (2) reduce operating current by adjustment of optical overlap, or (3) increase differential efficiency by adjusting the facet reflectivities.

4.3.5   Case study: QUASI-CW 1 cm-wide laser array bars

In addition to the myriad of scientific and commercial applications that use the semiconductor laser output directly, a very important application of semiconductor lasers is their utility as pumps for solid-state lasers. The advantages of semiconductor lasers over alternative pump sources such as flashlamps are their small size, high efficiency, low voltage, narrow wavelength band, and long operating lifetimes. Since spatial or temporal coherence are typically not required of solid-state laser pumps (with some fiber lasers and amplifiers representing one exception), semiconductor laser pumps are typically configured to emit as high a reliable output power as possible—hence design of the laser pump is driven significantly by packaging and thermal management issues.

The basic building block of the semiconductor laser pump is the 1-cm-wide laser bar, illustrated in Fig. 4.24. Diode laser arrays and bars have been the subject of extensive work over the past 15 years (Solarz *et al.*, 1994; Scifres and Kung, 1994). Each wide bar is subdivided into an array of current-pumped regions. In turn, each array element of the bar may also be subdivided into many parallel narrow stripes. The periodically spaced active elements of the bar constitute a percentage of the total chip area known as the "fill factor." The fill factor can be varied from as low as 10–20% to high values approaching 100%, depending upon the application requirement, primarily on required pulse length and duty factor. The bars are operated under quasi-cw conditions to pump Q-switched solid-state lasers, where high-energy pulses are required. The bars may also be operated under cw conditions (to lower peak power levels) for some applications.

In recent years, interest has evolved from pumping traditional solid-state lasers such as Nd:YAG to others such as Er:YAG and Cr:LiSrAlF$_6$. This interest has motivated rapid development of new semiconductor material systems operating at both longer and shorter wavelengths than GaAs:AlGaAs. Significant applications over a broad range of operating temperatures and duty cycles continue to exist at Nd:YAG and YLF pump

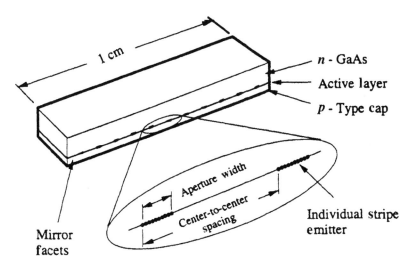

Figure 4.24: Geometry of 1-cm-wide laser diode array ("bar format") showing individual emitting apertures and center-to-center spacing ("duty factor"). Note each aperture is in turn typically a multistripe laser.

wavelengths of 798–810 nm. Erbium glass and Er:YAG lasers have created the need for pump wavelengths in the 920–980 nm range, which is met by InGaAs:AlGaAs semiconductors. At still longer wavelengths, Tm:YAG and Ho:YAG lasers require pump wavelengths near 1.8 μm, which is met by InGaAs:InGaAsP lasers. Finally, at the short wavelength extreme, solid-state materials such as Cr:LiCaAlF$_6$ and Cr:LiCaAlF$_6$ require pump lasers at near-680 nm wavelengths and are met by GaInP:AlGaInP laser diodes.

As mentioned above, high electrical-to-optical conversion efficiency is one of the fundamental advantages of diode laser arrays. In the past several years, operating efficiencies of quasi-cw bars have increased steadily from ~30% to >50%, primarily due to the advent of low-threshold, high-efficiency quantum well material. Figure 4.25 plots the *L–I* curve

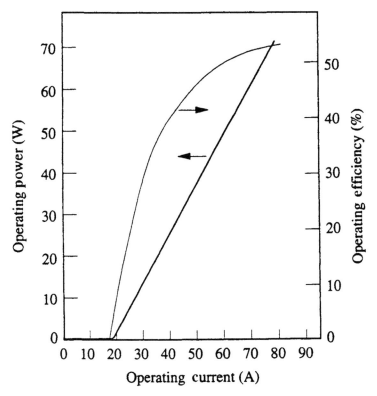

Figure 4.25: *L–I* curve of a quasi-cw laser array bar emitting near 807 nm wavelength, showing a >50% conversion efficiency (from Harnagel *et al.*, 1993).

and conversion efficiency of a quasi-cw 807 nm wavelength GaAs:AlGaAs bar of 1 cm width. The fill factor of the bar is ~90% and consists of 105-μm-wide emitting regions on 120-μm centers. The array is operated at $T=25$ °C in 200 μs pulses at a repetition rate of 50 Hz (duty factor = 1%). At its operating power of 60 W, the conversion efficiency approaches 54%. Such improvements in conversion efficiency significantly decrease dissipated heat, allowing comparable improvement in array operating lifetime. Alternatively, conversion efficiency increases allow operation of quasi-cw bars at higher temperatures or increased duty factors while maintaining a fixed "required" lifetime.

Improved material quality and design has been used to increase the T_o of quasi-cw bars. Figure 4.26 shows the L–I characteristic of such a bar

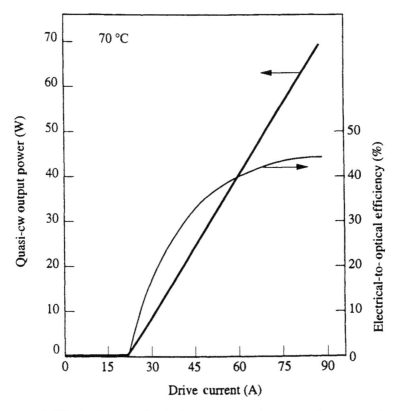

Figure 4.26: L–I curve of a high-T_o quasi-cw bar operating at an elevated temperature of 70 °C (from Harnagel *et al.*, 1993).

at $T=70$ °C ambient temperature. The conversion efficiency at operating power output of 60 W is approximately 43%. Due to the shift of operating wavelength with increased temperature, such a bar designed to operate at 807 nm at 70 °C emits near 793 nm at 25 °C.

An additional factor that enhances the lifetime of laser array bars is operating them significantly below their catastrophic damage level. Figure 4.27 presents the *L–I* curve of a 1-cm InGaAs/AlGaAs bar with a measured COD power limit of 247 W. Its operating efficiency peaks near 47%. These COD levels are comparable to those measured in the GaAs: AlGaAs materials. Operating life of the two material systems is nearly identical and is expected to exceed 10^{10} pulses of 200 μs duration. This projected lifetime corresponds to approximately 5000 hours of "on-time," which agrees well with reliability of 10 Watt cw bars that have a maximum output of 120 W cw (Sakamoto *et al.*, 1992).

There are many ways in which single 1-cm-wide laser bars, the building blocks of laser pumps, can be configured into pump modules. An

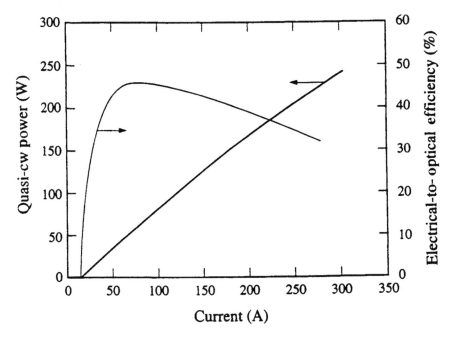

Figure 4.27: *L–I* curve under quasi-cw conditions of a 1-cm-wide strained-layer InGaAs/AlGaAs measured up to its COD limit of 247 W (from Harnagel *et al.*, 1993).

example method, practiced at SDL, is to mount each bar onto a submount, as shown in Fig. 4.28, which in turn may be combined together. Each submount consists of the laser bar, which is soldered p-side down onto a heat spreader and contacted on the n-side with an array of wire bonds that are bridged to an insulated n-contact. Such a part may be individually tested, burned-in, and yielded before being assembled into packages of larger arrays of bars.

Yielded arrays are in turn stacked and mounted within a variety of cooling subsystems, depending on the particular solid-state laser architecture for which it is designed. This architecture is not mated to any particular cooling scheme and therefore offers high flexibility. A few pump module geometries will be presented in this chapter; however, this represents but a small fraction of the commercially available designs. Figure 4.29 presents a cross-section of a stacked array of submounted bars spaced on 0.4-mm centers. If the submounts consist of 100-W bars, this configuration provides two-dimensional optical flux densities of 2.5 kW/cm^2.

Figure 4.30 shows how the basic submounted bar can be interleaved with copper heat spreaders to produce arrays for slab pumping spanning

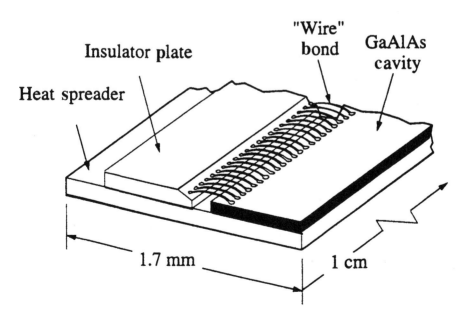

Figure 4.28: Schematic diagram of a modular submount used to heatsink 1-cm-wide laser bars (from Harnagel *et al.*, 1993).

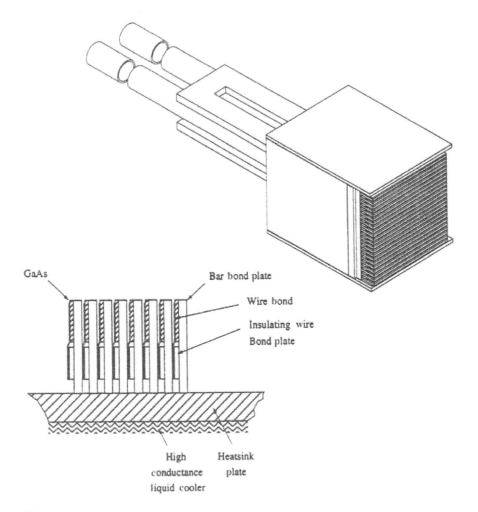

Figure 4.29: Schematic diagram of a stacked array of 1-cm-wide bars, commonly used for pumping solid-state lasers (from Harnagel *et al.*, 1993).

a broad range of required bar duty cycles. Figure 4.30a shows the high-density array of Fig. 4.29 mounted to a 1 cm × 1 cm cooler. This 25-bar, 1.5-kW peak power array can be operated at >3% duty cycle. By interleaving the basic submount with heat spreaders as in Figs. 4.30b and c, substantially higher duty cycles become possible. The slab array of Fig. 4.30b consists of 38 submounts interleaved with heat spreaders to produce a 1 cm × 3 cm array with >2.4 kW/cm^2 peak power (Harnagel

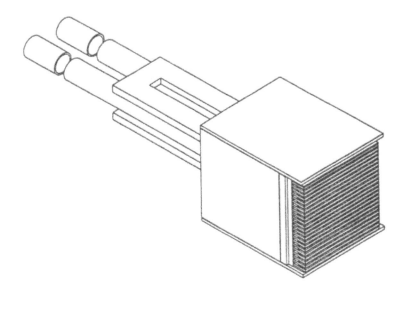

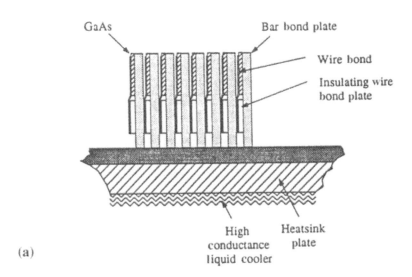

(a)

Figure 4.30: Schematic diagrams (a), (b), and (c) of successively lower duty-factor stacked arrays, interleaved with thicker heat spreaders, used to achieve higher-duty-cycle quasi-cw operation (from Harnagel *et al.,* 1993).

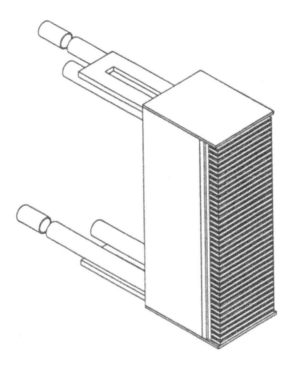

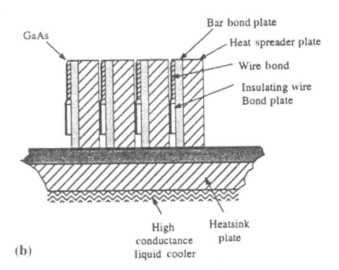

(b)

Figure 4.30: Continued

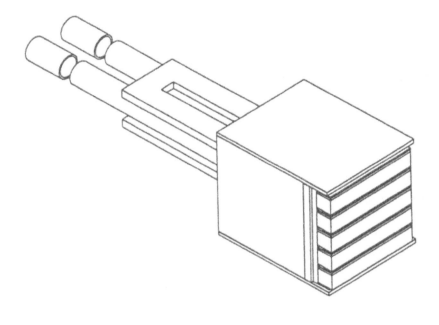

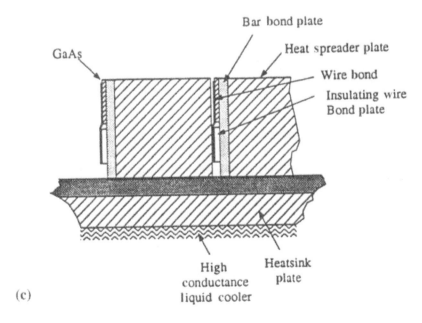

(c)

Figure 4.30: Continued

et al., 1993). This array is capable of operating at >6% duty cycle. Figure 4.30c shows the basic submount interleaved with a thicker heat spreader for a 2-mm bar spacing. This configuration allows array operation at 20% duty cycle. Arrays with *average* output power levels of 200 W/cm^2 or *peak* power levels of 5 kW/cm^2 are readily fabricated using this technique.

4.4   Conclusions

Within this chapter the framework for the analysis and design of high-power semiconductor lasers has been reviewed. These lasers have a variety of applications within the commercial and scientific marketplace, and the utility of semiconductor lasers lies within their wavelength span from the visible spectrum to the mid-infrared. High-power lasers are demanded by applications where the light is not acting as a passive probe but an active workhorse to pump lasers and amplifiers, activate photodynamic drugs, or act as laser tweezers and scissors. The issues reviewed in this chapter, namely (1) power limitations, (2) beam quality, and (3) thermal management are common to all semiconductor laser materials. As these issues have come to be understood more fully in the past several years, great advances in reliable output power have resulted, enabling semiconductor lasers to replace bulkier and less efficient gas and solid-state lasers, as well as opening up new applications. As new material systems mature, the future of semiconductor lasers continues to look bright.

Acknowledgments

The author would like to acknowledge the support of his colleagues at SDL Inc., especially D. Welch, R. Parke, R. Waarts, A. Hardy, R. Lang, R. Craig, and D. Scifres.

References

Adachi, S. (1985). *J. Appl. Phys.* **58**, R1–R29.

Agrawal, G., and Dutta, N. (1986). *Long-Wavelength Semiconductor Lasers.* New York: Van Nostrand Reinhold.

Arakawa, Y., and Yariv, A. (1985). *IEEE J. Quantum. Electron.* **21**, 1666–1674.

Botez, D. (1994). Monolithic phase-locked diode laser arrays, in *Diode laser arrays*, D. Botez and D. Scifres, eds., New York: Cambridge University Press.

Chand, N.; Becker, E. E.; van der Ziel, J. P.; Chu, S. N. G.; and Dutta, N. K. (1991). *Appl. Phys. Lett.* **58**, 1704–1706.

Chen, H. Z.; Ghaffari, A.; Morkoc, H.; and Yariv, A. (1987). *Electron. Lett.* **23**, 1334–1335.

Chinn, S. R. (1993). A review of edge-emitting coherent laser arrays, in *Surface-emitting semiconductor lasers and arrays*, G. Evans and J. Hammer, eds., New York: Academic Press.

Epperlein, P. W.; Buchmann, P.; and Jakubowicz, A. (1993). *Appl. Phys. Lett.* **62**, 455–457.

Goldberg, L.; Mehuys, D.; Surette, M. R.; and Hall, D. C. (1993). *IEEE J. Quantum. Electron.* **29**, 2028–2043.

Hadley, G. R.; Botez, D.; and Mawst, L. (1991). *IEEE J. Quantum. Electron.* **27**, 921–930.

Harnagel, G. L.; Vakili, M.; Anderson, K. R.; Worland, D. P.; Endriz, J. G.; and Scifres, D. R. (1993). *IEEE J. Quantum. Electron.* **28**, 952–965.

Henry, C. H. (1982). *IEEE J. Quantum. Electron.* **18**, 259–264.

Ikegami, T., IEEE J. Quantum. Electron. **8**, 470–475 (1972).

Jaeckel, H.; Bona, G.; Buchmann, P.; Meier, H.P.; Vettiger, P.; Kozlovsky, W.J.; and Lenth, W. (1991). *IEEE J. Quantum. Electron.* **27**, 1560–1567.

Joyce, W., and Dixon, R. (1975). *J. Appl. Phys.* **46**, 855–862.

Lang, R. (1979). *IEEE J. Quantum. Electron.* **15**, 718–726.

Lang, R. J.; Hardy, A.; Parke, R.; Mehuys, D.; O'Brien, S.; Major, J.; and Welch, D. (1993). *IEEE J. Quantum. Electron.* **29**, 2044–2051.

Lang, R. J.; Salzman, J.; and Yariv, A. (1986). *IEEE J. Quantum. Electron.* **22**, 463–468.

Lau, K., and Yariv, A. (1985). High-frequency current modulation of semiconductor lasers, in *Semiconductors and Semimetals*, Vol. 22, Part B. San Diego: Academic Press.

Leger, J. (1993). External methods of phase-locking and coherent beam addition of diode lasers, in *Surface-emitting semiconductor lasers and arrays*, G. Evans and J. Hammer, eds., New York: Academic Press.

Major, J. S., Jr. and Welch, D. F. (1993). *Electron. Lett.* **29**, 2121–2122.

Nash, F. (1973). *J. Appl. Phys.* **44**, 4696–4707.

Sakamoto, M.; Endriz, J. G.; and Scifres, D. R. (1992). *Electron. Lett.* **28**, 197–199.

Scifres, D., and Kung, H., "High power diode laser arrays and their reliability," in *Diode Laser Arrays*, D. Botez and D. Scifres, Eds., New York: Cambridge University Press (1994).

Scifres, D. R. (1993). *Reliability of high-power single-mode laser diodes.* SDL Inc. Product Information Release

Siegmann, A. E. (1990). New developments in laser resonators, *SPIE* vol. 1224 Optical Resonators, 2–14.

Solarz, R.; Beach, R.; Benett, B.; Freitas, B.; Emanuel, M.; Albrecht, G.; Comaskey, R.; Sutton, S.; and Krupke, W. (1994). High average power semiconductor laser arrays and laser array packaging with an emphasis on pumping solid state lasers, in *Diode Laser Arrays*, D. Botez and D. Scifres, eds. New York: Cambridge University Press

Streifer, W., and Kapon, E. (1979). *Appl. Optics* **18**, 3724–3725.

Suematsu, S.; Kishino, K.; Arai, S.; and Koyama, F. (1985). Dynamic single-mode semiconductor lasers with a distributed reflector, in *Semiconductors and semimetals*, Vol. 22, Part B, San Diego: Academic Press.

Susaki, W.; Tanaka, T.; Kan, H.; and Ishii, M. (1977). *IEEE J. Quantum. Electron.* **13**, 587–593.

Thompson, G. H. B. (1972). *Optoelectronics* **4**, 257–310.

Thompson, G. H. B. (1980). *Physics of semiconductor devices.* New York: John Wiley and Sons.

Vahala, K.; Chiu, L. C.; Margalit, S.; and Yariv, A. (1983). *Appl. Phys. Lett.* **42**, 631–633.

Walpole, J. N.; Kintzer, E. S.; Chinn, S. R.; Wang, C. A.; and Missaggia, L. J. (1992). *Appl. Phys. Lett.* **61**, 740–742.

Welch, D.; Craig, R.; Streifer, W.; and Scifres, D. (1990b). *Electron. Lett.* **26**, 1481–1482.

Welch, D.; Kung, H.; Sakamoto, M.; Zucker, E.; Streifer, W.; and Scifres, D. (1990c). Reliability characteristics of high power laser diodes, *SPIE Technical Digest,* vol. 1219, Laser Diode Technology and Applications.

Welch, D. F.; Parke, R.; Mehuys, D.; Hardy, A.; Lang, R.; O'Brien, S.; and Scifres, D. (1992). *Electron. Lett.* **28**, 2011–2013.

Welch, D.; Streifer, W.; Schaus, C.; Sun, S.; and Gourley, P. (1990a). *Electron. Lett.* **26**, 10–12.

Yariv, A. (1985). *Optical Electronics*, 3rd ed. New York: Holt, Rinehart, and Winston

Yariv, A. (1989). *Quantum electronics*, 3rd ed. New York: John Wiley and Sons

Yellen, S. L.; Shepard, A. H.; Dalby, R. J.; Baumann, J. A.; Serreze, H. B.; Guido, T. S.; Soltz, R.; Bystrom, K. J.; Harding, C. M.; and Waters, R. G. (1993). *IEEE J. Quantum. Electron.* **29**, 2058–2067.

Yonezu, H. O.; Ueno, M.; Kamejima, T.; and Hayashi, I. (1979). *IEEE J. Quantum. Electron.* **15**, 775–780.

Chapter 5

Surface-Emitting Lasers

Kenichi Iga and Fumio Koyama

Tokyo Institute of Technology
Nagatsuta 4259, Midori-ku, Yokohama 226-8503, Japan

Abstract

In this chapter, we review the device physics and technologies of surface-emitting (SE) lasers and related surface-operating optical devices. There have been developed four types of SE lasers, *i.e.,* (1) vertical-cavity surface-emitting lasers (VCSELs); (2) grating coupled type; (3) 45° deflecting mirror type; and (iv) Folded cavity type. In particular, we will feature the technology for vertical-cavity surface-emitting lasers based on GaAlAs/GaAs, GaInAsP/InP, and GaInAs/GaAs systems. Some lasing characteristics, device design, state-of-the-art performances, possible device characteristics, future prospects, and possible application will also be discussed.

5.1 Introduction

A new field of optoelectronics, including large-capacity parallel lightwave communications, multiaccess optical disks, optical computing, and optical interconnects, is becoming very important for future development of multimedia infrastructure. This circumstance is accelerating the importance of the surface-emitting (SE) laser. One of the authors (Iga) suggested a vertical-cavity surface-emitting (SE) laser, or VCSEL, in 1977. Figure 5.1 shows an image of arrayed vertical-cavity SE lasers. The cavity is formed by two surfaces of an epitaxial layer, and light output is taken vertically from one of the mirror surfaces (Soda *et al.*, 1979). On the other hand, other types of SE lasers employing in-plane cavity structures have been developed, which will be introduced in the next section.

Accordingly, the surface-emitting laser structures provide the following many advantages;

1. A huge number of laser devices can be fabricated by fully monolithic processes.

2. The initial probe test can be performed before separating devices into discrete chips.

3. The output beam can be taken out vertically from the wafer.

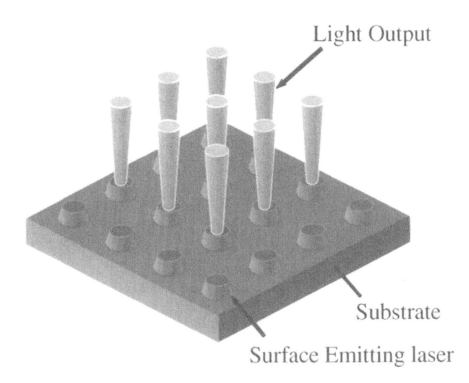

Figure 5.1: Surface-emitting laser and its array.

4. A densely packed two-dimensional laser array having phase locking possibility can be fabricated.

5. Very high power devices are realized by forming huge number of two-dimensional arrays.

5.2 Configurations of surface-emitting lasers

5.2.1 Categorization of surface-emitting lasers

A laser structure in which the emission is taken out perpendicular to the electrode was demonstrated by Melngailis (1965), with bulk InSb at 10 K under an intense magnetic field. After that, some studies on optically pumped platelet cavity lasers with CdSe or CdS film were made by several groups (Stillman *et al.*, 1966; Basov *et al.*, 1966; Packard *et al.*, 1969; Smiley *et al.*, 1971). The suggestion of a double-heterostructure SE laser as a device was given by Iga (1977). We obtained the first lasing operation of a GaInAsP/InP SE laser device in 1979, in which the threshold was 900 mA under pulsed condition at 77 K (Soda et al., 1979). Since then, the authors' group has been researching the possibility of vertical-cavity SE lasers with GaInAsP/InP and GaAlAs/GaAs systems (Soda *et al.*, 1979; Motegi *et al.*, 1982; Iga *et al.*, 1983; Ibaraki *et al.*, 1984; Iga *et al.*, 1988). We obtained room-temperature pulsed operation in a GaAlAs/GaAs SE laser in 1984 (Iga *et al.*, 1984). It is advisable for a long-haul system and network to use GaInAsP/InP SE single-mode lasers emitting 1.3 μm or 1.5 μm of wavelength region, if realized. On the other hand, GaAlAs SE lasers are attractive for optical interconnects, optical disks, optical sensing, and optical parallel processing.

There are several types of surface-emitting lasers where the light output is taken vertically from the substrate. We can classify those in terms of cavity structures, into the following four types

1. Vertical-Cavity Surface-Emitting Lasers (VCSEL).

2. Grating-Coupled Surface-Emitting Lasers.

3. 45° Deflecting-Mirror Surface-Emitting Lasers.

4. Folded-Cavity Surface-Emitting Lasers.

5.2.2   Vertical-cavity surface emitting lasers

The threshold current density of initial experimental SE lasers was rather high in comparison with conventional stripe lasers due to the short gain path, because the reflectivity of the mirrors was insufficient. This prevented room-temperature CW operation of SE lasers until 1988. To increase the reflectivity of the p-side (bonding side) reflector, we introduced a ring electrode in which the reflecting mirror is separated from the electrode (Uchiyama and Iga, 1984). In addition to this, we introduced an Au/SiO$_2$ mirror (Uchiyama *et al.*, 1986a), or dielectric multilayer reflector (Kinoshita *et al.*, 1987a), for improving the n-side (output side) reflectivity. For the purpose of effectively confining current in an active region, some types of current confining structures were introduced, *i.e.*, a round-low mesa, round-high mesa/polyimide buried, and circular buried heterostructure (Uchiyama *et al.*, 1986b). By introducing a circular buried heterostructure (CBH), the threshold was dramatically reduced, and low thresholds were obtained in a GaAlAs/GaAs system (Iga *et al.*, 1986, Kinoshita and Iga, 1987). In 1988, we achieved the first room-temperature CW operation of a GaAlAs/GaAs SE laser (Koyama *et al.*, 1988a, 1988b). After we demonstrated some good characteristics of CW SE lasers, much attention was focused on SE lasers, and many research groups started vertical-cavity SE laser research.

Vertical-cavity SE lasers utilizing semiconductor multilayer reflectors such as a DBR (Chailertvanitkul *et al.*, 1985; Sakaguchi *et al.*, 1986) or a DFB structure (Ogura *et al.*, 1984) may enable the integration of thin-film functional optical devices onto an SE laser by stacking them.

Jewell introduced these semiconductor multilayer reflectors and quantum wells into VCSELs and demonstrated 1~2mA threshold devices in 1989 (Jewell *et al.*, 1989). A lot of research and development of VCSELs has been performed since then, and some commercial applications have appeared since 1996. The VCSEL concept will open up a new scheme of three-dimensional integrated optics (Iga et al., 1982). In Fig. 5.2, we show a model of arrayed devices and SEM image of micro-cavities.

5.2.3   Grating coupled surface emitting lasers

The grating-coupled SE laser includes a distributed Bragg reflector (DBR) or distributed feedback (DFB) scheme using an in-plane higher-order coupling grating (Reinhart and Logan, 1975; Evans *et al.*, 1986; Scifres

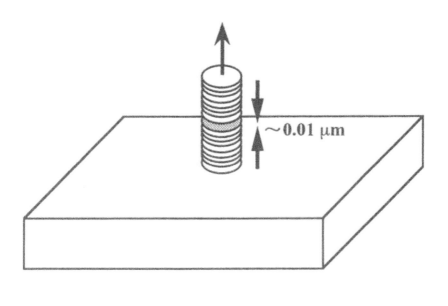

(a)

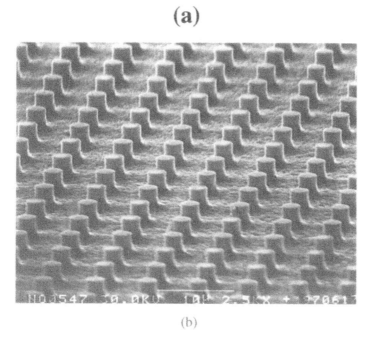

(b)

Figure 5.2: (a) A model of arrayed devices; (b) SEM image of micro-cavities.

et al., 1986). A fabrication technology developed for DFB lasers is automatically applicable in this scheme. The radiation length at the grating is several hundred microns, so that the output beam can be very narrow. An angle of 0.012° has been reported (Carlson *et al.*, 1988). By using an axial and lateral coupling configuration, a coherent array of multiple elements can be constructed (Evans *et al.*, 1989; Welch *et al.*, 1989). A three-row laser array was reported and emitted a pulsed 500 mW output (Kojima *et al.*, 1988).

One of the problems is that the output efficiency deteriorated due to the diffraction toward the substrate side. This was improved by using a highly reflective mirror, such as a gold metal, and semiconductor Bragg reflector prepared for the purpose of reflecting the substrate beam toward the output direction, and 400 mW pulsed and 250 mW CW were obtained (Macomber *et al.*, 1987). Output of 4 W of CW was also reported (Evans *et al.*, 1988; Welch *et al.*, 1989).

A circular grating surface-emitting laser was proposed to realize a circular beam from grating-coupled SE lasers by introducing a circularly symmetric grating (Wu *et al.*, 1991).

5.2.4   45° Deflecting-mirror surface-emitting lasers

An SE laser using a 45° deflector (Springthorpe, 1977; Liau and Walpole, 1985) was developed to take out the output beam from an etched mirror laser. The first device by Liau and Walpole. (1985) using GaInAsP/InP was fabricated by employing a mass-transport technique. Later, 164 devices were arrayed to get a very high power. Actually, more than one watt of CW power was obtained. Also, GaAlAs/GaAs devices were developed, and about 70 watts in pulsed condition with 66% of quantum efficiency was reported (Goodhue *et al.*, 1992).

5.2.5   Folded cavity semiconductor lasers

A folded cavity (or turn-up cavity) structure using a 45° deflecting intra-cavity mirror or a bending waveguide (Wu *et al.*, 1987; Yuasa *et al.*, 1988) was proposed. In a bent-waveguide structure, the active layer is grown on a structured substrate to have a bent cavity. The output light is radiated with an angle from the bent waveguide.

A folded cavity employing intra-cavity 45-degree reflectors similar to the former, but where the mirror is included in the cavity, was also conceived. The folded cavity together with a Bragg reflector can yield a very high power, and more than 50 W of CW output from a two-dimensional 16 × 94 array was achieved (Nam *et al.*, 1992).

5.3 Basics of vertical cavity surface-emitting lasers

5.3.1 Threshold current and quantum efficiency

We consider a model of a VCSEL with a heterostructure as shown in Fig. 5.3, in which the active region is buried in a material with smaller band gap or by an insulating material, and injected carriers are confined in

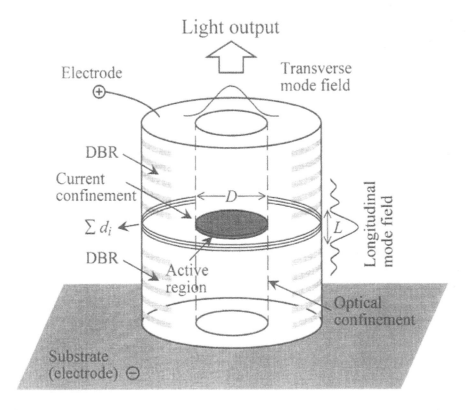

Figure 5.3: A model of AlAs oxide-confined VCSEL.

the circular active region with diameter D. The optical loss for the resonant mode must balance the mode gain ξg_{th} to reach the threshold:

$$\xi g_{th} = (\alpha_a + \alpha_d + \alpha_m). \tag{5.1}$$

where g_{th} is a material gain factor at threshold, α_a is absorption loss, α_d diffraction loss, and α_m is mirror loss. The equivalent diffraction loss α_d is averaged by cavity length.

The parameter ξ is the optical energy confinement factor, which is expressed by the product of the longitudinal confinement factor ξ_l and transverse one ξ_t, as follows:

$$\xi = \xi_l \cdot \xi_t . \tag{5.2}$$

ξ is defined by the overlap integral of the gain or loss and standing waves and expressed for a thick bulk medium as

$$\xi_l = \frac{d}{L}, \tag{5.3}$$

where d is the total active layer thickness and L is the effective cavity length. When a very thin layer ($\sim$100 Å) is placed at the maxima of standing waves, we have

$$\xi_l = \frac{2d}{L}. \tag{5.4}$$

This concept to reduce the threshold by placing the active layer at the maxima was suggested and is called "periodic gain" (Geels *et al.*, 1988; Raja *et al.*, 1988). The average absorption loss coefficient α_a (per unit length) is given by the equation

$$\alpha_a \cong \xi \alpha_{ac} + \alpha_{ex} , \tag{5.5}$$

where α_{ac} and α_{ex} are the absorption loss in the active and cladding or mirror layers, respectively. Here, we have assumed that $\alpha_{ac} > \alpha_{ex}$ and d $\ll$ L.

The mirror loss is defined by

$$\alpha_m = \frac{1}{L} \ln \frac{1}{\sqrt{R_f R_r}}, \tag{5.6}$$

where R_f and R_r are the reflectivities of the front and rear side reflector. We express the total internal cavity loss α as

$$\alpha = \alpha_a + a_d . \tag{5.7}$$

On the other hand, the threshold gain is expressed in terms of the threshold carrier density N_{th} as

$$g_{th} = A_0(N_{th} - N_t),$$
(5.8)

where N_t is the transparency carrier density for which the material turns, and A_0 is the differential gain dg/dN near the threshold. Thus, the threshold carrier density N_{th} is expressed by

$$N_{th} = \frac{g_{th}}{A_0} + N_t.$$
(5.9)

For the GaAlAs/GaAs system, N_{th} is about 2×10^{18} cm^{-3}.

From Eq. (5.1) and (5.9) we have

$$N_{th} = N_t + \frac{1}{\xi A_0}(\alpha_a + \alpha_d + \alpha_m).$$
(5.10)

On the other hand, the threshold carrier density N_{th} of injection lasers can be given by the driving current density at threshold J_{th} and is expressed as (Soda et $al.$, 1983)

$$N_{th} = \frac{\eta_i \tau_s}{ed} J_{th},$$
(5.11)

where η_i is the injection efficiency $i.e.$, the rate of generation of carriers by current injection.

We can assume that the carrier life time τ_s is expressed by

$$1/\tau_s = A + BN + CN^2,$$
(5.12)

where the first term is a nonradiative recombination independent of carrier density, the second linear term denotes a radiative recombination term, and the last one represents a nonlinear process such as Auger effect or the like (Asada et al, 1984). In case the radiative recombination is predominant, we can approximate

$$1/\tau_s = B_{eff}N/\eta_{spon},$$
(5.13)

where η_{spon} is spontaneous emission efficiency and B_{eff} is effective radiative recombination coefficient.

Then the threshold current density is given, from Eq. (5.11) by using Eq. (5.13), by

$$J_{th} = \frac{ed}{\eta_i \tau_s} N_{th} \cong \frac{edB_{eff}}{\eta_i \eta_{spon}} N_{th}^2.$$
(5.14)

When $R = 95\%$ and $d = 2 \sim 3$ μm, the threshold current density J_{th} is $25 \sim 30$ kA/cm^2 in early devices. This value is the same as that of high-radiance LEDs and not a surprisingly high level. In addition, quantum well structures can provide lower threshold by 40% (Uenohara *et al.*, 1988). By reducing the active layer thickness to 100 Å and increasing the reflectivity to 99.9%, a threshold current density can be as low as the lowest level of conventional stripe lasers, 200 A/cm^2 or even lower.

The threshold current of surface-emitting lasers can then be expressed by the equation reduced from Eq. (5.14);

$$I_{th} = \pi(D/2)^2 J_{th} = \frac{eVN_{th}}{\eta_i \tau_s} \cong \frac{eVB_{eff}}{\eta_i \eta_{spon}} N_{th}^2 , \qquad (5.15)$$

where e is electron charge, τ_s is recombination lifetime, and V is the volume of active region, which is given by

$$V = \pi(D/2)^2 d , \qquad (5.16)$$

where D is the diameter of the active region.

As seen from the equation, we recognize that it is essential to reduce the volume of active region in order to decrease the threshold current. Assume that the threshold carrier density does not change so much; if we reduce the active volume, we can decrease the threshold as we can make a small active region. We compare the dimensions of surface emitting lasers and conventional stripe geometry lasers in Table 5.1. It is noticeable that the volume of VCSELs could be $V = 0.07$ μm^3, whereas that for stripe lasers remains $V = 60$ μm^3. This directly reflects the threshold currents: the typical threshold of stripe lasers is the mA range

Parameters	Symbol	Stripe Lasers	VCSELs
Active Layer Thickness	d	100 Å–0.1 μm	80 Å–0.5 μm
Reflectivity	R_m	0.3	0.99–0.999
Cavity Length	L	300 μm	0.3 μm ($\lambda/2$-λ)
Optical Confinement Factor	ξ	0.5%	3% × 2
Area of Active Region	S	3 × 300 μm^2	5 × 5 μm^2
Photon Lifetime	τ_p	0.1 ps	0.1 ps
Relaxation Frequency	f_r	5 GHz	10 GHz

Table 5.1: Comparison of laser dimensions

or higher, but that for VCSELs can be reduced to submilliamperes by a simple carrier confinement scheme such as an air post structure. It could be even as low as μA by implementing carrier and optical confinement structures, as will be introduced later.

As has been seen from Fig. 5.4, which shows an early-stage estimation of threshold, it can be reduced proportional to the square of the active region diameter. However, there should be a minimum value originating from the decrease of the optical confinement factor defined by the overlap of optical mode field and gain region when the diameter is decreasing. In addition to this, the extreme minimization of volume, in particular, in the lateral direction, is limited by the optical and carrier losses due to optical scattering, diffraction of lightwave, nonradiative carrier recombination, and other technical imperfections. These topics will be discussed later.

Also, we discuss the differential quantum efficiency of the SE laser. If we use a nonabsorbing mirror for the front mirror, the differential quantum efficiency from the front mirror is expressed as (Kinoshita *et al.*, 1987):

$$\eta_d = \eta_i \frac{(1/L)\ln(1/\sqrt{R_f})}{\alpha + (1/L)\ln(1/\sqrt{R_f R_r})}, \tag{5.17}$$

where η_i is the injection efficiency and α is the total internal loss defined by Eq. (5.7).

5.4 Fundamental elements for vertical cavity surface-emitting lasers

The problems that should be seriously considered for making vertical cavity SE lasers will be:

1. Highly reflective and transparent DBRs.

2. Minimization of optical losses.

3. Maximization of optical-field and gain overlap.

4. Electrode formation to reduce the resistivity for high-efficiency operation.

5. Heat sinking for high-temperature and high-power operation.

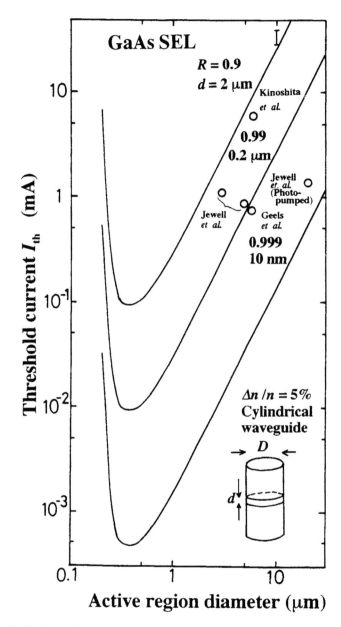

Figure 5.4: Threshold *vs.* active region diameter for GaAs surface-emitting lasers.

5.4.1 Dielectric multilayer mirrors

The dielectric multilayer mirrors have such advantages as (1) relatively large index difference, (2) almost no absorption loss, (3) small effective penetration depth, and so on. An electron-beam deposition is normally employed due to its high controllability. *In situ* optical path length monitoring has been introduced to accurately control the reflectivity and center wavelength (Oshikiri *et al.*, 1991). More than 99% of reflectivity for TiO$_2$/SiO$_2$ and SiO$_2$/Si reflectors is available.

5.4.2 Semiconductor multi-layer structures

Fine growth technologies such as metal organic chemical vapor deposition (MOCVD), molecular beam epitaxy (MBE), and chemical beam epitaxy (CBE) can provide superlattice structures that enable the fabrication of DFB- and DBR-type SE lasers introduced in the preceding section. For the purpose of demonstrating DBR SE lasers, Bragg reflectors composed of 30 layers of GaAlAs and AlAs stacks with quarter wavelength in medium were grown by the aforementioned MOCVD technique (Sakaguchi *et al.*, 1986). The period of the Bragg reflector is 1400 Å. The reflectivity of the multilayer Bragg reflector was measured from the top of the crystal surface. Reflectivity higher than 97% was obtained at a wavelength of 0.87 μm by an initial experiment, which corresponds to the lasing wavelength of the GaAlAs/GaAs SE laser. Also, we found that it is possible to inject carrier into an active region of 2~3 μm thickness through multilayers by appropriately doping them. We first succeeded in the oscillation of a GaAlAs SE laser that uses a multilayer reflector as one of the mirrors shown in Sakaguchi *et al.* (1986). Ibaraki *et al.* demonstrated a low-threshold room-temperature CW operation of DBR circular buried heterostructure SE laser with $I_{th} = 5.2$ mA (Ibaraki *et al.*, 1989).

Also, GaInAsP/InP semiconductor reflectors are grown by CBE. Reflectivities of 95~99% have been attained, but this system requires more than 40 pairs. By introducing such a periodic configuration, a reduction of the threshold current can be expected (Uchiyama *et al.*, 1986b). To fully activate a multilayered active region such as an MQW and DFB, we also proposed a transverse or interdigital injection scheme (Uchiyama, 1986, Uchiyama, *et al.*, 1986). A DBR or DFB structure without facet mirrors enables the integration of functional optical devices with SE lasers by stacking them. This may open up the way for a new three-dimensional integrated optics.

5.4.3   Quantum wells for gain medium

The quantum well laser exhibits many preferable characteristics, such as low threshold current (Lau and Yariv, 1988), high relaxation oscillation frequency (Arakawa and Yariv, 1985), and larger characteristic temperatures. Thus, an SE laser with a multiquantum well (MQW) for its active region is expected to provide not only a higher gain but also some other better performances. Lasing characteristics of an MQW SE laser by optical pumping were reported by Nomura *et al.*, (1985). We have reported a laser oscillation of an MQW SE laser by current injection (Uenohara *et al.*, 1989). Later, Jewell *et al.* demonstrated a quantum well VCSEL exhibiting a very low threshold, which will be introduced in a later section. Some other sophisticated quantum wells such as quantum wires (Arakawa *et al.*, Hatori *et al.*, 1996), quantum dots (Deppe *et al.*, 1997), and modulation doped quantum wells (Hatori *et al.*, 1997a,b) have been considered as highly performing active regions.

5.4.4   Periodic and matched gain structures

The periodic gain concept was proposed by Geels *et al.*, (1988), and Raja *et al.*, (1988). If the gain region is placed at the maxima of standing waves of a resonant mode, it is possible to increase the modal gain by a factor of two, and the reduction of threshold can be expected. Optically pumped CW operation of periodic gain SE lasers with a high-power conversion efficiency of more than 40% have been demonstrated (Schaus *et al.*, 1989; Gourley *et al.*, 1989). This concept is now commonly employed in most VCSELs.

5.4.5   Current and optical confinement

In order to effectively confine carriers in the active region some carrier confinement structure should be considered as in stripe lasers. In Fig. 5.5 we show some current confining schemes introduced into VCSELs. The ring electrode (a) is the simplest one and is easily fabricated, but the current diffuses out to the outer region. The proton-bombardment confinement (b) can limit the current effectively into the active region by producing insulating material at the surrounding region. The buried heterostructure (BH) (c) is one of the ideal confinement schemes, but it is technically difficult to reproducibly produce two-dimensional BH. The air-post structure (d) can be fabricated by etching to form a circular or

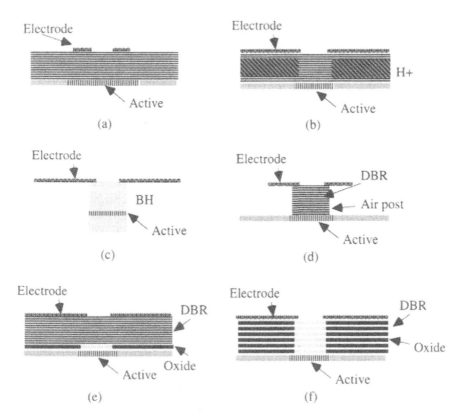

Figure 5.5: Some current confinement structures for VCSELs: (a) ring or circular electrode; (b) proton (H^+) bonbardment; (c) buried heterostructure (BH); (d) air-post; (e) oxide confinement; (f) oxide distributed Bragg reflector (DBR).

rectangular mesa to confine current into a small post. The oxide confinement structure (e) is very effective for making a very low threshold device by current and optical confinement at the same time.

Figure 5.5(e) shows a model of oxide confinement structure for confining an optical field to increase the overlap to gain region. The oxide is formed by oxidizing AlAs by a high-temperature water vapor (Holonyak *et al.*, 1990, Tsang, 1978). It has been considered that the oxidized AlAs is harmful for the reliability of optical devices. But it has been made clear that the high-temperature oxidation, such as under 400 °C, generates a stable Al_xO_y when the sample is cooled down to room temperature. This will be detailed below.

The oxidation velocity of AlGaAs strongly depends on Al content, and by using this nature we can preferentially oxidize AlAs or GaAlAs with higher Al content. This process was introduced by Deppe's group into the formation of VCSELs and about 200 μA of threshold was demonstrated (Deppe, 1994).

In addition to the carrier confinement effect of oxidized AlAs, the optical dielectric constant is about 2.4, which is much lower than the surrounding semiconductors, say GaAs or nonoxidized AlGaAs. This provides us with a kind of optical phase shift, and it has been recognized that there should be a lens effect to confine a beam along the optical axis of laser resonator (Coldren, 1995, Bissessur *et al.*, 1996). We can, therefore, solve the aforementioned problem which we face to reduce the lateral cavity size toward demensions of nearly an optical wavelength. In Fig. 5.6, we show a result of numerical simulation based on a realistic

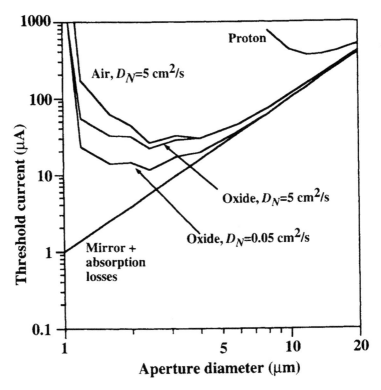

Figure 5.6: Threshold *vs.* aperture diameter of oxide-confined VCSELs.

model of VCSEL using the oxidation confinement structure (Bissessur *et al.*, 1997). It is noted that we could expect a threshold as low as 10 μA when we make an aperture size of 3 μm.

On the other hand, if we use a volume confinement of optical beam by oxidizing a whole structure of distributed Bragg reflector (DBR) consisting of AlAs/GaAs as shown in Fig. 5.5(f), a continuous optical guiding like a circularly symmetric optical fiber can be realized. A threshold of 70 μA at room temperature continuous wave operation has been achieved, as will be detailed later (Hayashi *et al.*, 1995).

5.5 Device design and technology

The materials for semiconductor lasers are now covering most of III~V compound alloys as shown in Fig. 5.7.

5.5.1 Long wavelength surface-emitting lasers

The importance of surface-emitting lasers emitting at 1.3 or 1.55 μm is currently increasing, since their application to parallel lightwave systems

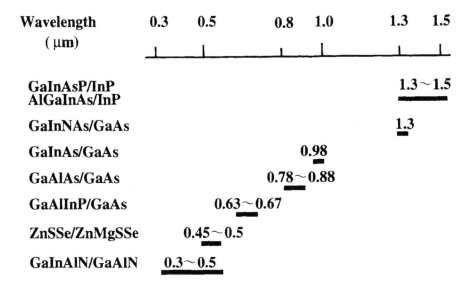

Figure 5.7: Materials for VCSELs in wide spectral bands.

and parallel optical interconnects are actually considered. The GaInAsP/ InP VCSEL is expected to be realized, but this material system has some substantial difficulties for making the VCSEL due to the following reasons:

1. The Auger nonradiative recombination is substantial, and this makes the threshold carrier density abnormally high near-room temperatures.

2. There is a resonant intravalence band absorption (IVBA), in particular, for the compositions near 1.55 μm.

3. The index difference between GaInAsP and InP is relatively small (10%), that requires more than 35 pairs in preparing distributed Bragg reflectors by stacking $\lambda/4$ pair of GaInAsP/InP heterostructure.

4. Moreover, the total thickness becomes several microns due simply to the wavelength being longer.

For these reasons, we did not succeed in achieving room-temperature continuous wave (CW) operation in this material until 1993. The key issue in this material for VCSEL is to have a high reflective mirror and build an effective current injection scheme. For this purpose, a GaInAsP/ InP SE laser with a circular buried heterostructure (CBH) (Kawasaki *et al.*, 1988) modified by using a flat-surface CBH (FCBH) was realized (Baba *et al.*, 1992). This laser can be grown by three-step epitaxies including regrowth processes followed by a successive fully monolithic fabrication process (Baba et al., 1992). The substrate is polished to 150 μm thick, and the *n*-side Au/Sn electrode is formed. Next, the substrate and etch-stop layer are preferentially etched off to form a short cavity ($= 7\ \mu$m). One of the favorable characteristics of this material system is that the chemical etch rate difference for GaInAsP and InP is relatively large, and preferential etch can be perfectly performed. Finally, SiO$_2$/Si reflectors are formed both on the bottom of an etched well and on the *p*-side epitaxial surface by an electron beam evaporation employing optical thickness monitoring. A current/light output (I~L) characteristic of an FCBH-VCSE laser device at 77K under CW condition was measured (Kawasaki *et al.*, 1988; Baba *et al.*, 1992). Pulsed operation has been obtained near-room temperature (Oshikiri *et al.*, 1989, Baba *et al.*, 1992) and room temperature (Kasukawa *et al.*, 1990). A polyimide buried structure device is fabricated by much simpler processes (Uchiyama *et al.*, 1986a) and pulsed operation at 66 °C has been obtained (Wada *et al.*, 1991).

Thermal problems for CW operation have now been extensively studied. A MgO/Si mirror with good thermal conductivity was demonstrated (Tanobe *et al.*, 1992). For realizing reliable devices, the buried heterostructure (BH) is crucial. We have fabricated a BH SE laser exhibiting a relatively low threshold at room temperature under pulsed operation (Baba *et al.*, 1992).

By improving the heat sinking using a diamond submount and highly reflective mirror, we have achieved room-temperature CW operation. We show its device structure in Fig. 5.8a with a typical current *vs.* light power output (I~L) characteristic as shown in Fig. 5.8b. The minimum threshold obtained was 22 mA at 14 °C for a 12-μm-diameter device size. The highest operation temperature is 34 °C (Uchiyama *et al.*, 1996).

Some hybrid mirror techniques for long wavelength VCSELs have been developed. One is to use a semiconductor/dielectric reflector, which is demonstrated by chemical beam epitaxy (CBE) (Miyamoto *et al.*, 1992) as shown in Fig. 5.9. The other is epitaxial bonding of quaternary/GaAs-AlAs mirror, where 144 °C pulsed operation was achieved by optical pumping (Babic *et al.*, 1995). Epitaxially bonded mirror made of GaAs/AlAs was introduced into surface-emitting lasers operating at 1.3 μm providing 3 mA of room-temperature pulsed threshold. The record including the CW threshold of 2.3 mA as well as the maximum CW operating temperature up to 32 °C have been reported for 1.5 μm SE lasers with epitaxially bonded mirrors (Babic *et al.*, 1995). The epitaxial bonding of GaInAsP-InP/GaAs-AlAs mirrors combined with the oxide confinement structure has been exhibiting CW operation up to 64 °C (Babic *et al.*, 1996). Further development may be necessary to compete with edge emitters employing DFB and spot sized transformer (SST) structures.

The introduction of GaInNAs/GaAs and AlGaInAs/InP may be a solution to the material bottleneck of the GaInAsP/InP system, *i.e.*, temperature characteristics, AlAs oxidation capability, the use of GaAs/AlAs DBRs, and the like.

It would be extremely important to develop GaInNAs/GaAs 1300 nm VCSELs for short distance optical links due to low-cost chip and moduling capability. In Fig. 5.10 we show the band gap *vs.* lattice-constant diagram featuring this system. It may readily replace any edge emitters for optical interconnects and links, if 1 mW devices can be realized. Some design consideration and fundamental growth have been performed (Miyamoto,

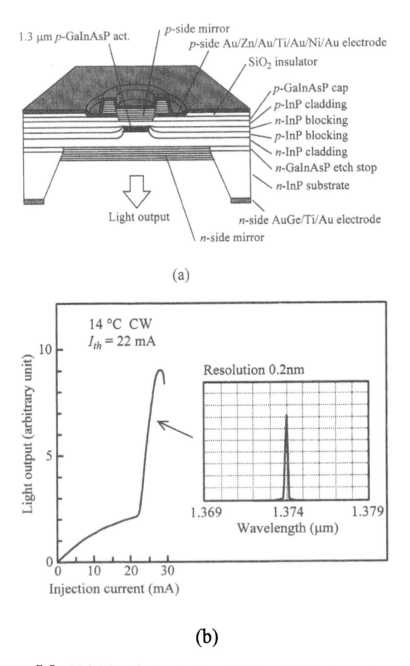

(a)

(b)

Figure 5.8: (a) A 1.3μm device structure that showed the first room-temperature CW operation; (b) a typical current *vs.* light power output (I~L) characteristic.

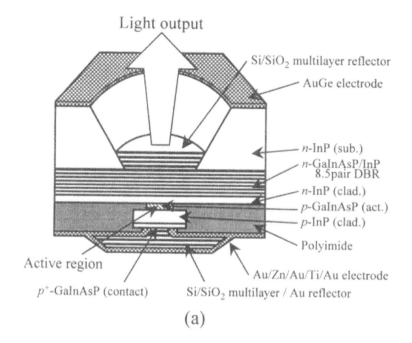

(a)

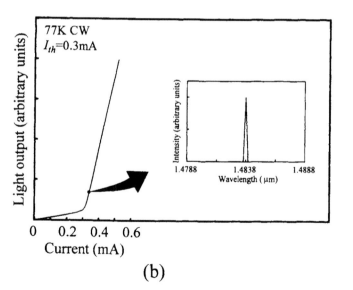

(b)

Figure 5.9: (a) Chemical beam epitaxy (CBE) grown VCSEL; (b) Its I~L characteristic and spectrum.

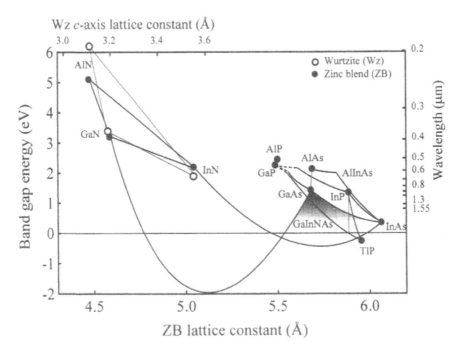

Figure 5.10: The band gap *vs.* lattice-constant diagram.

1997). A schematic idea of such a device is shown in Fig. 5.11. In late 1997, the first GaInNAs/GaAs VCSEL came out emitting 1.2 μm of wavelength (Kondow *et al.*, 1997).

5.5.2   GaAlAs/GaAs surface emitting lasers

A GaAlAs/GaAs laser can employ almost the same CBH structure as the GaInAsP/InP laser. In order to decrease the threshold, the active region is also constricted by the selective meltback method. In 1986, the threshold of 6 mA was demonstrated for the active region diameter ~6 μm under pulsed operation at 20.5 °C (Iga *et al.*, 1986). The threshold current density is about 200 μA/μm^2. It is noted that a micro-cavity of 7 μm long and 6 μm in diameter was realized.

The MOCVD-grown CBH SE laser of Fig. 5.12 (Koyama *et al.*, 1988 a,b; Koyama *et al.*, 1989) was demonstrated using a two-step MOCVD growth and fully monolithic technology. First, a GaAs/GaAlAs DH wafer

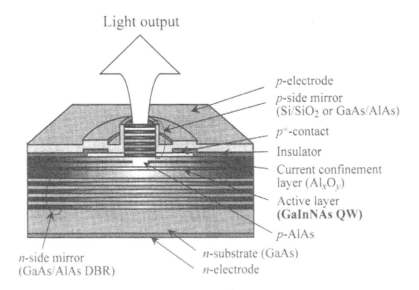

Figure 5.11: A schematic diagram of GaInNAs/GaAs VCSEL.

with an active layer thickness of 3 μm was grown by MOCVD at a temperature of 780 °C under atmospheric pressure. After the first growth, a silicon nitride (Si_3N_4) circular mask with a diameter of 10 μm was formed on the wafer for mesa etch and selective regrowth of current blocking layers. The p-cladding GaAlAs layer is lightly etched by a sulfuric acid (H_2SO_4):H_2O: hydrogen peroxide (H_2O_2) = (1:8:8) solution. Selective MOCVD regrowth of GaAs under atmospheric pressure is employed to form current blocking layers (0.7-μm-thick n-GaAs and 0.3-μm-thick p-GaAs). The growth condition is the same as that used in the double heterostructure (DH) wafer growth. There is no deposition on the top of a circular mesa covered with a Si_3N_4 mask. The short cavity structure with the cavity length of 5 μm is formed by selectively removing the GaAs substrate. A ring electrode with outer/inner diameter of 40 μm/10 μm was adopted and SiO_2/TiO_2 mirrors are applied to both surfaces.

The first room temperature CW operation has been achieved (Koyama *et al.*, 1988; Koyama *et al.*, 1989). The lowest CW threshold current was 20 mA $J_{th} \cong 260$ μA/μm^2). The differential quantum efficiency is typically 10%. The maximum CW output power is about 2 mW. The saturation of the output power is due to a temperature increase of the device. Stable

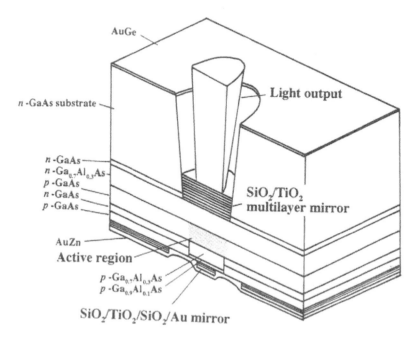

AuGe

n-GaAs substrate

Light output

n-GaAs
n-Ga$_{0.7}$Al$_{0.3}$As
p-GaAs
n-GaAs
p-GaAs

SiO$_2$/TiO$_2$
multilayer mirror

AuZn
Active region

p-Ga$_{0.7}$Al$_{0.3}$As
p-Ga$_{0.9}$Al$_{0.1}$As

SiO$_2$/TiO$_2$/SiO$_2$/Au mirror

Figure 5.12: A MOCVD-grown CBH SE laser that exhibited the first room-temperature CW operation.

single mode operation is observed with neither subtransverse modes nor other longitudinal modes. The spectral linewidth above the threshold is less than 1 Å, which is limited by the resolution of the spectrometer. The mode spacing of this device was 170 Å. The side-mode suppression-ratio (SMSR) of 35 dB is obtained at $I/I_{th} = 1.25$. This is comparable to that of well-designed DBR- or DFB-dynamic single-mode lasers.

Sub-mA thresholds and 10 mW outputs have been achieved. The power conversion efficiency of 57% has been demonstrated (Jager *et al.*, 1996). Some commercial optical links are already available in the market. The price of low-skew multimode fiber ribbons may be a key issue for inexpensive multimode-fiber-based data links.

5.5.3 GaInAs/GaAs surface-emitting lasers

The GaInAs/GaAs strained pseudomorphic system grown on a GaAs substrate emitting at 0.98 micron wavelength exhibits a high laser gain and has been introduced into surface-emitting lasers together using GaAlAs/ AlAs multi-layer reflectors. A low threshold ($\cong$ 1 mA at CW) has been demonstrated (Jewell *et al.*, 1989). The threshold current of vertical-cavity surface-emitting lasers has been reduced down to the submilliampere range in various institutions in the world. The record thresholds reported so far are 0.7 mA (Geels *et al.*, 1990), and 0.65 mA (Jager *et al.*, 1996), $\cong$ 0.2 mA (Numai *et al.*, 1993; Huffaker *et al.*, 1994). The material that exhibits the best performance for VCSEL is a pseudomorphic GaInAs/ GaAs system emitting 0.98 micron of wavelength. Moreover, a threshold of 91 micro-amperes at room-temperature CW operation was reported by introducing the oxide current and optical confinement (Huffaker, 1996a). The theoretical expectation was 10 microamperes or less, if some good current and optical confinement structure could be introduced.

In 1995, we developed a novel laser structure employing selective oxidizing process applied to AlAs that is one of the components of multilayer Bragg reflectors (Hayashi *et al.*, 1995). The schematic structure of the device now developed is shown in Fig. 5.13. The active region comprises three quantum wells consisting of 80-Angstrom GaInAs-

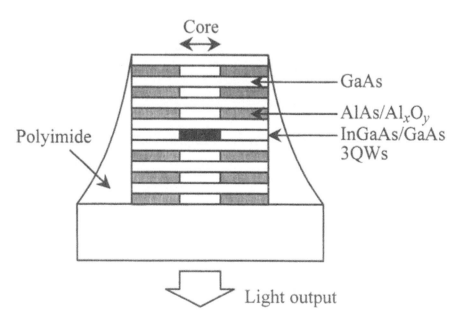

Figure 5.13: A schematic structure of oxide-confined InGaAs/GaAs VCSEL.

strained layers. The Bragg reflector consists of GaAs/AlAs quarter-wavelength stacks of 24.5 pairs. After etching the epitaxial layers, including the active layer and two Bragg reflectors, the sample was treated in the high-temperature oven with water vapor bubbled through nitrogen gas. The AlAs layers are oxidized preferentially with this process, and a native oxide of aluminum is formed at the periphery of etched mesas. It is recognized from the SEM picture that only AlAs layers in DBR have been oxidized, as shown in Fig. 5.14. The typical size is a 20-micron core starting from a 30-micron mesa diameter. We have achieved about 1 mW of power output and submilliampere threshold. The nominal lasing wavelength is 0.98 microns.

We have made a smaller-diameter device having 5 microns started from a 20 micron mesa. The minimum threshold achieved is 70 microamperes at room-temperature CW operation, as shown in Fig. 5.15.

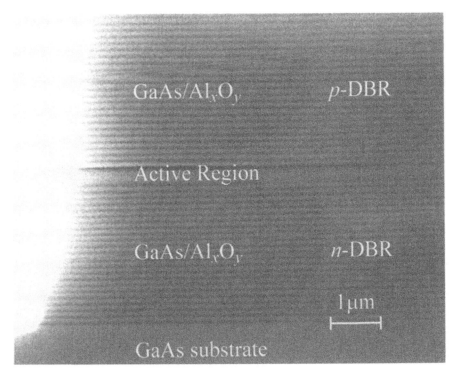

Figure 5.14: Oxidized AlAs layers in DBR.

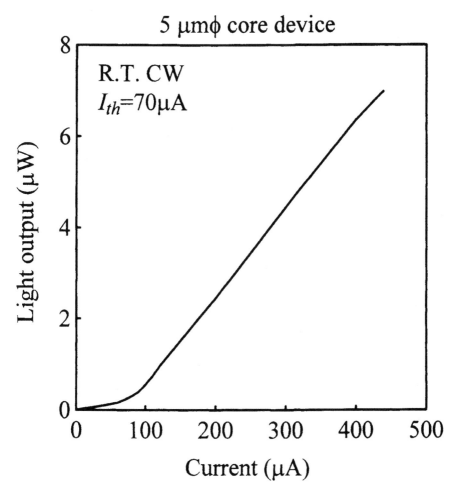

Figure 5.15: I~L characteristics of CW operation of an InGaAs/GaAs VCSEL.

As the theoretically predicted threshold is about 1 microampere, this record may soon be broken. Recently, the threshold of 8.5 μA (Yang *et al.*, 1995a, b) has been reported. It can thus be said that surface-emitting lasers are now meeting a new stage, exhibiting ultralow power consumption and real application to optical interconnects.

A relatively high power (as high as 50 mW) is becoming possible (Peters *et al.*, 1992). The power conversion efficiency of 50% was reported

(Lear *et al*, 1995). Also, near 200 mW has been demonstrated by a large device (Grabherr *et al.*, 1997).

However, due to the unavailability of low-cost detectors in this wavelength, it is difficult to find real applications.

5.5.4   Short-wavelength surface-emitting lasers

Visible surface-emitting lasers are extremely important for disk and display applications, in particular, red, green, and blue surface emitters may provide much wider technical applications, if realized. GaInAlP/GaAs VCSELs have been developed, and room-temperature operation exhibiting submilliamperes threshold has been obtained (Lott *et al.*, 1993; Huang *et al.*, 1993; Lott *et al.*, 1994). Materials yielding blue and green lasers are much more difficult than any other materials for considering SE lasers. Some design consideration and fundamental process technology for ZnSe and GaN systems are now being attempted in the author's laboratory (Honda *et al.*, 1994).

1. 780 nm: The device having 200 μA threshold and 1.1 mW output has been reported aiming at CD player application (Shin *et al.*, 1996). A VCSEL-loaded pickup is becoming commercialized.

2. 670~650 nm: AlGaInP/GaAs SE lasers are developed, and room-temperature operation exhibiting submilliamperes threshold and a few mW outputs have been obtained. The best application of red-color VCSELs may be laser printers, and the first generation DVDs (digital video disks), and plastic optical fiber systems.

3. 555 nm: The ZnSe system is still the only material to provide CW operation of green-blue semiconductor lasers operating beyond 1000 hours. Some trials for green surface emitters were attempted. It may be difficult to find a good application for such green VCSELs because GaN devices look more promising.

4. 470~340 nm: The GaN and related materials including Al, In, and B can cover wide spectral ranges from green to UV. The reported reliability of GaN-based LEDs and LDs (Nakamura *et al.*, 1996) seems to indicate a good material potential for VCSELs as well (Iga, 1996). The estimation of threshold for GaN/Al$_{0.1}$Ga$_{0.9}$N quantum well lasers is carried out using the density-matrix theory with intraband broadening (Honda *et al.*, 1997).

Attempts for realizing blue to UV VCSELs have started in 1997. Some optical pumping experiments have been reported. We have made a preliminary study to explore MOCVD selective area growth and dry etch a GaN system by a chlorine-based reactive ion beam etching. These may be crucial technologies for GaN-based devices (Honda *et al.*, 1997).

The application area of blue~UV VCSELs is huge. It may provide the next generation of DVDs, laser printers incorporating blue-sensitive drams, full color displays, high efficiency illuminations together with green and red devices, and so on.

5.6 Lasing characteristics of VCSELs

5.6.1 Ultimate performances

By overcoming various technical problems, such as making tiny structures, ohmic resistance of electrodes, and improving heat sinking, we believe that we can obtain a 1 μA device, as shown in Fig. 5.4 (Tamanuki *et al.*, 1992). A lot of efforts toward improving the characteristics of surface emitting lasers have been made, including surface passivation in the regrowth process for buried heterostructure, microfabrication, and fine epitaxies.

As has been introduced in sect. 5.4.4, very low thresholds of around 70 μA (Hayashi *et al.*, 1995), 40 μA (Deppe *et al.*, 1997) and about 10 μA (Yang *et al.*, 1996 a,b) were reported by employing aforementioned oxidation techniques. Therefore, by optimizing the device structures, we can expect thresholds lower than microamperes in the future.

The efficiency of devices is another important issue for various applications. By introducing the oxide confinement scheme, the power conversion efficiency has been drastically improved due to the effective current confinement and the reduction of optical losses. Also, the reduction of driving voltage by introducing model contacting technology helped a lot. As mentioned earlier, > 50% power conversion efficiency (sometime called wall-plug efficiency) has been realized (Weigl *et al.*, 1996b). The significant difference from the convenional stripe laser is that high efficiency can be obtained at relatively low driving ranges in the case of VCSELs. Therefore, further improvement may enable us to achieve very high efficiency arrayed devices, which has never been attained in any other types of lasers.

The high-speed modulation capability is very essential for communication applications. In VCSELs, 10 Gbits/s or higher modulation experiments have been reported. It is a big advantage for VCSEL systems that over 10 Gbits/s modulation is possible at around 1 mA driving levels. In Fig. 5.16, we show an example of small signal analog modulation of an oxide-confined InGaAs/GaAs VCSEL (Hatori *et al.*, 1997a). Also, in Fig. 5.17, we show an eye diagram for a 10 Gbits/s transmission experiment through a 100 m multimode fiber (Hatori *et al.*, 1997b). This characteristic is very preferable for low power consumption optical interconnect applications. Actually, transmission experiments at 10 Gbits/s and zero-bias transmission at 1 Gbit/s have been reported (Ebeling *et al.*, 1996).

The reliability of devices is a final screening of applicability of any components and systems. A high-temperature acceleration life test of protoninplanted VCSELs showed an expected room-temperature lifetime of over 10^7 hours (Guenter *et al.*, 1996). There is no reason why we cannot have very long-life VCSEL devices, because the active region is completely

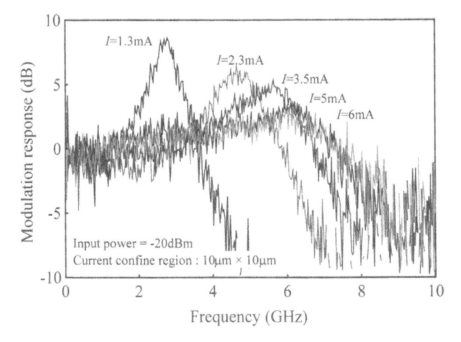

Figure 5.16: An example of small signal analog modulation of an oxide-confined InGaAs/GaAs VCSEL.

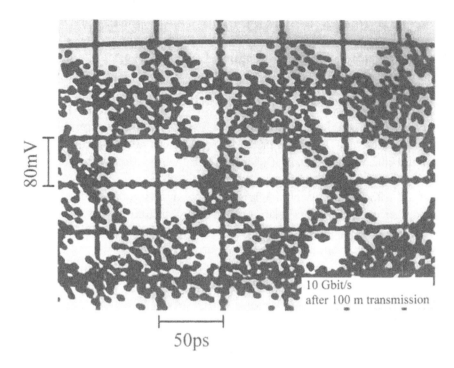

Figure 5.17: Eye diagram at 10 Gbit/s after 100 m in multimode fiber transmission.

embedded in wide-gap semiconductor materials and the mirrors are already passivated.

5.6.2 Single-mode behavior

The relative intensity noise was measured by a standard-intensity noise-measurement method (Koyama *et al.,*). Similar to other single longitudinal mode lasers, a VCSEL behaves as a low-noise laser exhibiting about $-140\sim-150$ dB/Hz (Koyama *et al.,* 1989). A side-mode suppression ratio of 35 dB was obtained at $I/I_{th} = 1.25$, which is comparable to that of a well-designed DBR or DFB dynamic single-mode laser. The temperature dependence of lasing wavelength was 0.07 nm/K. Single-mode operation was maintained in a temperature range of more than 50 K. This originated from the large mode spacing between neighboring longitudinal modes

($\sim$160 Å). The temperature characteristic of the threshold current looks different from that of conventional lasers. The increase in the threshold is caused by gain detuning and heating.

5.6.3   Polarization characteristics

For polarization-sensitive applications such as magneto-optic disks and coherent detection, the polarization state of lasers must be well defined. The polarization characteristic of several SE laser samples was measured by detecting the output through a rotating Glan-Thompson prism (Shimizu *et al.*, 1988). No noticeable change in polarization directions was observed with varying injection currents in the particular device. The output light was linearly polarized along the (011) or (01$\bar{1}$) direction. We consider that the polarization direction is determined by the anisotropy of the crystal surface and the evaporated mirror, the irregularity of the mesa, and so on. In order to investigate the polarization selectivity, we introduce a theoretical model and calculated the relative intensities of modes along two different directions by using rate equations. The loss difference between polarization states of the SE laser is much smaller than that of a conventional laser. Therefore, a polarization control mechanism is needed for polarization-sensitive applications. Some trials have been made to control the polarization of VCSEL. One approach consists of an SE laser with a grating terminator formed on a DBR (Mukaihara *et al.*, 1994).

The VCSEL prepared on GaAs (311) substrates is attractive due to a stable polarization operation based upon intrinsic anisotropic optical gain in quantum wells (Ohtoshi *et al.*, 1986), which is essential for low-noise applications such as in high-speed optical interconnects and optical disk memories. So far, several polarization-controlled VCSELs on GaAs (311) substrates have been demonstrated (Tateno *et al.*, 1997, Mizutani *et al.*, 1997, Kaneko *et al.*, 1995, Takahashi *et al.*, 1996). A very-low-threshold current density of 180A/cm^2 in VCSELs on GaAs (311)A substrates grown by molecular beam epitaxy (MBE) was reported as shown in Fig. 5.18(a) (Takahashi *et al.*, 1996). A very stable polarization was demonstrated in a wide range of driving current and temperature variation as shown in Fig. 5.18(b).

In 1997, NTT and the authors' group independently realized MOCVD-grown VCSELs on GaAs (311)B substrates (Tateno *et al.*, 1997, Mizutani *et al.*, 1997). However, there have been no reports on (311)B substrate VCSELs having both the low threshold and the low electrical resistance

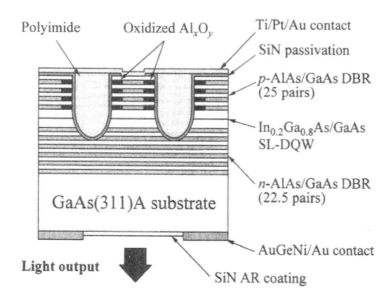

(a)

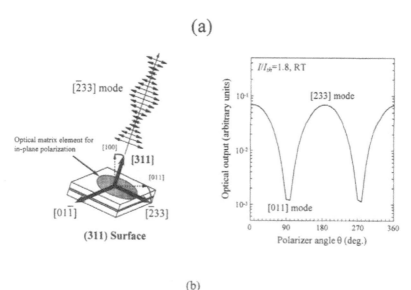

(b)

Figure 5.18: (a) Structure of an oxide-confined VCSEL grown on a (311) a substrate; (b) stable polarization demonstrated in the wide range of driving current and temperature variations.

that are required for low power consumption. The difficulty in highly *p*-type doping of AlAs on GaAs (311)B substrates by MOCVD has been removed by using a carbon auto-doping technique for reducing the electrical resistance of *p*-type distributed Bragg reflectors (DBRs) (Mizutani *et al.*, 1997). In this work, a low threshold current (600 μA) and polarization-controlled VCSEL on a GaAs (311)B substrate by MOCVD was realized. The fabricated vertical cavity surface-emitting laser (VCSEL) exhibited a low specific resistance of $2 \times 10^{-4}\Omega$ cm^2 and a stable polarization state along [$\bar{2}$33].

5.6.4   Quantum noise

The linewidth is measured by a standard delayed self-homodyne method with a 4-km-long single-mode fiber (Tanobe *et al.*, 1989). Two optical isolators with a total isolation of 60 dB are used to eliminate the effect of external optical feedback. The optical coupling to single-mode fiber is rather easy, because the modes can be almost matched with each other. From another group, a very high coupling efficiency (90%) was reported (Tai *et al.*, 1990). This is not surprising when we consider a well-matched mode profile. The spectral linewidth of 50 MHz is obtained with an output power of 1.4 mW (Tanobe *et al.*, 1989). Even in such an ultrashort cavity device with cavity length of less than 10 μm, a relatively narrow spectral linewidth can be attained. This is due to a high reflectivity of the mirrors employed. We can expect much narrower laser linewidth by increasing the output power and reducing the cavity loss.

5.6.5   Spontaneous emission control

Spontaneous emission control is considered by taking advantage of the microcavity structure. A possibility of no distinct threshold devices has been suggested (Yamamoto, 1988). The spontaneous emission factor has been estimated on the basis of three-dimentional mode density analysis (Baba *et al.*, 1991; Baba *et al.*, 1992). In Fig. 5.19 we show the estimated spontaneous emission factor of microcavity VCSEL (Baba *et al.*, 1991).

By overcoming technical problems to make tiny structures and to improve the thermal resistance, we believe that we can obtain a 1 μA threshold device (Tamanuki *et al.*, 1991). A lot of efforts toward improving the characteristics of surface emitting lasers have been made, including surface passivation in the regrowth process for buried heterostructure,

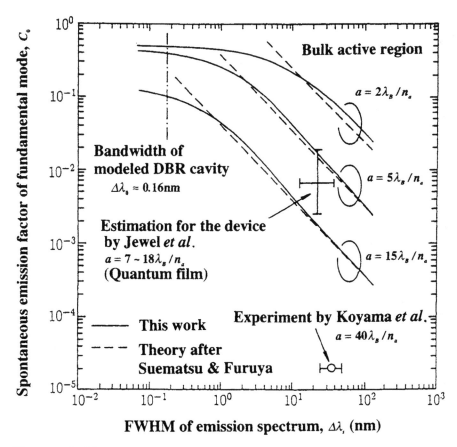

Figure 5.19: The estimated spontaneous emission factor of microcavity VCSEL.

micro-fabrication, and fine epitaxies (Tamanuki *et al.*, 1991). The ohmic resistance of semiconductor DBRs is reduced down to the order of low $10^{-5}\,\mathrm{Wcm}^2$ (Lear *et al.*, 1992). Spontaneous emission control is considered by taking the advantage of microcavity structures.

5.6.6   Photon recycling

One another interesting topic for microcavity SE lasers is photon recycling. By covering the side-bounding surfaces of the cavity with a highly reflective materials, some amount of spontaneously wasted photons can be recycled. It has been demonstrated that the photon cycling SE laser device appears to have no distinct threshold (Numai *et al.*, 1991). The efficiency of photon recycling has been estimated (Tamanuki *et al.*, 1992). The quantum noise characteristics, such as relative intensity noise (RIN) (Koyama *et al.*, 1992) and linewidths (Tanobe *et al.*, 1990) are being studied.

5.7   Integrations and applied optical subsystems

5.7.1   Photonic-stack integration

A stacked planar optics (Iga *et al.*, 1982) consists of planar optical components and their stack. All components must have the same two-dimensional spatial relationship, which can be achieved with planar technology with the help of photolithographic fabrication as used in electronics. Once we align the optical axis and adhere all of the stacked components, two-dimensional arrayed components are realized; the mass production of axially aligned discrete components is also possible if we separate individual components. To realize the stacked planar optics, all of the employed optical devices must have a planar structure. An array of microlenses on a planar substrate is required to focus and collimate the light in optical circuits. A planar microlens (Iga *et al.*, 1982) was developed, and 10 × 10-cm-square array samples are now available. As summarized in Fig. 5.20, we can consider a variety of integration based upon VCSELs.

5.7.2   Surface operating devices and monolithic integrations

In addition to surface-emitting lasers, surface-emitting laser-type optical devices such as optical switches, frequency tuners (Yokouchi *et al.*, 1990; Chang-Hasnain *et al.*, 1991), optical filters (Koyama *et al.*, 1991 a,b), and ultra-minute structures such as quantum wells and superlattices have become very active.

Some important technologies for integration related to surface-emitting lasers are progressing, *i.e.*, photonic integration, array formation, and OEIC. A wide variety of functions, such as frequency tuning, amplification, and filtering can be integrated along with surface emitting lasers by

(1) Lateral integration
Monitoring/receiver detector
Photothyristor
/phototransistor
OEIC
On-Si ULSI

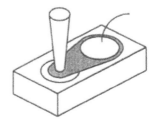

(2) Stack integration
Wavelength tuner
Saturable absorber
Modulator/switch
Functional element

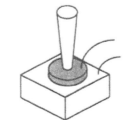

(3) Array integration
Individually accessible
Incoherent high power
Coherent high power
Beam steering
Multi-wavelength
Display
Light signal

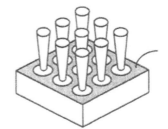

(4) Hybrid integration
Fiber array module
Optical interconnect
Disk pick-up
Optical sensor

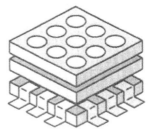

Figure 5.20: A variety of integration schemes based upon VCSELs.

stacking. Moreover, a two-dimensional parallel optical logic system (Ichioka and Tanida, 1984) can deal with a large amount of image information at high speed. Vertical optical interconnection of LSI chips and circuit boards may be another interesting issue.

Further development of the SE laser may open up various applications and accelerate the integration of optical devices and optical circuits with the possibility of two-dimensional arrays.

5.7.3   Two-dimensional arrays of surface-emitting lasers

Densely packed arrays have also been demonstrated for the purpose of making high-power lasers and coherent arrays. The coherent coupling of these arrayed lasers has been tried by using a Talbot cavity (Ho *et al.*, 1990) and diffraction coupling. It is pointed out that two-dimensional arrays are more suitable for making a coherent array than a linear configuration, since we can take the advantage of two-dimensional symmetry (Orenstein, 1989). The multi-wavelength emission from two-dimensional arrayed SELs was demonstrated by Chang-Hasnain *et al.*, (1991). This kind of device should be used in wavelength domain multiplexing (WDM) lightwave systems. A novel method of MOCVD grown multi-wavelength VCSELs was demonstrated as shown in Fig. 5.21 (Koyama *et al.*, 1996).

5.7.4   Applied subsystems

If we can have two-dimensional arrayed devices available for actual uses, we can expect to simultaneously align a tremendous number of optical components, as in parallel multiplexing lightwave systems. A wide variety of functions, such as frequency tuning, amplification, and filtering can be integrated along with surface-emitting lasers by stacking. Moreover, a two-dimensional parallel optical logic system (Numai *et al.*, 1991) can deal with a large amount of image information at high speed. For this purpose, the surface-emitting (SE) laser will be a key device. Its technical aspects and future prospect are detailed in the textbook published by Iga and Koyama (1990). Vertical optical interconnection of LSI chips and circuit boards may be another interesting issue.

Lastly, we consider some possible applications including optical interconnects, parallel fiber-optic subsystems, and so on. In optical fiber communication systems, ultrafine semiconductor dealing with a scale of

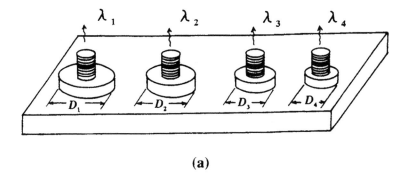

(a)

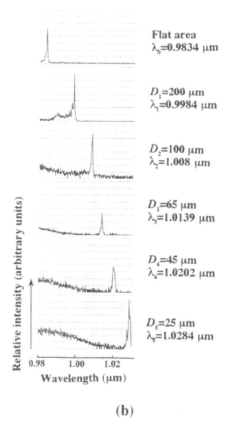

(b)

Figure 5.21: (a) Conceptional schematic structure of multi-wavelength vertical cavity surface-emitting laser array fabricated by MOCVD; (b) multi-wavelength operation.

0.1 or less. On the other hand, longer than 10,000 km of fibers are considered to be directly connected without any electrical repeaters. It could be said then that optoelectronics can treat the dimensions of 10^{-11} to 10^7 m. Another important advance is parallel lightwave systems including more than 4000 fibers, for example, broad-band ISDN systems.

By taking advantage of wide band and small volume transmission capability, the optical interconnect is considered to be inevitable in the computer technology. Some parallel interconnect scheme is desirable, and new concepts are being researched, as shown in Fig. 5.22. Vertical optical

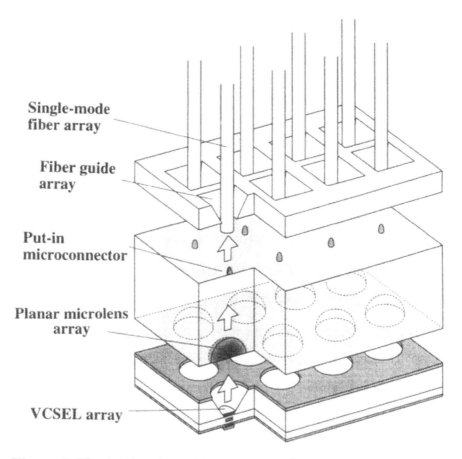

Single-mode
fiber array

Fiber guide
array

Put-in
microconnector

Planar microlens
array

VCSEL array

Figure 5.22: An idea of parallel interconnect scheme.

interconnect of LSI chips and circuit boards may be another interesting issue. In any case, the two-dimensional arrayed configuration of surface-emitting lasers and planar optics will open up a new era of optoelectronics. A new architecture for 64-channel interconnect has been proposed, and a modeling experiment was performed using GaAlAs VCSEL arrays (Redmond and Schenfeld, 1994).

The application of VCSELs to optical disk systems was suggested by Iga (1992) as shown in Fig. 5.23. Commercialization was started in 1997. Several schemes for optical computings have been considered, but one of the bottlenecks may be a lack of suitable optical devices, in particular, two-dimensional surface-emitting lasers and surface-operating switches. Fortunately, very-low-threshold surface-emitting lasers have been developed, and stack integration together with two-dimensional photonic devices is actually considered.

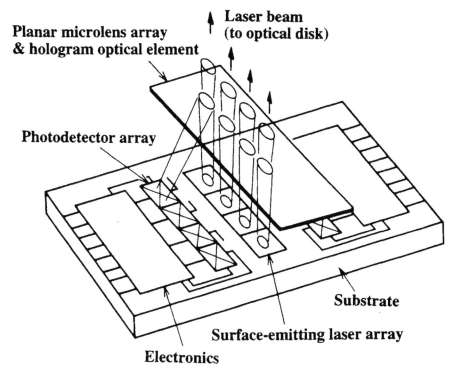

Figure 5.23: The application of VCSELs to optical disk systems.

5.8   Summary

In order to further achieve substantial innovations in VCSEL performances, the following technical issues remain unsolved or are still nonoptimized:

1. AlAs oxidation and its application to current confinement and optical beam focusing, in particular, for long wavelength devices.

2. Modulation doping, p-type and n-type modulation doping to quantum wells/barriers.

3. Quantum wires and dots for active regions.

4. Strained quantum wells and strain-compensation.

5. Angled substrates such as (311)A, (311)B, (411), etc., for polarization control (Takahashi *et al.*, 1996 a,b).

6. New material combinations such as GaInNAs/GaAs or AlGaInAs/InP for long wavelength emission.

7. Reproducible wafer fusion techniques to achieve optimum combination of active region and mirrors.

8. Transparent mirrors to increase quantum efficiency and output power.

9. Multi-quantum barriers (MQBs) to prevent carrier leakage to a p-cladding layer.

10. Wavelength controlling technologies such as tuning, steering, and trimming.

11. Arrays and high-power technologies

Vertical optical interconnects of LSI chips and circuit boards and multiple fiber systems may be the most interesting fields related to surface-emitting lasers. From this point of view, the device should be as small as possible. The future process technology for it, including epitaxy and etching, will drastically change the situation of surface-emitting lasers. Some optical technologies have already been introduced in various subsystems, but the arrayed micro-optic technology would be very helpful for advanced systems.

References

Arai, S.; Asada, M.; Suematsu, Y.; Itaya, Y.; Tanbun-ek, T.; and Kishino, K. (1980). *Electron. Lett.* **16**, 349.

Arakawa, Y., and Yariv, A. (1985). *IEEE J. Quantum Electron.* WE-21, 1666.

Asada, M.; Kameyama, A; and Suematsu, Y. (1984). *IEEE J. Quantum Electron.* QE-20, 745.

Baba, T.; Hamano, T.; Koyama, F.; and Iga, K. (1991). *IEEE J. Quant. Electron.* **27**, 1347.

Baba, T.; Hamano, T.; Koyama, F.; and Iga, K. (1992). *IEEE J. Quant. Electron,* **28**, 1310.

Baba, T.; Yogo, Y.; Suzuki, K.; Koyama, F.; and Iga, K. (1993). *Electron. Lett.* **29**, 913–914.

Baba, T.; Yogo, Y.; Suzuki, K.; Koyama, F.; and Iga, K. (1993). *IEICE Trans. Electronics.* E76-C, 1423.

Babic, D. I.; Dudley, J. J.; Streubel, K.; Milin, R. P.; Margalit, N. M.; Bowers, J. E.; and Hu, E. L. (1995). *7th Int'l Conf. on Indium Phosphide and Related Materials,* SB1.2.

Babic, D. I.; Streubel, K.; Mirin, R. P.; Margalit, N. M.; Bowers, J. E.; Hu, E. L.; Mars, D. E.; Yang, L.; and Carey K. (1995). *IOOC'95,* Postdeadline papers, PD1-5, Hong Kong.

Babic, D. I.; Streubel, K.; Mirin, R. P.; Pirek, J.; Margalit, N. M.; Bowers, J. E.; Hu, E. L.; Mars, D. E.; Yang, L.; and Carey, K. (1996). *IPRM'96,* ThA1-2.

Basov, N. G.; Bogdankevich, O. V.; and Grasyuk, A. Z. (1966). *IEEE J. Quantum Electron.* QE-2, 594.

Bissessur, H.; Koyama F.; and Iga, K. (1996). *IEICE,* C-320, 320

Bissessur, H.; Koyama, F.; and Iga, *K.* (1997). *IEEE J. Quantum Electron.* **3**, 2.

Botez, D.; Zinkiewicz, L. M.; Roth, T. J.; Mawst, L. J.; and Peterson, G. (1989). *IEEE Photonics Technology Letters,* **1**, 8, 205.

Carlson, N. W.; Evans, G. A.; Hammer, J. M.; Carr, L. A.; and Hawrylo, F. Z. (1988). *Appl. Phys. Lett.* **52**, 939.

Chailertvanitkul, A., Iga, K., and Moriki, K. (1985). *Electron Lett.* **21**, 303.

Chang-Hasnain, C. J.; Harbison, J. P.; Zah, C. E.; Florez, L. T.; and Andreadakis, N. C. (1991). *Electron. Lett.* **27**, 1002.

Coldren, L. A. (1995). *10th International Confer. on Integrated Optics and Optical Fibre Commun.,* TuB1-1, 26.

Dallesasse, J. M.; Gavrilovic, P.; Holonyak, Jr. N.; Kaloski, R. W.; Nam, D. W.; and Vesely, E. J. (1990). *Appl. Phys. Lett.* **56**, 2436.

Dallesasse, J. M.; Holonyak, Jr. N. ; Sugg, A. R.; Richard, T. A.; (1990). *Appl. Phys. Lett.* **57**, 2844.

Deppe, D. G.; Huffaker, D. L.; Deng, Q.; Oh, T.-H.; and Graham, L. A. (1997). *IEEE/LEOS'9,* ThA 1.

Deppe, D. G.; Huffaker, D. L.; Shin, J.; and Deng, Q. (1995). *IEEE Photon. Tech. Lett.* **7**, 965.

Donnely, J. P.; Bailey, R. J.; Goodhue, W. D.; Wang, C. J.; Lincoln, G. A.; Johnson, G. D. (1992). *CLEO'92,* CWN7.

Ebeling, K. J.; Fiedler, U.; Michalzik, R.; Reiner, G.; and Weigl, B. (1996). *22nd European Conf. on Optical Commun.* TuC2.1, 2.81.

Evans, G. A.; Carlson, N. W.; Hammer, J. M.; Lurie, J. M.; Butler, M.; Palfrey, J. K.; Amantea, S. L.; Carr, R.; Hawrylo, L. A.; James, F. Z.; Kaiser, E. A.; Kirk, C. J.; Reichert, J. B.; Reichert, W. F. (1988). *Appl. Phys. Lett.* **53**, 2123.

Evans, G. A.; Hammer, J. M.; Carlson, N. W.; Elia, F. R.; James, E. A.; and Kirk, J. B. (1986). *Conference on Lasers and Electro-Optics,* Postdeadline Paper, ThU3.

Geels, R. S.; and Coldren, L. A. (1990). *48th Device Research Conference,* VIIIA-1.

Geels, R. S.; and Coldren, L. A. (1990). *12th IEEE International Semiconductor-laser Conference.*

Geels, R. S.; Corzine, S. W.; Scott, J. W.; Young, D. B.; and Coldren, L. A. (1990). *Photonics Lett.* **2**, (4), 2345–2346.

Geels, R.; Yan, R. H.; Scott, J. W.; Corzine, S. W.; Simes, R. J.; and Coldren, L. A. (1988). *The Conf. on Lasers and Electro-Optics,* paper WM-1.

Goodfellow, R. C.; Carter, A. C.; Rees, G. J.; and Davis, R. (1981). *IEEE Trans. on Electron Devices,* ED-28, 365.

Gourley, P. L.; Brennan, T. M.; and Hammons, B. E. (1989). *Appl. Phys. Lett.* **54**, 1209.

Grabherr, M.; Weigl, B.; Reiner, G.; Miller, M.; and Ebeling, K. J. (1996). *Electron. Lett.* **32**, 1723–1724.

Guenter, J. K.; Hawthorne, R. A.; and Granville, D. N.; Hibbs-Brenner, M. K.; and Morgan, R. A. (1996). *Proc. of SPIE,* 2683, 1.

Hatori, N.; Mukaihara, T.; Abe, M.; Ohnoki, N.; Mizutani, A.; Matsutani, A.; Koyama, F.; and Iga, K. (1996). *Jpn. J. Appl. Phys.* **35**, 12A, 6108–6109.

Hatori, N.; Mizutani, A.; Nishiyama, N.; Koyama, F.; and Iga, K. (1998). *PTL,* **10**, (2), 194–196.

Hatori, N.; Mizutani, A.; Nishiyama, N.; Koyama, F.; and Iga, K. (1997). *Electron. Lett.* **33**, (12), 1096–1097.

Hayashi, Y.; Mukaihara, T.; Hatori, N.; Ohnoki, N.; Matsutani, A.; Koyama, F.; and Iga, K. (1995). *Electron. Lett.* **31**, 560–561.

Henry, C. H. (1983). *IEEE J. Quantum Electron.* QE-18, 259.

Ho, E.; Koyama, F.; and Iga, K. (1990). *Appl. Opt.,* **29**, (34), 5080–5085.

Honda, T.; Koyama, F.; and Iga, K. (1997). *MRS'96, MRS 1996 Fall Meeting Symposia Proceedings, Material Research Society,* Pittsburgh.

Honda, T.; Sakaguchi, T.; Katsube, A.; Koyama, F.; and Iga, K. (1994). *13th Symposium Record on Alloy Semiconductor Physics and Electronics,* S-15, 231–234.

Honda, T.; Yanashima, K.; Koyama, F.; Kukimoto, H.; and Iga, K. (1994). *Jpn. J. Appl. Phys.,* **33**, (2), 1211–1212.

Huang, K. F.; Tai, K.; Wu, C. C.; and Wynn, J. D. (1993). *Device Res. Conf.* III B-7.

Huffaker, D. L.; Deppe, D. G.; Kumar, K.; and Rogers, T. J. (1994). *Appl. Phys. Lett.* **65**, 1.

Huffaker, D. L.; Deppe, D. G.; Lei, C.; and Hodge, L. A. (1996). *CLEO'96* (Anaheim), JTuH5.

Huffaker, D. L.; Shin, J.; and Deppe, D. G. (1994). *Electron. Lett.* **31**, 1946.

Ibaraki, A.; Ishikawa, S.; Ohkouchi, S.; and Iga, K. (1984). *Electron. Lett.* **20**, 420.

Ibaraki, A.; Kawashima, K.; Furusawa, K.; Ishikawa, T.; Yamaguchi, T.; Niina, T. (1989). *Jpn. J. Appl. Phys.,* **28**, L667.

Ichioka, Y.; and Tanida, J. (1984). *Proc. IEEE,* **72**, 787.

Iga, K. (1996). *International Symposium on Blue Laser and Light Emitting Diodes,* Th-11.

Iga, K. (1993). *Plenary Talk at Integrated Photonics Research.*

Iga, K. (1994). *Optoelectronics—Device and Technologies—9*, (2), 167.

Iga, K. (1996). *Gakujutu-Geppo (Monthly J. of MECSS),* **49** ,42–46.

Iga, K.; Ishikawa, S.; Ohkouchi, S.; and Nishimura, T. (1984). *Appl. Phys. Lett.* **45**, 348.

Iga, K.; Kinoshita, S.; and Koyama, F. (1986). *10th IEEE Int'l Semiconductor Laser Conf.,* PD-4, 12.

Iga, K.; Kinoshita, S.; and Koyama, F. (1987). *Electron. Lett.* **23**, 3, 134–136.

Iga, K.; Koyama, F. (1990). *Surface emitting lasers,* (in Japanese). Tokyo: Ohmsha Pub. Co. Ltd., Tokyo.

Iga K., Koyama F., and Kinoshita S. (1988). *IEEE J. Quant. Electron.,* QE-24, 9, 1845.

Iga, K.; Oikawa, M.; Misawa, S.; Banno, J.; and Kokubun, K. (1982). *Appl. Opt.* **21**, 3456.

Iga, K.; Soda, H.; Terakado, T.; and Shimizu, S. (1983). *Electron. Lett.* **19**, 457.

Ikegami, T.; and Suematsu, Y. (1968). *IEEE J. Quantum Electron.* QE-4, 148.

Imajo, Y.; Kasukawa, A.; Kashiwa, S.; Okamoto, H.; (1990). *Jpn. J. Appl. Phys.* **29**, (7), L1130.

Jager, R.; Grabherr, M.; Jung, C.; Michalzik, R.; Reiner, G.; Weigl, B.; and Ebeling, K. J. (1997). *Electron. Lett.* **33**, (4), 330–331.

Jewell, J. L.; Lee, Y. H.; Tucker, R. S.; Burrus, C. A.; Scherer, A.; Harbison, J. P.; Florez, L. T.; and Sandroff, C. J. (1990). *CLEO'90* (Anaheim), CFF1, 500.

Kaneko, Y.; Nakagawa, S.; Takeuchi, T.; Mars, D. E.; Yamada, N.; and Mikoshiba, N. (1995). *Electron. Lett.* **31**, 805.

Kasukawa, A.; Imajo, Y.; Fukushima, T.; Okamoto, H. (1990). *48th device Research Conf.* Post Deadline Paper, VB-2.

Kawasaki, H., Koyama, F. and Iga, K. (1988). *Jpn. J. Appl. Phys.* **27**, 1548.

Kinoshita, S.; and Iga, K. (1987b). *IEEE J. Quantum Electron.* QE-23, 882.

Kinoshita, S.; Sakaguchi, T.; Odagawa, T.; and Iga, K. (1987a). *Jpn. J. Appl. Phys.* **26**, 410.

Kishino, K.; Kinoshita, S.; Konno, S.; and Tako, T. (1983). *Jpn. J. Appl. Phys.* **22**, L473.

Kojima, K.; Morgan, R. A.; Mullally, T.; Guth, G. D.; Focht, M. W.; Leibenguth, R. E.; Asom, M. T.; and Gault, W. A. (1992). *13th IEEE Semiconductor Laser Conf.* PD-2.

Kojima, K.; Noda, S.; Mitsunaga, K.; Kyuma, K.; Hamanaka, K.; Nakayama, T. (1987). *IEEE J. Quant. Electron.* **26**, (12), 227.

Kondow, M.; Uomi, K.; Niwa, A.; Kitatani, T.; Watahiki, S.; and Yazawa, Y. (1996). *Jpn. J. Appl. Phys.* **35**, 1273.

Kondow, M.; Uomi, K.; Niwa, A.; Kitatani, T.; Watahiki, S.; and Yazawa, Y. (1997). *IEEE/LEOS'97*, ThE3.

Koyama, F.; and Iga, K. (1987). *Trans. of IEICE of Japan* E70, 455.

Koyama, F.; Kubota, S.; and Iga, K. (1991). *Electron. Lett.* **27**, (12), 1093.

Koyama, F.; Kinoshita, S.; and Iga, K. (1988b). *Trans. of IEICE of Japan* E71, 1089.

Koyama, K.; Kinoshita, S.; and Iga, K. (1989). *Appl. Phys. Lett.* **55**, 221.

Koyama, F.; Morito, D.; and Iga, K. (1991). *IEEE J. of Quant. Electron.* **27**, (6), 1410–1416.

Koyama, F.; Morito, K.; and Iga, K. (1992). *IEEE J. Quant. Electron.* **27**, 1419.

Koyama, F.; Tomomatsu, K.; and Iga, K. (1988a). *Appl. Phys. Lett.* **52**, 528.

Koyama, F.; Uenohara, H.; Sakaguchi, T.; and Iga, K. (1987). *Jpn. J. Appl. Phys.* **26**, 1077.

Lau, K. Y.; Derry, P. L.; and Yariv, A. (1988). *Appl. Phys. Lett.* **52**, 88.

Lau, K. Y.; and Yariv, A. (1985) *IEEE J. Quantum Electron.* QE-21, 121.

Lear, K. L.; Choquette, K. D.; Schneider, Jr. R. P.; and Kilcoyne, S. P. (1995). *Appl. Phys. Lett.* **66**, 2616

Lear, K. L.; Chalmers, S. A.; and Kileen, K. P. (1992). *LEOS Annual,* PD.

Lear, K. L.; Schneider, Jr. R. P.; Choquette, K. D.; Kilcoyne, S. P.; and Geib, K. M. (1995). *Electron. Lett.* **31**, 208–209.

Leger, J. R.; Scott, M. L.; and Veldkamp, W. B. (1988). *Appl. Phys. Lett.,* **52**, 1771.

Liau, Z. L.; and Walpole, J. N. (1985). *Appl. Phys. Lett.* **46**, 115.

Lott, J. A.; and Schneider, Jr. R. P. (1993) *Electron. Lett.* **29**, 830.

Lott, J. A.; Schneider, Jr. R. P.; Mallo, K. J.; Kilcoyne, S. P.; and Choquette, K. D. (1993). *14th IEEE International Semiconductor Laser Conf.* (Hawaii), W1.1.

Macomber, S. H.; Mott, J. S.; Noll, R. J.; Gallatin, G. M.; Gratrix, E. J.; O'Dwyer, S. L.; and Lambert, S. A. (1987). *Appl. Phys. Lett.* **51**, 472.

Maeda, W.; Chang-Hasnain, C. J.; Lin, C.; Patel, J. S.; Johnson, H. A.; and Walker, J. A. (1991). *IEEE Photonics Tech. Letts.* **3**, (3), 268.

Margalit, N. M.; Babic, D. I.; Streubel, K.; Mirin, R. P.; Naone, L.; Bowers, J. E.; and Hu, E. L. (1996). *Electron. Lett.* **32**, 1675.

Melngailis, I. (1965). *Appl. Phys. Lett.* **6**, 59.

Misawa, S.; Oikawa, M.; and Iga, K. (1984). *Appl. Opt.,* **23**, (11), 1784.

Miyamoto, T. (1997). *IEEE/LEOS'97,* ThE2.

Miyamoto, T.; Takeuchi, K.; Koyama, F.; Iga, K. (1997). *IEEE Photonics Technology Letters,* **9**, 11.

Miyamoto, T.; Uchida, T.; Yokouchi, N.; Inaba, Y.; Koyama F.; and Iga, K. (1992). *LEOS Annual,* DLTA13.2.

Mizutani, A.; Hatori, N.; Nishiyama, N.; Koyama, F.; and Iga, K. (1997). *Electron. Lett.* **33**, 1877.

Mizutani, A.; Hatori, N.; Ohnoki, N.; Nishiyama, N.; Ohtake, N.; Koyama, F.; and Iga, K. (1997). *Jpn. J. Appl. Phys.* **36**, 6728–6729.

Moriki, K.; Nakahara, H.; Hattori, K.; and Iga, K. (1987). *The Trans. of IEICE of Japan* J70-C, 501.

Motegi, Y.; Soda, H.; and Iga, K. (1982). *Electron. Lett.* **18**, 461.

Mukaihara, T.; Koyama, F.; and Iga, K. (1992). *Electron. Lett.* **28**, (6), 555–556.

Mukaihara, R.; Ohnoki, N.; Baba, T.; Koyama, F.; and Iga K. (1994). *14th IEEE International Semiconductor Laser Conf.* (Hawaii), W1.6.

Nakamura, S.; Senoh, M.; Nagahama, S.; Iwasa, N.; Yamada, T.; Matsushita, T.; Kiyoku, H.; and Sugimoto, Y. (1996). *Jpn. J. Appl. Phys.* **35**, L14–L16, 15 J.

Nam, D. W.; Waarts, R. G.; Welch, D. F.; Scifres, D. R.; (1992). *CLEO'92,* CWN8.

Nomura, Y.; Shinozuka, K.; Asakawa, K.; and Ishii, M. (1985). *Extended Abstracts of 17th Conf. on Solid State Devices and Materials,* 71.

Numai, T.; Kawakami, T.; Yoshikawa, T.; Sugimoto, M.; Sugimoto, Y.; Yokoyama, H.; Kasahara, K.; and Asakawa, K. (1993). *Jpn. J. Appl. Phys.* **32**, 10B, L1533–1534.

Numai, T.; Ogura, I.; Kosaka, H.; Kunihara, K.; Sugimoto, M.; and Kasahara, K. (1991). *LEOS'91,* OE7.5.

Numai, T.; Sugimoto, M.; Ogura, I.; Kosaka, H.; Kasahara, K. (1991). *J. J. Appl. Phys.* **30**, (4A), L602, April.

Ogura, M.; Hata, T.; and Yao, T. (1984). *Jpn. J. Appl. Phys.* **23**, L512.

Ohkoshi, T.; Kikuchi, K.; and Nakamura A. (1980). *Electron. Lett.* **16**, 630.

Ohnoki, N.; Mukaihara, T.; Hatori, N.; Mizutani, A.; Koyama, F.; and Iga, K. (1996). *Extended Abstracts of 1996 International Conference on Sokid State Devices Materials,* 595.

Ohtoshi, T.; Kuroda, T.; Niwa, A.; and Tsuji, S. (1994). *Appl. Phys. Lett.* **65**, 1886.

Okuda, H.; Soda, H.; Moriki, K.; Motegi, Y.; and Iga, K. (1981). *Jpn. J. Appl. Phys.* **20**, 563.

Olshansky, R.; and Su, C. B. (1985). *Electron. Lett.* **21**, 721.

Orenstein, M.; Kapon, E.; Stoffel, N. G.; Harbison, J. P.; Florez, L. T.; Wullert, J. (1991). *Appl. Phys Lett.* **58**, (8), 804, Feb.

Oshikiri, M.; Koyama, F.; and Iga, K. (1991). *Electron. Lett.* **27**, 22, 2038–2039.

Packard, J. R.; Tait, W. C.; and Campbell, D. A. (1969). *IEEE J. Quantum Electron.* QE-5, 44.

Peters, F. H.; Peters, M. G.; Young, D. B.; Scott, J. W.; Tibeault, B. J.; Corzine, S. W.; and Coldren, L. A. (1992). *13th IEEE Semiconductor Laser Conf.* PD–1.

Raja, M. Y. A.; Brueck, S. R. J.; Osinski, M.; Schaus, C. F.; McInerney, J. G.; Brennan, T. M.; and Hammon, B. E. (1988). *Electron. Lett.* **24**, 1140.

Redmond, I.; and Schenfeld, E. (1994). *Optical Computing '94,* Scottland.

Reinhart, F. K.; and Logan, R. A. (1975). *Appl. Phys. Lett.* **26**, 516.

Sakaguchi, T.; Koyama, F.; and Iga, K. (1986). *Electron. Lett.* **24**, 928.

Sakaguchi, T.; Koyama, F.; and Iga, K. (1990). *Laser Review,* **18**, 137.

Schaus, C. F.; Sun, S.; Schaus, H. E.; Raja, M. Y. A.; McInerney, J. G.; and Brueck, S. R. J. (1989). *The Conf. on Lasers and Electro-Optics,* PD-13.

Shimizu, M.; Koyama, F.; and Iga, K. (1988). *Japan. J. Appl. Phys.* **27**, 1774.

Shin, H. E.; Zoo, Y. G.; and Le, E. H. (1996). *16th Conference on Lasers and Electro-Optics* (Anaheim), JTuH7.

Smiley, V. N.; Tayler, H. F.; and Lewis, A. L. (1971). *J. Appl. Phys.* **42**, 5859.

Soda, H.; Iga, K.; Kitahara, C.; and Suematsu, Y. (1979). *Jpn. J. Appl. Phys.* **18**, 2329.

Soda, H.; Motegi, Y.; and Iga, K.; (1983). *IEEE J. Quantum Electron.* QE-19, 1035.

SpringThorpe, A. J. (1977). *Appl. Phys. Lett.* **31**, 524.

Stillman, G. E.; Sirkis, M. D.; Rossi, J. A.; Johnson, M. R.; and Holonyak, N. (1966). *Appl. Phys. Lett.* **9**, (7), 268–269.

Suematsu, Y.; and Furuya, K. (1977). *Trans. IECE of Japan,* E-60, 467.

Suematsu, Y.; Arai, S.; and Kishino, K. (1983). *IEEE J. Lightwave. Tech.* LT-1, 161.

Tai, K.; Hasnain, G.; Wynn, J. D.; Fischer, R. J.; Wang, Y. H.; Weir, B.; Gamelin, J.; Cho, A. Y. (1990). *Electronics Letter* **26**, (19), 1628–1629.

Takahashi, K.; Vaccaro, P. O.; Watanabe, T.; Mukaihara, T.; Koyama, F.; and Iga, K. (1996). *Jpn J. Appl. Phys.* **35**, 6102–6107.

Takahashi, M.; Vaccaro, P.; Fujita, K.; Watanabe, T.; Mukaihara, T.; Koyama, F.; and Iga, K. (1996). *IEEE Photon. Technol. Lett.* **8**, 737.

Tamanuki, T.; Koyama, F.; and Iga, K. (1992). *Jpn. J. Appl. Phys.* **31**, 1810.

Tanobe, H.; Oshikiri, M.; Araki, M; Koyama, F.; and Iga, K. (1992). *LEOS Annual,* DLTA12.2.

Tanobe, H.; Koyama, F.; and Iga, K. (1989). *Electron. Lett.* **25**, 1444.

Tateno, K.; Ohiso, Y.; Amano, C.; Wakatsuki, A.; and Kurokawa, T. (1997). *Appl. Phys. Lett.* **70**, 3395.

Tsang, W. (1978). *Appl. Phys. Lett.* **33** (5) 426–429.

Uchiyama, S.; and Iga, K. (1984). *IEEE J. Quantum Electron.* QE-20, 1117.

Uchiyama, S.; and Iga, K. (1985). *Electron. Lett.* **21**, 162.

Uchiyama, S. (1986). PhD Dissertation (Tokyo Institute of Technology).

Uchiyama, S.; Iga, K. (1986) *IEEE J.Q.E.,* **22**, 2, 302–309.

Uchiyama, S.; Iga, K.; and Kokubun, Y. (1986b). *The 12th European Conf. on Optical Commun.* 37.

Uchiyama, S.; and Ninomiya T. (1996). *1st Optoelectronics and Communications Conference,* 19D1-3.

Uchiyama, S.; Ohmae, Y.; Shimizu, S.; and Iga, K. (1986a). *IEEE / OSA Lightwave Tech.* LT-4, 846.

Uchiyama, S.; Yokouchi, N.; and Ninomiya, T. (1996). *The 43th Spring Meeting of Jpn. Soc. Appl. Phys.* 26p-C-7.

Uenohara, H.; Koyama, F.; and Iga, K. (1989). *Jpn. J. Appl. Phys.* **28**, 740.

Wada, H.; Babic, D. I.; Crawford, D. L.; Dudley, J. J.; Bowers, J. E.; Hu, E. L.; Merz, J. L. (1991). *Device Research Conference, Post Deadline Paper,* A-8.

Warren, M. E.; Gourley, P. L.; Hadley, G. R.; Vawter, G. A.; Brennan, T. M.; Hammons, B. E.; and Lear, K. L. (1992). *CLEO'92,* JThA5.

Watanabe, I.; Koyama, F.; and Iga, K. (1988). *Jpn. J. Appl. Phys.* **16**, 1598.

Weigl, B.; Grabherr, M.; Jager, R.; Reiner, G.; and Ebeling, K. J. (1996). *15th IEEE International Semicon. Laser Conf.* Post deadline paper, PDP2.

Weigl, B.; Reiner, G.; Grabherr, M.; and Ebeling, K. J. (1996). *CLEO'96* (Anaheim), JTuH2.

Welch, D. F.; Parke, R.; Hardy, A.; Streifer, W.; Scifres, D. R. (1989). *Electron. Lett.* **25**, 819.

Wipiejewski, T.; Panzlaf, K.; Zeeb, E.; and Ebeling, K. J. (1992). *18th European Cof. on Opt. Comm. ECOC'92,* PDII-4.

Wu, C.; Svilans, M.; Fallhi, M.; Makino, T.; Glinsk, J.; Maritan, C.; and Blaauw, C. (1991). *ECOC'91* (Paris), Postdeadline, 25.

Wu, M. C.; Ogura, M.; Hsin, W.; Whinnery, J. R.; and Wang, S. (1987). *Conference on Lasers and Electro-Optics,* WG4.

Yamamoto, Y.; Machida, S.; Igeta, K.; and Horikoshi, Y. (1988). *XVI Int. Conf. Quantum Electron.* WB-2.

Yang, G. M.; MacDougal, M. H.; and Dapkus, P. D. (1995). *CLEO'95,* Post Deadline Papers, CPD4-1, Baltimore.

Yang, G. M.; MacDougal, M. H.; and Dapkus, P. D. (1995). *Electron. Lett.* **31**, 886

Yokouchi, N.; Koyama, F.; and Iga, K. (1990). *The Trans. of IEICE,* E73, 1473.

Yokouchi, N.; Koyama, F.; and Iga, K. (1992). *Photonics Tech. Lett.* **7**, 701.

Yoo, H. J.; Scherer, A.; Harbison, J. P.; Florez, L. T.; Paek, E. G.; Van der Gaag, B. P.; Hayes, J. R.; Lehmen, E.; Kapon, E.; and Kwon, Y. S. (1990). *Appl. Phys. Lett.* **56**, 1198.

Young, D. B.; Kapila, A.; Scott, J. W.; Malhotra, V.; and Coldren, L. A. (1994). *Electron. Lett.* **30**, 233.

Yuasa, T.; Hamano, N.; Sugimoto, M.; Takado, N.; Ueno, M.; Iwata, H.; Tashiro, T.; Onabe, K.; and Asakawa, K. (1988). *Conference on Lasers and Electro-Optics,* WO6.

Index

Optics and Photonics
(Formerly Quantum Electronics)

Edited by   Paul F. Liao, *Bell Communications Research, Inc., Red Bank, New Jersey*
Paul L. Kelley, *Tufts University, Medford, Massachusetts*
Ivan P. Kaminow, *AT&T Bell Laboratories, Holmdel, New Jersey*
Gorvind P. Agrawal, *University of Rochester, Rochester, New York*

N. S. Kapany and J. J. Burke, *Optical Waveguides*
Dietrich Marcuse, *Theory of Dielectric Optical Waveguides*
Benjamin Chu, *Laser Light Scattering*
Bruno Crosignani, Paolo DiPorto and Mario Bertolotti, *Statistical Properties of Scattered Light*
John D. Anderson, Ir, *Gasdynamic Lasers: An Introduction*
W. W. Duly, *CO_2 Lasers: Effects and Applications*
Henry Kressel and J. K. Butler, *Semiconductor Lasers and Heterofunction LEDs*
H. C. Casey and M. B. Panish, *Heterostructure Lasers: Part A. Fundamental Principles; Part B.*
    Materials and Operating Characteristics
Robert K. Erf, editor, *Speckle Metrology*
Marc D. Levenson, *Introduction to Nonlinear Laser Spectroscopy*
David S. Kliger, editor, *Ultrasensitive Laser Spectroscopy*
Robert A. Fisher, editor, *Optical Phase Conjugation*
John F. Reintjes, *Nonlinear Optical Parametric Processes in Liquids and Gases*
S. H. Lin, Y. Fujimura, H. J. Neusser and E. W. Schlag, *Multiphoton Spectroscopy of Molecules*
Hyatt M. Gibbs, *Optical Bistability: Controlling Light with Light*
D. S. Chemla and J. Zyss, editors, *Nonlinear Optical Properties of Organic Molecules and Crystals.*
    Volume 1, Volume 2
Marc D. Levenson and Saturo Kano, *Introduction to Nonlinear Laser Spectroscopy, Revised Edition*
Govind P. Agrawal, *Nonlinear Fiber Optics*
F. J. Duarte and Lloyd W. Hillman, editors, *Dye Laser Principles: With Applications*
Dietrich Marcuse, *Theory of Dielectric Optical Waveguides, 2nd Edition*
Govind P. Agrawal and Robert W. Boyd, editors, *Contemporary Nonlinear Optics*
Peter S. Zory, Jr., editor, *Quantum Well Lasers*
Gary A. Evans and Jacob M. Hammer, editors, *Surface Emitting Semiconductor Lasers and Arrays*
John E. Midwinter, editor, *Photonics in Switching, Volume I, Background and Components*
John E. Midwinter, editor, *Photonics in Switching, Volume II, Systems*
Joseph Zyss, editor, *Molecular Nonlinear Optics: Materials, Physics, and Devices*
F. J. Duarte, editor, *Tunable Lasers Handbook*
Jean-Claude Diels and Wolfgang Rudolph, *Ultrashort Laser Pulse Phenomena: Fundamentals,*
    Techniques, and Applications on a Femtosecond Time Scale
Eli Kapon, editor, *Semiconductor Lasers I: Fundamentals*
Eli Kapon, editor, *Semiconductor Lasers II: Materials and Structures*

Yoh-Han Pao, Case Western Reserve University, Cleveland, Ohio, Founding Editor 1972–1979

Printed and bound by CPI Group (UK) Ltd, Croydon, CR0 4YY

08/05/2025

01864900-0002